The Role of Halo Substructure in Gamma-Ray Dark Matter Searches

The Role of Halo Substructure in Gamma-Ray Dark Matter Searches

Special Issue Editors

Miguel A. Sánchez-Conde
Michele Doro

MDPI • Basel • Beijing • Wuhan • Barcelona • Belgrade • Manchester • Tokyo • Cluj • Tianjin

Special Issue Editors

Miguel A. Sánchez-Conde
Institute for Theoretical Physics
(IFT UAM/CSIC) and
Department of Theoretical
Physics, Universidad Autónoma
de Madrid
Spain

Michele Doro
Department of Physics and
Astronomy G. Galilei
Italy

Editorial Office
MDPI
St. Alban-Anlage 66
4052 Basel, Switzerland

This is a reprint of articles from the Special Issue published online in the open access journal *Galaxies* (ISSN 2075-4434) (available at: https://www.mdpi.com/journal/galaxies/special_issues/Gamma-RayDMS).

For citation purposes, cite each article independently as indicated on the article page online and as indicated below:

LastName, A.A.; LastName, B.B.; LastName, C.C. Article Title. *Journal Name* **Year**, *Article Number*, Page Range.

ISBN 978-3-03936-044-4 (Pbk)
ISBN 978-3-03936-045-1 (PDF)

Cover image courtesy of Diemand et al., Nature, Volume 454, Issue 7205, pp. 735-738 (2008).

© 2020 by the authors. Articles in this book are Open Access and distributed under the Creative Commons Attribution (CC BY) license, which allows users to download, copy and build upon published articles, as long as the author and publisher are properly credited, which ensures maximum dissemination and a wider impact of our publications.

The book as a whole is distributed by MDPI under the terms and conditions of the Creative Commons license CC BY-NC-ND.

Contents

About the Special Issue Editors vii

Preface to "The Role of Halo Substructure in Gamma-Ray Dark Matter Searches" ix

Jesús Zavala and Carlos S. Frenk
Dark Matter Haloes and Subhaloes
Reprinted from: *Galaxies* **2019**, *7*, 81, doi:10.3390/galaxies7040081 1

Shin'ichiro Ando, Tomoaki Ishiyama and Nagisa Hiroshima
Halo Substructure Boosts to the Signatures of Dark Matter Annihilation
Reprinted from: *Galaxies* **2019**, *7*, 68, doi:10.3390/galaxies7030068 55

Javier Rico
Gamma-Ray Dark Matter Searches in Milky Way Satellites—A Comparative Review of Data Analysis Methods and Current Results
Reprinted from: *Galaxies* **2020**, *8*, 25, doi:10.3390/galaxies8010025 85

Geoff Beck
Radio-Frequency Searches for Dark Matter in Dwarf Galaxies
Reprinted from: *Galaxies* **2019**, *7*, 16, doi:10.3390/galaxies7010016 113

Martin Stref, Thomas Lacroix and Julien Lavalle
Remnants of Galactic Subhalos and Their Impact on Indirect Dark-Matter Searches
Reprinted from: *Galaxies* **2019**, *7*, 65, doi:10.3390/galaxies7020065 127

Moritz Hütten, Martin Stref, Céline Combet, Julien Lavalle and David Maurin
γ-ray and ν Searches for Dark-Matter Subhalos in the Milky Way with a Baryonic Potential
Reprinted from: *Galaxies* **2019**, *7*, 60, doi:10.3390/galaxies7020060 145

Francesca Calore, Moritz Hütten and Martin Stref
Gamma-Ray Sensitivity to Dark Matter Subhalo Modelling at High Latitudes
Reprinted from: *Galaxies* **2019**, *7*, 90, doi:10.3390/galaxies7040090 165

Javier Coronado-Blázquez and Miguel A. Sánchez-Conde
Constraints to Dark Matter Annihilation from High-Latitude HAWC Unidentified Sources
Reprinted from: *Galaxies* **2020**, *8*, 5, doi:10.3390/galaxies8010005 179

Ángeles Moliné, Jascha A. Schewtschenko, Miguel A. Sánchez-Conde, Alejandra Aguirre-Santaella, Sofía A. Cora and Mario G. Abadi
Properties of Subhalos in the Interacting Dark Matter Scenario
Reprinted from: *Galaxies* **2019**, *7*, 80, doi:10.3390/galaxies7040080 193

About the Special Issue Editors

Miguel A. Sánchez-Conde, Obtained his PhD in physics at the Institute of Astrophysics of Andalusia (IAA-CSIC) and University of Granada in 2009 with a thesis titled "The Nature of Dark Matter: Gamma-Ray Searches and the Formation of CDM Halos". He held appointments as a postdoctoral researcher at the Institute of Astrophysics of Canary Islands (2009–2011); at the Kavli Institute for Particle Astrophysics and Cosmology (KIPAC/SLAC, Stanford University, 2011–2014); and at the Oskar Klein Centre of Stockholm University (2014–2017). In March 2017, he joined the Institute for Theoretical Physics and the Department of Theoretical Physics of the Madrid Autonomous University as a senior researcher. Dr. Sánchez-Conde's main research activities focus on shedding light on the fundamental nature of dark matter with gamma rays. He is a former member of the MAGIC Collaboration and a current member of both the Fermi-LAT Collaboration and the Cherenkov Telescope Array Consortium. In addition to astroparticle physics, his research interests include N-body cosmological simulations, dark matter halo formation and evolution, large-scale structure, and cosmology.

Michele Doro, Graduated with a PhD in physics at the University of Padova in 2004 with the thesis titled "Commissioning of the Calibration System of the MAGIC Telescope", presented at the Institut de Fisica d'Altes Energies (IFAE, Barcelona, Spain). He obtained a PhD in Physics in 2009 at the University of Padova with the thesis titled "Novel Reflective Elements and Indirect Dark Matter Searches with MAGIC and CTA" on the technology of mirror facets for the Cherenkov telescopes (MAGIC and CTA) and indirect dark matter searches with gamma rays from annihilation or decay events in astrophysical environments. He was a research fellow at IFAE, at the Universitat Autònoma de Barcelona (UAB, Spain 2010–2013), and at the Max Planck Institut for Physics (MPI-Munich, Germany 2015). He is currently an associate professor in fundamental physics interactions at the Department of Physics and Astronomy of the University of Padova and is associated with the Italian Nuclear Physics Institute (INFN). His research expertise focuses on dark matter and fundamental physics searches with TeV gamma rays, atmospheric calibration through remote sensing instrumentations, and technology for mirror facets.

Preface to "The Role of Halo Substructure in Gamma-Ray Dark Matter Searches"

This is an era in which a tremendous international effort, both theoretical and experimental, is being made to unveil the ultimate nature of particle dark matter in the universe. Yet, not only is its microscopic nature unknown, but key aspects regarding its distribution and clustering also remain to be fully comprehended, especially at the smallest scales, where both observations and simulations are challenging. This lack of knowledge also has a great impact on the search for dark matter. As a prominent example, an important, open topic of research is to understand the precise role that the dark matter halo substructure may have for ongoing, or yet-to-come, dark matter search strategies and observational campaigns.

Halo substructure represents a natural expectation of the way structure formation works within the standard ΛCDM cosmological model: small structures or halos collapse and form first, and then, by accretion and merging, larger structures come into existence. In this bottom-up formation scenario, halo substructures, or subhalos, are predicted to be largely abundant within larger halos, for instance galaxies such as ours. Dwarf satellite galaxies are thought to be the most massive exponents of halo substructure in the Milky Way, while a very numerous population of lighter subhalos, not massive enough to possess a visible baryonic counterpart, is also expected to fill the galactic volume up to distances much larger than the size of the bulge and the disk of stars and gas.

When it comes to gamma-ray dark matter searches—those that look for gamma-ray photons generated from the decay or annihilation of dark matter particles—it is currently evident that halo substructure plays a critical role. Both dwarf galaxies and dark subhalos are known to be excellent targets per se for current or future gamma-ray observatories. Furthermore, the clumpy distribution of subhalos residing in larger halos boosts the predicted dark matter signals at Earth considerably with respect to a scenario in which no subhalos are accounted for in the computations. In this regard, a precise characterization of the statistical and structural properties of subhalos becomes critical. Intriguingly, as said, many questions remain open on the subhalo population that need to be answered should we want to make further progress. Indeed, current and planned gamma-ray experiments possess, for the first time, the exciting potential to reach the most relevant regions of the dark matter parameter space, but, in order to achieve this potential, a more profound knowledge of the most promising dark matter targets and scenarios is mandatory. Only in doing so will we be able to obtain accurate predictions of dark matter-induced photon fluxes to properly motivate and invest a significant fraction of telescopes' observing time on selected targets and to derive robust conclusions from these observations and from this whole dark matter search effort. Certainly, the field is now mature enough to demand precision work on every front.

This Special Issue was partially conceived during the workshop "Halo Substructure and Dark Matter Searches" held at the Institute for Theoretical Physics (IFT UAM-CSIC) in Madrid in June 2018[1], in which an overview of the current knowledge about dark matter subhalos was laid out. In written form now, building upon the discussions of the workshop, we aim to summarize where we stand today with respect to our knowledge of dark matter halo substructure from different viewpoints, to identify what are the remaining big questions, to discuss how we could address them in the near future, and, by doing so, to find new avenues for further research. With this goal in mind, the present volume includes a series of comprehensive reviews, written by some of the most

---

[1] https://eventos.uam.es/16257/detail/halo-substructure-and-dark-matter-searches.html

renowned experts, covering some of the key topics in the field: what we know about halo and subhalo formation and evolution from N-body cosmological simulations (by J. Zavala and C. Frenk), our current understanding of the annihilation signal boost due to the presence of subhalos (S. Ando et al.), and the present status of dark matter searches in dwarf galaxies using both gamma-ray (J. Rico) and radio frequencies (G. Beck). These reviews are then complemented with a number of research articles focusing on specific hot topics, e.g., dark subhalo searches (M. Stref et al.), subhalo annihilation boosts (M. Hütten et al.), subhalo survival (F. Calore et al.), subhalo properties in alternative cosmological scenarios (A. Moliné et al.), and subhalo searches with the HAWC TeV gamma-ray observatory (J. Coronado and M. Sánchez-Conde).

Our collection is not complete. We had planned for further contributions that would have completed the Special Issue. These included, e.g., gravitational lensing as a test of halo substructure, the dark matter content of Milky Way dwarf satellite galaxies from observations, stellar streams as a test of halo substructure, and the search for low-mass galactic subhalos with gamma rays, among others. We could not get such papers in time for this Special Issue. However, the interested reader has access to specific talks on these and related topics in the video recording of the Madrid workshop[2].

All in all, this Special Issue will act as a necessary and long-awaited bridge between the N-body simulation and gamma-ray dark matter search communities. In summary, we hope that this Special Issue will provide a fairly balanced and useful summary of the state of the art in halo substructure, thereby serving as a starting point for further research on this topic.

Madrid and Venice: 7 April 2020

Miguel A. Sánchez-Conde, Michele Doro
Special Issue Editors

---

[2] https://eventos.uam.es/16257/section/13616/halo-substructure-and-dark-matter-searches.html

Review

Dark Matter Haloes and Subhaloes

Jesús Zavala [1,*] and Carlos S. Frenk [2]

[1] Center for Astrophysics and Cosmology, Science Institute, University of Iceland, Dunhagi 5, 107 Reykjavík, Iceland
[2] Institute of Computational Cosmology, Department of Physics, Durham University, South Road, Durham DH1 3LE, UK; c.s.frenk@durham.ac.uk
* Correspondence: jzavala@hi.is

Received: 29 July 2019; Accepted: 20 September 2019; Published: 25 September 2019

Abstract: The development of methods and algorithms to solve the N-body problem for classical, collisionless, non-relativistic particles has made it possible to follow the growth and evolution of cosmic dark matter structures over most of the universe's history. In the best-studied case—the cold dark matter or CDM model—the dark matter is assumed to consist of elementary particles that had negligible thermal velocities at early times. Progress over the past three decades has led to a nearly complete description of the assembly, structure, and spatial distribution of dark matter haloes, and their substructure in this model, over almost the entire mass range of astronomical objects. On scales of galaxies and above, predictions from this standard CDM model have been shown to provide a remarkably good match to a wide variety of astronomical data over a large range of epochs, from the temperature structure of the cosmic background radiation to the large-scale distribution of galaxies. The frontier in this field has shifted to the relatively unexplored subgalactic scales, the domain of the central regions of massive haloes, and that of low-mass haloes and subhaloes, where potentially fundamental questions remain. Answering them may require: (i) the effect of known but uncertain *baryonic* processes (involving gas and stars), and/or (ii) alternative models with new dark matter physics. Here we present a review of the field, focusing on our current understanding of dark matter structure from N-body simulations and on the challenges ahead.

Keywords: dark matter; structure formation; cosmological N-body simulations

Contents

1 Introduction   2

2 Formation of Dark Matter Haloes   3
    2.1 Initial Conditions: The Primordial Power Spectrum in the Linear Regime 3
    2.2 The Non-Linear Regime: N-Body Simulation Methods 5
    2.3 The Non-Linear Regime: Initial Conditions and The Emergence of the Cosmic Web . . . 7
    2.4 The Structural Properties of Dark Matter Haloes 11

3 Halo Mergers and the Emergence of Subhaloes   18
    3.1 Halo Mass Assembly: Smooth Accretion vs Mergers 18
    3.2 Evolution of Subhaloes: Initial Conditions 19
    3.3 Dynamics of Subhaloes 21
    3.4 The Abundance, Spatial Distribution and Internal Structure of Dark Matter Subhaloes . 28
    3.5 The Impact of the Nature of the Dark Matter 32

4   **Outlook**   **33**
     4.1 The Impact of Baryonic Physics on Dark Matter Structure 34
     4.2 Astrophysical Tests of the Nature of the Dark Matter 36

References   **39**

1. Introduction

The current theory of the formation and evolution of cosmic structure in the universe is based on the dark matter hypothesis in which ∼84% of the mass-energy density of the universe [1] is in the form of a new type of particle, or particles, with negligible electromagnetic interactions. The evidence for the existence of dark matter is varied and compelling. It comes from cosmic structures on all scales and across all epochs: from the smallest, dark-matter-dominated dwarf galaxies (e.g., [2]), through the largest clusters of galaxies (e.g., [3]), to the large-scale structure of the universe (e.g., [4]) and back to the very seeds of cosmic structure reflected in the temperature of the cosmic background radiation (CMB; e.g., [1]). This body of evidence, accumulated over the past three decades, can be accounted for within a coherent theory of structure formation in which the gravity of the dark matter amplifies primordial density perturbations imprinted during an early period of cosmic inflation [5,6]. Empirical evidence for the existence of dark matter comes purely from its gravitational effect: despite significant efforts, experimental searches for dark matter particles in accelerators (e.g., for a review in LHC searches see [7]), and dedicated detectors on Earth (e.g., [8,9]) and in space (e.g., [10,11]) so far remain unsuccessful. Until the particles are discovered, dark matter will remain a hypothesis, albeit one with strong empirical support.

In addition to the dark matter hypothesis, the standard theory of structure formation makes a specific assumption about the nature of dark matter, which is only partially supported by observations. This is that the dark matter consists of classical, non-relativistic, collisionless particles which had negligible thermal velocities at early times. This "cold dark matter" (CDM) is assumed to behave as a fluid throughout most of the universe's history, except at very early times when this assumption breaks down in different ways depending on the specific mechanism of dark matter production. The most common hypothesis is that the dark matter particles are thermal relics from the Big Bang (e.g., [12]). In this case, dark matter was symmetric[1] and in thermal equilibrium with the photon-baryon plasma through interactions with standard model particles. As the universe cools down, dark matter decouples from the standard model particles, its creation annihilation stops and its co-moving density *freezes out*. If the strength of the interactions is assumed to be on the scale of the weak force, then the thermal-relic abundance of these weakly interacting massive particles (WIMPs) is quite close to the observed abundance of dark matter. This remarkable coincidence, haplessly known as the *WIMP miracle*, has enshrined WIMPs as the most popular dark matter candidates, especially since new physics at the weak scale (and with them the emergence of WIMP-like particles) was anticipated by Supersymmetric theories in order to solve the hierarchy problem (e.g., [13]). Moreover, WIMPs are the quintessential CDM candidate because once they decouple, they are nearly collisionless and, since they are massive (∼10 GeV–1 TeV), they behave as a classical (non-quantum) fluid that becomes non-relativistic very early on.

The combination of the *WIMP miracle* with the success of the CDM model in explaining the observed large-scale structure of the universe in the mid-1980s [14] established the current paradigm of structure formation in which gravity is the only dark matter interaction. This model has been widely adopted by the community working on galaxy formation and evolution and, as a result, most of our understanding on how cosmic structure emerges comes from studies that assume the CDM model. This is a relevant remark in the context of this review because the properties of dark matter haloes and

---

[1]   Equal amounts of dark matter and anti-dark matter.

their substructure depend on the nature of dark matter (see Section 2). In reality, the range of allowed dark matter models, motivated to varying degrees by particle physics considerations, is vast. In this landscape of models, only a fraction fall in the CDM category alongside WIMPs, e.g., the QCD axion (motivated by a proposed solution to the strong CP problem in particle physics [15]).

Dark matter could become non-relativistic at sufficiently late times to suppress, by free-streaming, the formation of low-mass galactic-scale haloes. This case is, in fact, one of the best-studied alternatives to CDM, known as warm dark matter (WDM). In contrast to WIMPs, these particles have masses of $\mathcal{O}$ (1 keV). A sterile neutrino, included as a part of a model that accounts for neutrino masses and for the baryon asymmetry of the universe, is the favorite WDM candidate (for a recent review see [16]). Another possibility is that dark matter is made of extremely light bosons with a $\mathcal{O}$ (1 kpc) de Broglie wavelength, in which case quantum effects would be relevant on galactic scales (such possibility falls in the category of "fuzzy dark matter"; for a review see [17].)

Although the interactions between dark matter and standard model particles are severely constrained, the interactions among the dark matter particles themselves are not. It is possible that dark matter may have its own rich phenomenology hidden from the ordinary matter. This *hidden dark matter sector* might possess new forces and particles, some of which could be viable dark matter particles that are strongly self-interacting[2]. These collisional particles fall under the category of self-interacting dark matter (SIDM; for a review see [18])[3]. Some of the hidden particles might be light enough that they effectively act as *dark radiation* that prevents the gravitational collapse of dark matter on subgalactic scales[4] (e.g., [20,21]). As mentioned earlier, the CDM hypothesis is only supported to some extent: astronomical data allow a variety of models in which dark matter behaves significantly differently from CDM.

The goal of this paper is to provide a review of the formation, evolution and dynamics of dark matter haloes and subhaloes, as revealed primarily by N-body simulations. Although no account of the properties of haloes based purely on gravitational dynamics can be complete since baryonic processes play a significant role in galaxy formation, and new dark matter physics could also do so, we focus on the standard CDM paradigm of structure formation in part because the subfield of cosmological N-body simulations has historically been developed in this context, and also because the emergence and properties of dark matter structures are most simply understood in the context of CDM. Alternative dark matter models with additional physical ingredients to gravity, albeit appealing, are more complicated. In various parts of this review, we will explore how different assumptions for the nature of dark matter can lead to different predictions from CDM.

2. Formation of Dark Matter Haloes

2.1. Initial Conditions: The Primordial Power Spectrum in the Linear Regime

A theory of structure formation aims to explain the evolution of the universe from a nearly homogeneous initial state, with tiny matter density perturbations, $\delta\rho/\rho$, seeded by inflation, which grow to leave an imprint on the CMB (emitted at the time of recombination, $z \sim 1100$, when $\delta\rho/\rho \sim 10^{-3}$), through the emergence of the self-gravitating dark matter haloes where galaxies form ($\delta\rho/\rho \gg 1$), to the universe we observe today characterized by a web of filamentary large-scale structure ($\delta\rho/\rho \sim 1$).

The starting point is the end of cosmic inflation when dark matter perturbations are predicted to have a nearly scale-invariant power spectrum, $\Delta^2 \propto k^{3+n_s}$, where $\Delta^2(k) = k^3 P(k)/2\pi^2$ is the

---

[2] By strong, we mean that the cross-section for self-interaction is of the order of the nuclear cross-section for visible matter (set by the strong force).
[3] Some SIDM models are motivated by the baryon asymmetry; in these models, dark matter, unlike traditional WIMPs, shares this asymmetry (for a review of asymmetric dark matter see [19]).
[4] In contrast to WDM, the damping of small structures is not due to free-streaming, but to a collisional, Silk-like, damping.

dimensionless power spectrum, and the spectral index $n_s = 0.965$ [1]. The growth of dark matter perturbations in the expanding universe is driven by self-gravity. As long as the perturbations are small, $\delta\rho/\rho \ll 1$ (the *linear regime*), this growth can be described by linear perturbation theory in which each perturbation evolves independently of all others.

Two important processes occur in the linear regime, which modify the primordial power spectrum. The first (known as the Mészáros effect [22]) operates during the period when the energy density in the universe is dominated by radiation: the growth of dark matter perturbations on scales smaller than the horizon stagnates, while super-horizon scales continue to grow. This situation pertains until matter overcomes radiation as the dominant component of the energy density, after which all perturbations grow at the same rate. The transition introduces a characteristic scale in the power spectrum, the size of the horizon at the time of matter-radiation equality. On scales smaller than this, the power spectrum flattens. The second important scale, a cutoff in the power spectrum, is of non-gravitational origin and reflects the particle nature of dark matter. The physical mechanism that imposes this cutoff is model dependent. For thermal relics (like many WIMP models and certain types of WDM), the mechanism is free-streaming, a form of collisionless (Landau) damping, whose scale is given by the horizon size at the epoch when the dark matter particles become non-relativistic; the more massive the particle, the earlier this epoch, and thus the smaller the (co-moving) free-streaming scale is [5], $k_{fs} = 2\pi/l_{fs}$. This is the best-known cutoff mechanism, which has been traditionally used to classify dark matter into three categories (where m_χ denotes the mass of the particle): cold [6] ($m_\chi \sim 100$ GeV, $k_{fs} \sim 2.5 \times 10^6$ h/Mpc); warm ($m_\chi \sim 1$ keV, $k_{fs} \sim 3.8$ h/Mpc); and hot ($m_\chi \sim 30$ eV, $k_{fs} \sim 0.3$ h/Mpc).

A different type of damping is collisional damping, which prevents the gravitational collapse of small structures, resulting in an effective cutoff in the power spectrum. An example is kinetic coupling of WIMPs, which effectively keeps dark matter coupled to the photon-baryon plasma until the universe cools enough that the interactions become inefficient, damping perturbations beyond a scale in the range $(2.6 \times 10^5$–$1.2 \times 10^8)$ h/Mpc [24]. Another example is collisional damping due to interactions between dark matter and relativistic particles in the early universe (either photons or neutrinos, e.g., [25,26], or, in non-standard models, dark radiation in *hidden dark sector* models, e.g., [20,21]). The relativistic particles create an effective radiation pressure that counteracts the gravitational collapse, driving oscillations in the density perturbations, akin to the well-known baryon acoustic oscillations (BAOs), but on much smaller scales; by analogy they are called dark acoustic oscillations, DAOs [7]. Once the universe cools down, the dark matter decouples from the relativistic particles, imprinting a characteristic scale (the size of the sound horizon at the time of decoupling) in the power spectrum, followed by a Silk-like damping cutoff.

The main features of the clustered dark matter distribution during the linear regime are illustrated in Figure 1. On the largest scales, not affected by the Mézáros effect, the power spectrum is nearly scale-invariant, $\Delta^2 \propto k^{3+n_s}$; on smaller scales it bends to increasingly shallower slopes. For CDM (black line), the power spectrum remains featureless well below galactic scales. For reference, a dark matter halo today hosting a typical dwarf galaxy would have a mass $\sim 10^{10}$ M$_\odot$, roughly corresponding to a (co-moving) wavenumber ~ 12 h/Mpc[8]. Measurements of galaxy clustering on scales larger than individual galaxies, together with constraints from the flux spectrum of the Ly-α forest (e.g., [28]) constrain the power spectrum to be like CDM to the left of the hashed area in Figure 1. On smaller

---

[5] The (co-moving) free-streaming scale is given by: $l_{fs} = 2ct_{nr}/a_{nr}\left[1 + \ln(a_{eq}/a_{nr})\right]$, where t_{nr} is the age of the universe at the time when the dark matter particles become non-relativistic (at a temperature $3k_B T_{nr} \sim m_\chi c^2$); $a_{nr} = 1/(1 + z_{nr})$ is the scale factor at t_{nr} ($a \propto t^{1/2}$ in the radiation-dominated era); and a_{eq} is the scale factor at the time of matter-radiation equality.
[6] For cold particles, we have assumed CDM WIMPs, which requires taking into account the kinetic decoupling temperature and epoch; specifically, we took Equation (43) of [23].
[7] Please note that acoustic oscillations are also present in WIMP-CDM models (e.g., [27]), but they occur at much smaller scales than in relevant hidden dark sector models where they can be of galactic scale.
[8] We use $M = 4\pi/3\bar{\rho}(\pi/k)^3$, where $\bar{\rho}$ is the mean dark matter density today.

scales the power spectrum could have a damping cutoff due to either collisionless (as in WDM models; red line) or to collisional (as in models with DAOs; blue line) processes. We have not included the cutoff characteristic of *fuzzy dark matter* models, but we note that it is also oscillatory like the DAOs models (but due to quantum rather than collisional effects; see e.g., Figure 2 of [29]).

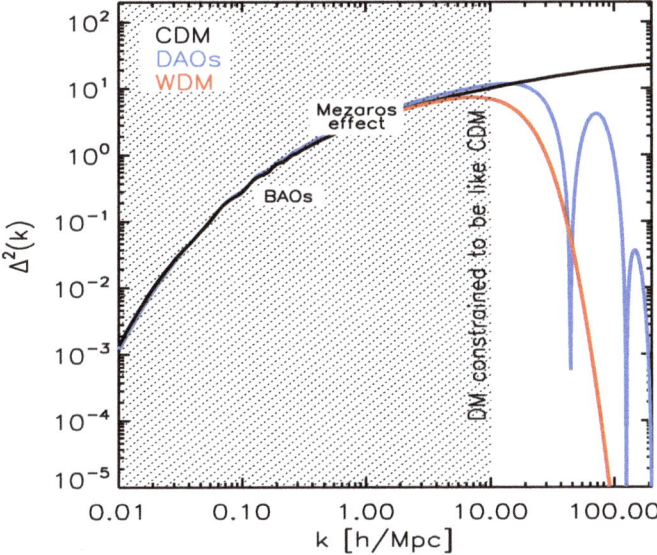

Figure 1. Dimensionless linear dark matter power spectrum in different dark matter models. In the current paradigm, cold dark matter (CDM), the power spectrum keeps on rising to well below subgalactic scales. Alternative models such as warm dark matter (WDM) or interacting dark matter (DAOs) have a cutoff at or slightly below galactic scales, which determines the abundance and structure of small-mass dark matter haloes and subhaloes and the galaxies within. In the black hashed area, the dark matter is constrained by the observed large-scale distribution of galaxies (e.g., [30,31]) and the Ly-α forest constraints on WDM [28] to behave as CDM. Figure adapted from [32].

As long as the dark matter perturbations remain linear ($\delta\rho/\rho \ll 1$), they grow at a rate that does not depend on their co-moving scale, $\Delta^2(k;t) \propto D^2(t)$, where $D(t)$ is the growth factor, which depends only on the mean density of matter and dark energy (see e.g., [33]). Once the density contrast is no longer small ($\delta\rho/\rho \sim 0.1$), perturbation theory breaks down since gravity couples perturbations on different scales and their evolution can no longer be calculated as independent modes.

2.2. The Non-Linear Regime: N-Body Simulation Methods

To follow the evolution of dark matter density perturbations beyond the linear regime, several approaches are possible depending on the problem of interest. (i) High order perturbation theory which can be used to study the quasi-linear regime ($\delta\rho/\rho \lesssim 1$), particularly in a modern reformulation such as the *Effective field theory of large-scale structure* [34,35]. (ii) Analytical models with simplified assumptions for the growth, turnaround (i.e., decoupling from the expansion of the universe), collapse and virialization (i.e., the formation of a gravitationally self-bound structure) of individual perturbations. The best-known examples are the *Spherical collapse* [36] and *Ellipsoidal collapse* [37] models which link a primordial perturbation to the final equilibrium configuration: the dark matter halo. (iii) The *halo model* (for a review see [38]), which combines the analytical models in (ii) with the assumption of a Gaussian density field and can be used to compute the abundance of virialized haloes as a function of their mass (the halo mass function); together with a model for the dark matter distribution within haloes, it can be used to model the non-linear dark matter power spectrum on all scales. (iv) Models based on the *Stable clustering hypothesis* ([39,40]), which assumes that the number of neighboring dark matter particles within a fixed physical separation remains constant, and can be used to study the deeply non-linear regime; a recent reformulation in phase space has been shown to be a promising alternative to the halo model [41–43]. (v) Numerical *N-body simulations*, which solve *ab initio* the gravitational evolution in phase space of a distribution of N particles sampled from an initial power spectrum. This is the most general and powerful approach to study the clustering evolution on all scales and is the focus of this review. (vi) Techniques that avoid the particle discretization inherent in N-body simulations by following the phase-space distribution function directly [44]. These are particularly useful to study evolution from truncated power spectra such as for hot or warm dark matter for which standard N-body techniques suffer from artificial fragmentation [45].

In the case of classical, non-relativistic, collisionless particles, i.e., CDM, N-body simulations follow the evolution of the dark matter phase-space distribution function, $f(\vec{x},\vec{v};t)$, which in principle is given by the collisionless Boltzmann equation coupled with the Poisson equation for the gravitational field, $\Phi(\vec{x})$ (a combination known as the Vlasov-Poisson equation):

$$\frac{df}{dt} = \frac{\partial f}{\partial t} + \sum_i v_i \frac{\partial f}{\partial x_i} - \sum_i \frac{\partial \Phi}{\partial x_i} \frac{\partial f}{\partial v_i} = 0 \quad (1)$$

$$\rho_\chi(\vec{x};t) = \int f(\vec{x},\vec{v};t) d^3\vec{v} \quad (2)$$

$$\nabla^2 \Phi(\vec{x}) = 4\pi G \rho_\chi(\vec{x}) \quad (3)$$

where d/dt is the Lagrangian derivative. Cosmological N-body simulations[9] solve this equation in an expanding universe using a co-moving reference frame (with the expansion included explicitly through the solution of the Friedmann equations for the scale factor), discretizing the distribution function as an ensemble of N phase-space elements or "particles", $\{\vec{x}_i,\vec{v}_i\}$, with $i=1,...,N$. Since the collisionless Boltzmann equation implies that the phase-space distribution remains constant in time along any trajectory $\{\vec{x}(t),\vec{v}(t)\}$, the distribution obtained by following the N particles from initial conditions sampled from the phase-space distribution at $t=0$, constitute a representative Monte-Carlo sampling of the distribution function at any subsequent time, t. The N particles are thus a statistical representation of the coarse-grained[10] distribution function:

$$\tilde{f}(\vec{x},\vec{v}) \sim \sum_i m_i W(|\vec{x}-\vec{x}_i|;\varepsilon)\delta^3(\vec{v}-\vec{v}_i); \quad \frac{d\tilde{f}}{dt}=0 \quad (4)$$

$$\tilde{\rho}(\vec{x}) = \int \tilde{f}(\vec{x},\vec{v})d^3\vec{v} \sim \sum_i m_i W(|\vec{x}-\vec{x}_i|;\varepsilon) \quad (5)$$

$$\tilde{\Phi}(\vec{x}) = \int g(\vec{x}-\vec{x}')\tilde{\rho}(\vec{x}')d^3\vec{x}' \quad (6)$$

where m_i is the mass of the simulation particle, $\delta^3(\vec{v}-\vec{v}_i)$ is the DiRAC delta function in 3D, W is a *kernel density* with a softening length ε[11], introduced to obtain a smooth density field from the set of N discrete particles; i.e., the kernel effectively models each simulation particle as an extended mass distribution[12]; finally, the last equation for the potential is the general solution to Poisson's equation as

---

[9] For a review see e.g., Section 3 of [46].
[10] By this we mean an average of the fine-grained distribution function in the collisionless Boltzmann equation over the scales resolved in the simulation, typically several times the interparticle separation.
[11] In principle, each particle can have an individual softening, see e.g., Section 4 of [47].
[12] The introduction of a softening scale in the density (or potential) suppresses gravitational two-body large-angle scatterings which are artificial for an approximately continuous dark matter density distribution.

a convolution of the density field with a suitable Green's function[13]. Since each simulation particle represents a region of phase space containing a very large number, m_i/m_χ, of real dark matter particles, the information in an N-body simulation is always incomplete, limited by the phase-space resolution and the softening length.

With the discretization method employed in an N-body simulation, calculating the evolution of the phase-space distribution is reduced to following self-consistently the dynamics of a system of N particles (usually in terms of the Hamiltonian of the system in the co-moving frame) according to the potential derived from the particle distribution. Modern codes used to solve this problem employ efficient methods for computing the gravitational potential and integrating the Hamiltonian system forward in time. Early cosmological simulation codes used the particle-mesh (PM) technique in Fourier space (e.g., [48,49]) or direct integration of the N^2 interactions (e.g., [50]). The former is limited in resolution by the size of the mesh while the latter is limited by speed. These two shortcomings can be overcome by combining both techniques in the P^3M method (e.g., [51,52]), in which the long-range forces acting on a particle are calculated on a PM grid and the short-range forces by direct N^2 summation. An alternative approach is the hierarchical tree method [53] in which an octree is used to divide the volume recursively into cubic cells and increasingly coarse cells are used to compute the forces on a particle at increasingly large distances. The most widely used cosmological simulation code is GADGET-2 [54], which uses the treePM algorithm, whereby short-range forces are computed with the tree method and long-range forces with Fourier techniques[14].

If dark matter cannot be treated as CDM, then the fundamental equations may need to be modified. For models that only deviate from CDM because of a cutoff in the initial power spectrum (such as hot or warm dark matter and certain DAO models), the N-body Equations (1)–(6) and methods used for CDM are still valid as long as the dark matter behaves as a collisionless, classical system, and the simulation starts well after the dark matter particles have become non-relativistic; all that is needed is a modification of the initial conditions (see Section 2.3 below). On the other hand, if dark matter is non-relativistic but no longer collisionless, like in SIDM, then the collisionless Boltzmann equation needs to be replaced by the full collisional Boltzmann equation, which has an extra term (the collisional operator) in the right-hand-side of Equation (1), to account for the effect of dark matter collisions according to a self-scattering cross-section. It is possible to incorporate this new term within the Monte-Carlo approach of traditional N-body simulations by adding "collisions" between each simulation particle and its immediate neighbors in a probabilistic way that reflects the effective scattering rate given by the cross-section (e.g., [55–59]; see Appendix A of [58] for a detailed derivation). An alternative to the N-body approach is the "gravothermal fluid" approximation [60], which considers an SIDM dark matter halo as a self-gravitating, spherically symmetric, ideal gas with an effective thermal conductivity (related to the self-scattering cross-section, see e.g., [61]). Although this approach is restricted, it provides physical insight into the evolution of SIDM haloes, and a degree of validation of SIDM N-body simulations. Finally, if quantum effects are important for the dark matter fluid, then the Vlasov-Poisson equation needs to be replaced by the Schrödinger–Poisson equation, whose solution requires numerical methods quite distinct from the N-body approach (e.g., [62,63]).

2.3. The Non-Linear Regime: Initial Conditions and The Emergence of the Cosmic Web

The techniques of Section 2.2 can be used to integrate forward in time a particle distribution starting from an initial state, the initial conditions, usually taken to be in the linear regime described by perturbation theory. The basic techniques for generating general initial conditions were laid out in [14,64] and have been refined over the years (e.g., [65,66]; for a review see [67] or Appendix C1.1.4

---

[13] In Fourier space, Equation (6) is simply a multiplication $\hat{\Phi}(\vec{k}) = \hat{g}(\vec{k})\hat{\rho}(\vec{k})$.
[14] For a review of the force computation methods see Section 3.5 of [46].

of [68]). They provide a particle realization with the statistical properties of the linear dark matter density field described by the power spectrum. In general the procedure can be divided into two steps:

(i) create a realization of an unperturbed cube of side L by distributing N particles homogeneously in a lattice or in a glass-like configuration[15] to avoid imprinting a grid-like pattern in the simulation.
(ii) perturbations of wavelength λ down to the Nyquist frequency of the particle distribution are represented by plane waves of spatial frequency in Fourier space, $k = 2\pi/\lambda$, whose amplitudes and phases are drawn at random from a Gaussian distribution with variance proportional to the desired linear power spectrum. The density field and its gravitational potential in real space are then obtained by an inverse Fourier transform. Using the Zel'dovich approximation [70], or the more accurate second-order Lagrangian perturbation theory (e.g., [71]), these fields are used to compute the displacements needed to transform the uniform N-particle distribution in part (i) into a distribution that has the desired power spectrum.

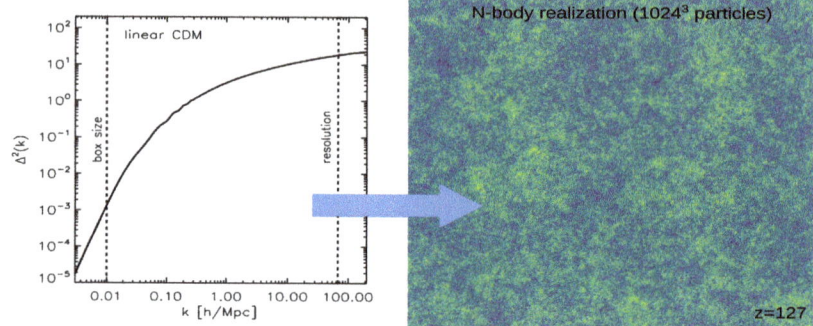

Figure 2. Illustration of the initial conditions for an N-body simulation. *Left:* the dimensionless linear CDM power spectrum. The vertical dashed lines mark the modes corresponding to the maximum and minimum scales that can be represented in the initial conditions: the fundamental mode, $2\pi/L$, and the Nyquist mode, π/d, where L and d are the cube length and interparticle separation, respectively. *Right:* a realization of the dark matter density field generated from the power spectrum on the left at redshift $z = 127$ using $N = 1024^3$ particles in a cosmological cube of co-moving side, $L = 40$ Mpc/h. The code MUSIC [65] was used to generate the particle distribution and the Pynbody package [72] to create the image.

An illustration of the end result of this procedure is shown in Figure 2. The initial conditions generator, MUSIC [65], was used to construct the particle distribution on the right, which is a statistical realization of the CDM linear power spectrum shown on the left. The main limitations for a cosmological simulation are already set in the initial conditions: the maximum length scale that can be simulated is determined by the (co-moving) side of the computational cube[16], and the minimum length scale that can represented in the initial conditions is set by the Nyquist frequency of the particle distribution[17]. The choice of cube length and particle number depends on the science goal of the simulation and on the computing resources available. We will come back to this point below.

---

[15] The particles are initially placed at random in the simulation cube and then left to evolve under a repulsive force by reversing the sign of the gravitational force until they reach an equilibrium configuration that has no discernible grid pattern [69].
[16] A sufficiently large volume is needed to sample large-scale modes that remain approximately linear during the simulation where power is transferred from large to small scales; without appropriate large-scale sampling, the clustering is no longer accurate once perturbations on the scale of the cube become non-linear.
[17] In practice, power below the Nyquist frequency is generated non-linearly so the resolution of the simulation is not limited by the Nyquist frequency but rather by the gravitational softening scale, ε.

The procedure illustrated in Figure 2 for CDM can be readily applied to other dark matter models with different initial power spectra. In fact, in models where dark matter only behaves differently from CDM at very early times, e.g., in thermal-relic WDM models, it is the different initial conditions (the lack of power on small scales in WDM relative to CDM in the linear regime) that gives rise to the main differences between these models (since the residual thermal motions in WDM models of interest are negligible (see e.g., [73]). In models with a truncated initial power spectrum, the subsequent evolution is affected by particle discreteness in the reconstruction of the density field which introduces an irreducible (shot-noise) power. This results in spurious clustering on scales close to the cutoff length [74] that requires careful treatment to either remove small-scale artificial clumps [75] or avoid their formation altogether by using non-standard simulation techniques [45,76].

Once the initial conditions are generated, an N-body simulation is performed, most commonly in a computational cube with periodic boundary conditions, to follow the evolution of the density and velocity fields into the non-linear regime across all resolved scales. An example, the Millennium II simulation [77], is illustrated in Figure 3. The left set of panels shows the projected dark matter density distribution at various *snapshots* corresponding to the redshifts shown at the top right of each panel. The emergence of the cosmic web, the result of gravitational clustering, is apparent, with its now familiar pattern of filaments over a range of scales surrounding voids. The right panel shows the evolution of the power spectrum at the same snapshots (solid lines). The hierarchical onset of non-linear structure, from small to large scales is clearly apparent by reference to the linear power spectrum (grey lines).

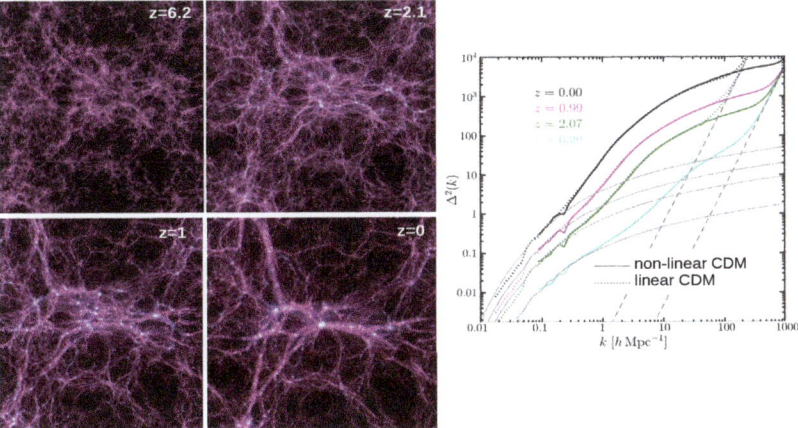

Figure 3. Emergence of the cosmic web. *Left:* evolution of the (projected) dark matter density field in a slab of length $L = 100$ Mpc/h and thickness 15 Mpc/h from the Millennium-II simulation [77]. The redshift corresponding to each snapshot is shown on the top right. *Right:* The dimensionless dark matter power spectrum (solid lines) at the redshifts shown on the left. For comparison, also shown are: the linear power spectrum (thin grey lines) and the non-linear power spectrum for the lower resolution but larger scale (500 Mpc/h) Millennium I simulation (in dotted lines; [4]). The dashed lines show the Poisson noise limit for the Millennium I (left) and Millennium-II (right) simulations. Figure adapted from [77][18].

---

[18] Reproduced from Michael Boylan-Kolchin et al. Resolving cosmic structure formation with the Millennium-II Simulation. MNRAS (2009) 398 (3): 1150-1164, doi: 10.1111/j.1365-2966.2009.15191.x. By permission of Oxford University Press on behalf of the Royal Astronomical Society. For the original article, please visit the following u. This figure is not included under the CC-BY license of this publication. For permissions, please email: journals.permissions@oup.com.

The first CDM cosmological N-body simulations in the 1980s [14,78] already contained all the relevant physical processes of gravitational clustering for collisionless dark matter, but were computationally limited; they could follow the evolution of only $\mathcal{O}(10^4)$ particles. In the decades since then, the tremendous improvement in computational capabilities has been such that cosmological ($L \gtrsim 100$ Mpc/h) simulations with $\mathcal{O}(10^9)$ particles are routinely performed[19], and the most expensive simulations to date have reached the one trillion particle milestone [80].

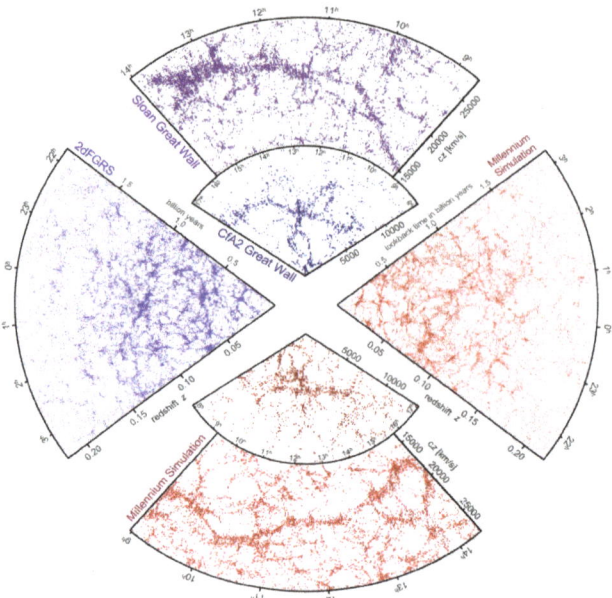

Figure 4. The galaxy distribution in various redshift surveys and in mock catalogues constructed from the Millennium simulation [4]. The small slice at the top shows the CfA2 "Great Wall" [81], with the Coma cluster at the center. Just above is a section of the Sloan Digital Sky Survey in which the "Sloan Great Wall" [82] is visible. The wedge on the left shows one half of the 2-degree-field galaxy redshift survey [83]. At the bottom and on the right, mock galaxy surveys constructed using a semi-analytic model applied to the simulation [84] are shown, selected to have geometry and magnitude limits matching the corresponding real surveys. Adapted from [85].

To compare the simulations to astronomical data it is necessary to make a correspondence between dark matter haloes and the galaxies that would form within them. In the earliest simulations, galaxies were identified with high peaks of the suitably filtered density field, an assumption known as "biased galaxy formation" [14,86]. Physically based models of galaxy formation that could be grafted onto N-body simulations were developed in the early 1990s [87]. These are known as "semi-analytic models" because they encapsulate the relevant physical models in a set of coupled differential equations that are solved numerically. These equations assume spherical symmetry and describe the cooling of gas, the formation of stars, chemical evolution, the growth and merging of central supermassive black holes and feedback effects arising from energy injected into the gas during the course of stellar evolution and by active galactic nuclei triggered by accretion of gas onto the central black hole. The model is applied at every stage of the gravitational evolution of the merging hierarchy of haloes, described by a merger tree (see Section 3.1). Semi-analytic models have been extremely successful in linking the distribution

---

[19] For a review of the state of cosmological simulations circa 2012 see [79].

of dark matter computed in an *N*-body simulation to the observed universe [88–91] and have become
very sophisticated in predicting visible galaxy properties over a large range of wavelengths (e.g., [92]).
An example based on the Millennium simulation is shown in Figure 4.

Dark matter *N*-body simulations are the cornerstone of the current understanding of how galaxies form and evolve and, as illustrated in Figure 4, have been very successful in explaining the large-scale structure of the universe [85]. The latter accomplishment is non-trivial and demands certain conditions about the nature of dark matter. For instance, already in the 1980s, light neutrinos were ruled out as the dominant component of dark matter by their incompatibility with the observed large-scale structure [93], thus demonstrating the potential of *N*-body simulations to test models for the nature of the dark matter. By contrast, the fact that CDM matched the observations available at the time remarkably well contributed greatly to its establishment as the standard model of cosmogony [14]. By now, it is firmly established that whatever the dark matter is, it must behave as CDM on large scales (see Figures 1 and 4). It is important to recognize, however, that a wide range of dark matter candidates behave just as CDM on large scales and thus are also allowed by the large-scale structure data, as we discussed in Sections 1 and 2.1. In this sense, the success of the CDM model in explaining the large-scale structure of the universe is shared by allowed WDM, SIDM and *fuzzy dark matter* models.

2.4. The Structural Properties of Dark Matter Haloes

As a consequence of gravitational instability, the tiny density perturbations present when the CMB was emitted grow in time, eventually separating from the expansion of the universe and becoming self-gravitating bound structures known as dark matter haloes. This process of forming *virialized* haloes can be understood from the simple spherical collapse model [36]. Haloes become increasingly more massive with time, *smoothly* by accreting mass from their surroundings or *merging* with other, smaller haloes. The latter thus become subhaloes, which is the topic of Section 3. Although the large-scale environment and spatial clustering of dark matter haloes are clearly relevant, here we focus on the abundance (halo mass function) and internal structure of dark matter haloes. These properties are the most useful when attempting to differentiate dark matter models. Currently, however, the halo mass function and halo structure on the key subgalactic scales are only weakly constrained observationally. We should also note that not all dark matter is contained within haloes. The fraction of *unclustered* dark matter is naturally a strong function of time reflecting the growth of collapsed objects by hierarchical clustering. Even today at most $\sim 20\%$ is expected to be unclustered in the CDM model [94]. A recent update of this work (applied also to WDM) puts the fraction even lower, at the percent level [95].

Definition of a halo—Because of the dynamic nature of haloes and their lack of spherical symmetry, precisely defining the boundary of a halo, and thus its mass, is, to some extent, arbitrary [14,96,97]. A variety of definitions exist in the literature with the most common ones being: (i) the FOF (friends-of-friends) mass, defined as the mass of the set of particles that are linked together by a percolation scale, defined by a linking length, $b \sim 0.2$ in units of the mean interparticle separation [14]; (ii) a spherical overdensity mass, M_Δ, contained within a sphere centered on the halo (with the center placed at the minimum of the gravitational potential of the halo), with a radius given by the spherical collapse model, whereby the collapsed region that defines a halo contains an average density $\Delta(z)$ times the critical density for closure [98]. The overdensity, $\Delta(z)$, is a redshift-dependent function of cosmology [99,100], but for the Einstein-de Sitter cosmology, $\Delta \sim 178$ at all times; (iii) the viral mass, defined with $\Delta = 200$, which early simulations identified as the radius that separates the region of the halo that is in dynamical equilibrium from the surrounding region that is still collapsing [98]. Given the simplicity of the latter, its relation to dynamical equilibrium, and its connection with the Einstein-de Sitter spherical collapse overdensity, the radius, r_{200}, and the enclosed mass, M_{200}, are widely used in the field as the boundary and mass of dark matter haloes, respectively.

The halo mass function—The mass function of dark matter haloes, i.e., the number density of haloes of different mass, has been characterized quite precisely in the last couple of decades by *N*-body simulations (e.g., [77,101–106]), and is now well determined over the full range of epochs and masses relevant to galaxy formation, from $\mathcal{O}(10^8\ M_\odot)$ dwarf-size haloes to $\mathcal{O}(10^{15}\ M_\odot)$ cluster-size haloes. The number density of haloes per unit mass scales as:

$$\frac{dn}{dM} \propto M^\alpha, \text{ where } \alpha \sim -1.9 \text{ for low masses,} \tag{7}$$

with an overall normalization that correlates with the large- scale environment, with denser environments having a larger halo abundance [78,107].

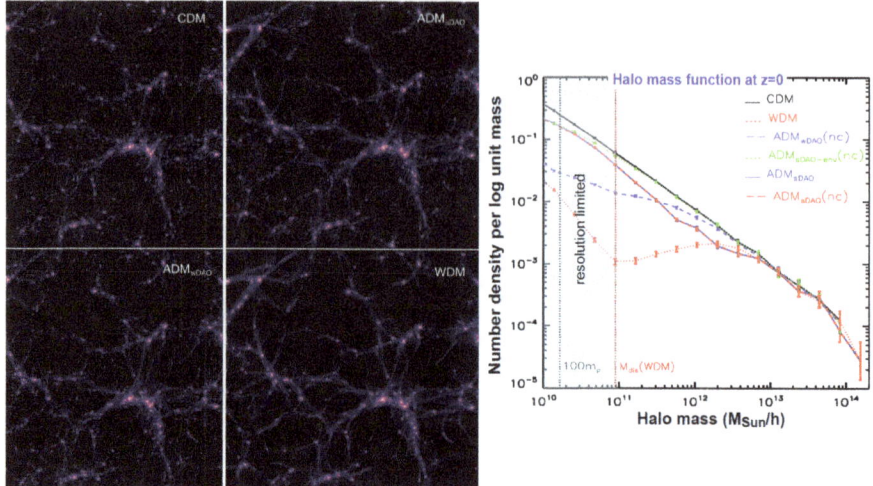

Figure 5. Halo mass function for different dark matter models (adapted from [20]). *Left:* The large-scale dark matter distribution in a slab of a 64 Mpc/h cube in different cosmologies: CDM and WDM in the top left and bottom right, respectively; two interacting dark matter models in the other two panels. *Right:* The halo mass function at $z = 0$ for the models on the left. The transparent light blue region marks the resolution limit of the simulations. The cutoff in the primordial linear power spectrum of the non-CDM models results in a lower abundance of low-mass haloes, visible in the panels on the left and quantified in the halo mass function on the right.

The shape of the halo mass function is reasonably well understood from statistical arguments based on the properties of the initial Gaussian density field (described by the power spectrum) and the gravitational collapse of density perturbations into virialized haloes as modeled by the spherical collapse model. These arguments are the basis of the Press–Schechter [108] and extended Press–Schechter (EPS) formalisms [109,110], which provide a good fit to the simulation results, particularly if the assumption of spherical symmetry for the collapse of overdensities is replaced by the assumption of ellipsoidal collapse [37]. At the small-mass end, the power-law form of the mass function is broken at a mass that depends on the nature of the dark matter. For example, a cutoff in the power spectrum, whether due to relativistic, collisional, or quantum effects, introduces a corresponding cutoff in the halo mass function. The mass function for WDM (e.g., [45,75,111,112]) and interacting dark matter (e.g., [20,113,114]) models have now been fairly well characterized with N-body simulations (with appropriate corrections for spurious fragmentation due to particle discreteness near the cutoff [74,75]). The Press–Schechter approach can be readily extended to provide a reasonable approximation to the halo mass function in these models as well (e.g., [115–117]).

Figure 5 provides an example of the effect of a cutoff in the primordial power spectrum on the halo mass function relative to CDM. The "atomic dark matter model", ADM$_{sDAO}$ [118], is an example of a model with dark acoustic oscillations, while WDM is a well-known example of the free-streaming effect (see Figure 1). These two models give rise to qualitatively different types of suppression in the abundance of low-mass haloes. The halo mass function thus contains a signature of the type of primordial power spectrum cutoff.

The inner structure of dark matter haloes—One of the remarkable findings of the past few decades is that the spherically averaged mass density profiles of dark matter haloes in dynamical equilibrium have a nearly universal form which is independent of halo mass, initial conditions[20] and cosmological parameters. These profiles are quite well described by a very simple functional form with just two parameters, the so-called Navarro-Frenk-White (NFW) profile [119,120]:

$$\rho_{\text{NFW}}(x) = \frac{\rho_s(c)}{cx(1+cx)^2}, \tag{8}$$

where $x = r/r_{200}$, and $c = r_{200}/r_s$ is the concentration of the halo; r_s is the scale length, which, for the NFW profile, coincides with the radius, r_{-2}, at which the logarithmic slope of the profile is equal to -2; finally $\rho_s(c) = \delta_c \rho_{\text{crit}}$, where $\rho_{\text{crit}} = 3H^2/8\pi G$ is the critical density of the universe, and:

$$\delta_c = \frac{\Delta c^3}{3K_c(c)} \tag{9}$$

where $K_c(c) = \ln(1+c) - c/(1+c)$. Although recent simulations have shown that a different profile, the so-called Einasto profile, which has three parameters, is a slightly better fit to simulations [121], the asymptotic behavior of the NFW profile for $\rho(r \to 0) \sim r^{-1}$ remains a remarkably good approximation to the inner structure of CDM haloes (see top left panel of Figure 6). The physical origin of this divergent *cusp* and the remarkably universal profile shape are not fully understood. It has been argued from N-body simulations of the early stages of structure formation that the first CDM haloes to form, i.e., those near the free-streaming scale of CDM have a steeper cusp than NFW, $\sim r^{-1.5}$, which is subsequently flattened after a few mergers to $\sim r^{-1}$ [122–125]. More recent simulations which follow the growth of the first mini-haloes all the way to the present, seem to confirm this, suggesting that the ubiquitous $\sim r^{-1}$ slope develops at some point after the formation of the halo and remains until $z = 0$.

Halo concentration—The remarkable simplicity of the NFW halo density profile goes beyond Equation (8): the profile is, in fact, fully specified by a single parameter, halo mass, because the concentration (or scale radius) correlates with mass, with lower-mass haloes generally being more concentrated than higher-mass haloes [120,126–136]. This correlation is ultimately a consequence of the hierarchical nature of structure formation by gravitational instability from a primordial power spectrum that as in CDM, grows monotonically towards small scales (see Figure 1). Lower-mass haloes form earlier, when the mean density of the universe is larger, and larger-mass haloes form later when the mean density of the universe is lower. The inner regions of haloes collapse first [137] and their density reflects the mean density of the universe at that time. Hence, smaller-mass haloes are more concentrated than larger-mass haloes. Furthermore, since (at least for CDM), the power spectrum, $\Delta^2(k)$, becomes flatter at larger k (due to the Mészáros effect), haloes with a wide range of masses collapse in a short time interval and this flattens the concentration-mass relation at low masses. Models based on these simple arguments explain, at some level, the concentration-mass relation measured in simulations [120,131,135], and also provide a natural connection between the mass assembly of haloes in time and their structure as a function of radius: each radial shell has the characteristic density of the cosmic background density at the time when it collapses [132]. Random deviations around the mean collapse time expected for haloes of a fixed mass and a stochastic merger history introduce scatter in the concentration-mass relation.

---

[20] This is only true if, on the scales of interest, the primordial power spectrum grows monotonically towards large k.

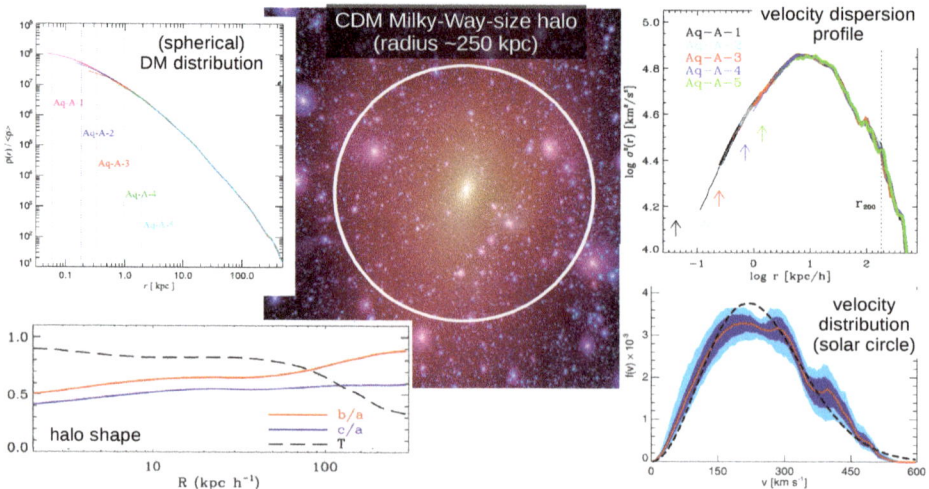

Figure 6. The structure of CDM haloes. The different panels show several characteristics of the spatial (left) and dynamical (right) structure of a Milky Way-size CDM halo ($M_{200} \sim 1.8 \times 10^{12}$ M$_\odot$; $r_{200} \sim 250$ kpc) from the Aquarius project [138]. The top panels show the spherically averaged radial density (left; [138][21]) and velocity dispersion (right; [121][22]) profiles, which are nearly universal for haloes in dynamical equilibrium. The bottom panels show the halo shape (left: moment of inertia axis ratios, and triaxiality: $T = (a^2 - b^2)/(a^2 - c^2)$; [139][23]) and local dark matter velocity distribution near the solar circle: 2 kpc $< r <$ 9 kpc (right; [140][24]).

Halo velocity distribution—For a spherical, self-gravitating, collisionless system in dynamical equilibrium, with radial density profile, $\rho(r)$, the Boltzmann equation reduces to the well-known Jeans equation [141]:

$$\frac{d(\rho \sigma_r^2)}{dr} + 2\frac{\beta}{r}\rho\sigma_r^2 = -\rho \frac{d\Phi}{dr}, \tag{10}$$

where $\Phi(r)$ is the gravitational potential related to the density by Poisson's equation, $\sigma_r(r)$ is the radial velocity dispersion profile, and $\beta(r) = 1 - (\sigma_\theta^2 + \sigma_\phi^2)/\sigma_r^2$ is the velocity anisotropy profile, which quantifies the degree of anisotropy of the particle orbits in the dark matter halo. Haloes tend to be isotropic only in their centers and are radially anisotropic in their outskirts [121,142], with a velocity anisotropy that is related to the logarithmic slope of the density profile [143]. The velocity structure of dark matter haloes in equilibrium is thus intimately linked to their spatial distribution (see top right panel of Figure 6), which is strikingly evident in the so-called *pseudo-phase-space density*, $Q \equiv \rho/\sigma^3$, where $\sigma^2 = \sigma_r^2 + \sigma_\theta^2 + \sigma_\phi^2$ is the square of the 3D velocity dispersion. This quantity has been found to be an almost perfect power law, $Q \propto r^{-1.875}$, over several orders of magnitude [121,144], and is in

---

[21] Reproduced from Volker Springel et al. The Aquarius Project: the subhaloes of galactic haloes. MNRAS (2008) 391 (4): 1685–1711, doi: 10.1111/j.1365-2966.2008.14066.x. By permission of Oxford University Press on behalf of the Royal Astronomical Society. For the original article, please visit the following https://academic.oup.com/mnras/article/391/4/1685/1747035.

[22] Reproduced from Julio Navarro et al. The diversity and similarity of simulated cold dark matter haloes. MNRAS (2010) 402 (1): 21–34, doi: 10.1111/j.1365-2966.2009.15878.x. By permission of Oxford University Press on behalf of the Royal Astronomical Society. For the original article, please visit the following https://academic.oup.com/mnras/article/402/1/21/1028856.

[23] Reproduced from Carlos Vera-Ciro et al. The shape of dark matter haloes in the Aquarius simulations: evolution and memory. MNRAS (2011) 416 (2): 1377–1391, doi: 10.1111/j.1365-2966.2011.19134.x. By permission of Oxford University Press on behalf of the Royal Astronomical Society. For the original article, please visit the following https://academic.oup.com/mnras/article/416/2/1377/1061105.

[24] Reproduced from Mark Vogelsberger et al. Phase-space structure in the local dark matter distribution and its signature in direct detection experiments. MNRAS (2009) 395 (2): 797–811, doi: 10.1111/j.1365-2966.2009.14630.x. By permission of Oxford University Press on behalf of the Royal Astronomical Society. For the original article, please visit the following https://academic.oup.com/mnras/article/395/2/797/1747020. The figures mentioned in footnotes 21–24 are not included under the CC-BY license of this publication. For permissions, please email: journals.permissions@oup.com

remarkable agreement with the self-similar solution for infall onto a point mass in an Einstein-de Sitter universe [145]. The radial behavior of Q is a manifestation of the nearly universal structure of dark matter haloes, which connects both their spatial and kinematical distributions.

Besides having anisotropic particle orbits, CDM haloes have a non-Maxwellian velocity distribution. This may be seen in a simple way by noting that for a purely isotropic spherical system ($\beta = 0$), the full velocity distribution function of the halo depends only on the specific energy, $f(\mathcal{E})$, and is fully given by the Eddington formula [146]:

$$f(\mathcal{E}) = \frac{1}{\sqrt{8}\pi^2} \int_0^{\sqrt{\mathcal{E}}} du \frac{d^2\rho}{d\Psi^2}(r(\Psi(u))), \quad (11)$$

where $u = \sqrt{\mathcal{E} - \Psi}$ and $\mathcal{E}$ and $\Psi(r)$ are the (negative) specific energy and gravitational potential, respectively. If haloes were spherical and isotropic, their velocity distribution would be given purely by the NFW density profile through Equation (11). Even at this level of approximation we can see that haloes would not be described by a Maxwellian distribution function (only the singular isothermal sphere, $\rho \propto r^{-2}$, results in a Maxwellian distribution in Equation (11)). In other words, haloes are non-Maxwellian simply by virtue of their NFW density profiles. In fact, CDM haloes in simulations have local[25] velocity distributions that show significant departures from Maxwellian, related to the dynamical assembly of the halo. The features that appear in the local velocity distribution are unique for a particular halo, and retain the memory of its assembly history ([140]; see bottom right panel of Figure 6).

Halo shapes—Although to first order, CDM haloes are well described by the spherical NFW profile, they are, in fact, triaxial [78]. In general, CDM haloes are prolate in the inner parts and oblate in the outskirts ([78,139,147,148]; see bottom left panel of Figure 6). This radial dependence of halo shape seems to be related to the assembly history of the halo within the cosmic web: the central regions, being assembled at earlier times through accretion along narrow filaments, end up being more prolate, while the outskirts, more recently assembled by less anisotropic accretion end up more oblate [139,149]. Thus, the halo shape profile at $z = 0$ carries some memory of its assembly history. Overall, more massive haloes are more aspherical than lower mass haloes [150,151] because in the hierarchical CDM model, the more massive haloes form more recently and thus their shapes retain memory of the most recent accretion event [152].

Dependence on the nature of the dark matter—There are significant changes in the structure of dark matter haloes if the dark matter particles do not behave as CDM. In currently allowed models, these deviations are mostly confined to the central regions, i.e., is within the scale radius, r_s. By introducing a new scale in the process of structure formation, be it in the initial conditions (e.g., a cutoff in the linear power spectrum), or during the non-linear evolution phase (e.g., a subgalactic scale mean free path due to self-interactions), these models break the near universality of CDM haloes.

---

[25] Given the limited resolution of simulations, local in this sense refers to regions of at least $\mathcal{O}$ (10 kpc^3) as in [140].

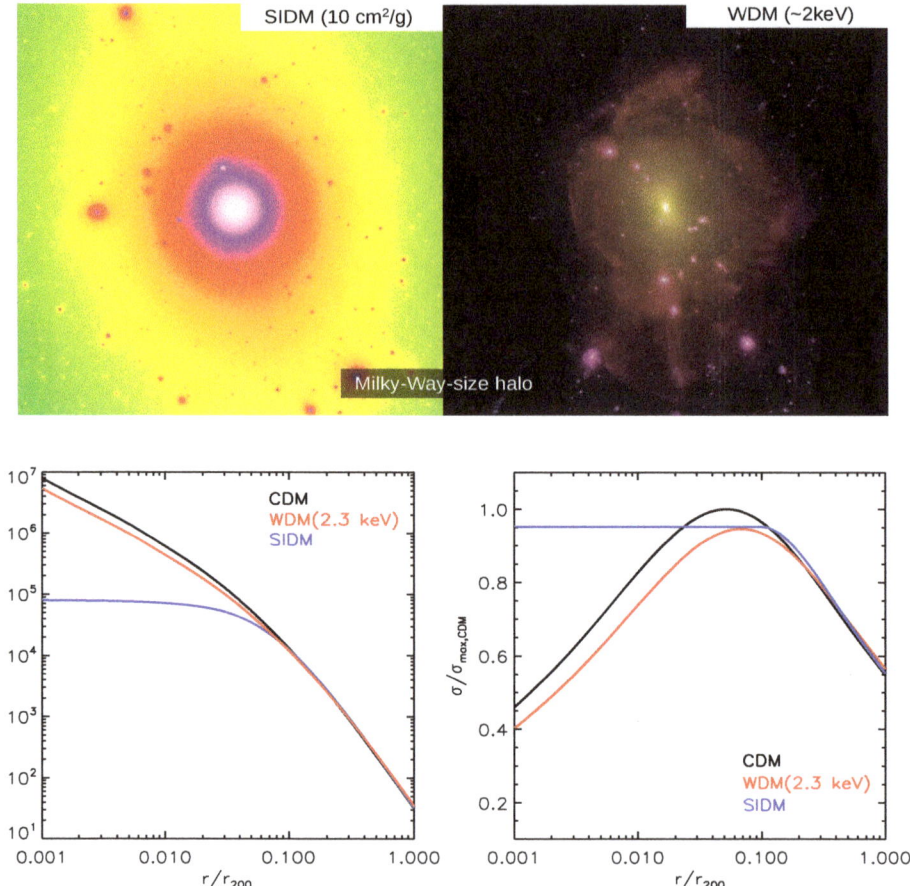

Figure 7. Structure of haloes in models with different types of dark matter: collisional (SIDM; $\sigma_T/m_\chi \gtrsim 1$ cm^2/g) and with a galactic-scale free-streaming cutoff (WDM; $m_\chi \sim 2.3$ keV). *Upper panels:* projected dark matter distribution of a Milky Way-size halo in the SIDM model (left panel; [57][26]) and in the WDM model (right panel; [153][27]). *Bottom left:* spherically averaged density profiles. WDM haloes are well described by an NFW profile, but have lower concentrations than their CDM counterparts of the same mass; SIDM haloes develop flat density cores during a transient stage that inevitably ends with the collapse of the core once the gravothermal catastrophe is triggered. *Bottom right:* spherically averaged velocity dispersion profiles. WDM haloes still obey the universal scaling for the pseudo-phase-space density, $\rho/\sigma^3 \propto r^{-1.875}$, at most radii, except in the very center, which results from a similar velocity dispersion profile to that in CDM but shifted downwards and to the right as a result of the lower concentration. SIDM haloes develop isothermal density cores of size of the order of the scale radius.

---

[26] Reproduced from Mark Vogelsberger et al. Subhaloes in self-interacting galactic dark matter haloes. MNRAS (2012) 423 (4): 3740–3752, doi: 10.1111/j.1365-2966.2012.21182.x. By permission of Oxford University Press on behalf of the Royal Astronomical Society. For the original article, please visit the following https://academic.oup.com/mnras/article/423/4/3740/1749150. This figure is not included under the CC-BY license of this publication. For permissions, please email: journals.permissions@oup.com.

[27] Reproduced from Mark Lovell et al. The haloes of bright satellite galaxies in a warm dark matter universe. MNRAS (2012) 420 (3): 2318–2324, doi: 10.1111/j.1365-2966.2011.20200.x. By permission of Oxford University Press on behalf of the Royal Astronomical Society. For the original article, please visit the following https://academic.oup.com/mnras/article/420/3/2318/979379. This figure is not included under the CC-BY license of this publication. For permissions, please email: journals.permissions@oup.com.

In models with a *galactic-scale cutoff in the primordial power spectrum*, such as WDM and interacting DM, the main changes can be understood from the later collapse of the first generation of haloes in these models compared to CDM. In contrast to CDM, these galactic-size haloes are not formed hierarchically from the assembly of smaller haloes but, instead, by *monolithic* collapse. Their characteristic density therefore reflects the mean background density at the time of the monolithic collapse. By contrast, a CDM halo of the same mass forms from the assembly of smaller fragments that typically formed earlier and are therefore denser. Simulations of WDM models have characterized the spatial structure of WDM haloes quite accurately, showing that the density profiles of allowed models are, in fact, well described by the NFW profile but with a lower concentration at a given mass [75,112,154–157]. The concentration-mass relation for WDM haloes can then be modeled in an analogous way to CDM, but taking into account the cutoff in the power spectrum [158] which is therefore reflected in the concentration of the haloes. An example of this is shown in the bottom left panel of Figure 7, where the density profile of a CDM halo is mapped into that of a 2.3 keV thermal relic WDM halo by simply scaling down the concentration using Equation (39) of Reference [157], which connects the concentration to the cutoff scale in the power spectrum[28]. This lower concentration is also reflected in the velocity dispersion profile (see bottom-right panel of Figure 7).

It is interesting to note that the pseudo-phase-space density, $Q \equiv \rho/\sigma^3$, in WDM haloes scales with radius in the same way as in CDM, $Q \propto r^{-1.9}$. In principle, Q can never exceed its primordial value, $Q = Q_{\max}$, determined by the *thermal* velocities of the unclustered dark matter particles [159]. This is because for a collisionless system, Liouville's theorem requires conservation of the fine-grained phase-space density and the coarse-grained density, approximated by Q, can never exceed this value. Thus, the central regions of WDM haloes cannot exceed a maximum density, i.e., they form a central density core. However, the value of $Q_{\max}$ is so large in allowed WDM models that the core size is tiny, $\mathcal{O}$ (10 pc) for $\sim$keV thermal WDM relics in dwarf-size haloes [160,161]; WDM cores are thus irrelevant in practice. Finally, WDM haloes are slightly less triaxial than CDM haloes as a whole for a fixed mass, at masses near the cutoff scale [112].

In SIDM, collisions between the dark matter particles have an on impact in the inner structure of haloes once the timescale for collisional relaxation at the characteristic radius of the halo r_s, $t_{\text{rel},s} \sim (\rho(r_s) \langle v_{\text{rel},s} \rangle \sigma_T / m_\chi)^{-1}$, where $\langle v_{\text{rel},s} \rangle$ is the characteristic local relative velocity, becomes comparable to the age of the inner halo. The original CDM density cusp turns into a core within the region where this condition is satisfied. The interaction cross-section thus introduces a new scale in structure formation—the mean free-path for particle collisions—which breaks the near universality of CDM haloes. The transformation of the cusp into a core due to elastic collisions at the halo center is a transitory phase that leads to a quasi-equilibrium configuration once the core has acquired its maximum size, which is approximately the radius at which the velocity dispersion profile peaks (see bottom panels in Figure 7). Prior to this, the transfer of energy during elastic collisions proceeds from the outside in since the velocity dispersion profile has a positive gradient in the inner regions and so there is a net "heat flux" from the regions close to the maximum of the velocity dispersion to the center (e.g., [162]). Once the core reaches its maximum size, subsequent collisions can only result in a net heat flux from the inside out since the velocity dispersion profile has a negative slope in the outer regions. This condition triggers the gravothermal collapse of the central parts of the SIDM halo, which results in the contraction of the core to form a new cusp, ultimately collapsing into a black hole [61,163][29].

---

[28] This functional form has been corroborated by [112], but the parameters in the two studies are different. The formula is nevertheless a good approximation to the general behavior.

[29] The gravothermal collapse [164] is a familiar process in globular clusters, where the inner regions have negative specific heat that is smaller than the positive specific heat in the outer regions. In the case of globular clusters, the collapse can be prevented by the formation of binary stars at the center. In the case of a SIDM halo, since the interactions are purely elastic, the process is expected to continue until a black hole forms. The black hole efficiently accretes the inner core of the SIDM halo (e.g., [165]). This discussion refers strictly to elastic self-scattering. If collisions are inelastic, then the energy released needs to be taken into account and, in fact, it could prevent the gravothermal collapse; see [166].

For $0.1 \lesssim \sigma_T/m_\chi \lesssim 10$ cm^2/g, cosmological N-body simulations have shown that SIDM haloes today should have cores of size of the order of the scale radius [56–58,162,167]. At the lower end of this range of cross-sections, SIDM cores are small, $\sim 0.2 r_s$ [58], making SIDM haloes only slightly different from their CDM counterparts. This is why below this cross-section, SIDM is essentially indistinguishable from CDM as a theory of structure formation [168]. At the higher end of the cross-section range, the core sizes are slightly larger than the scale radius and approach the full thermalization of the core, with a maximum size bounded by the radius at which the velocity dispersion peaks (a case such as this is shown in the bottom panels of Figure 7). Within the thermalized region, the orbits of dark matter particles are isotropized by collisions, erasing most of the memory of the assembly of the central regions [169]. This makes haloes centrally rounder than their CDM counterparts [170] and causes them to have local velocity distributions that are close to Maxwellian [171]. Since the onset of the gravothermal collapse phase is expected to be $\sim (250$–$400) t_{\rm rel,s}$ [165], the core phase of SIDM haloes in this range of cross-sections is relatively long-lived.

3. Halo Mergers and the Emergence of Subhaloes

In the previous section we reviewed the structural properties of dark matter haloes in CDM and in well-known alternatives such as WDM and SIDM. In this section, we focus on dark matter subhaloes. Since these exhibit similar structural properties to haloes, modified by a few relevant physical processes, we draw extensively on the results of the previous section. Our goal now is to describe these processes and how they affect the abundance and structure of subhaloes.

3.1. Halo Mass Assembly: Smooth Accretion vs Mergers

Haloes grow by accreting dark matter, either through mergers with smaller haloes or by accretion of diffuse, smooth material. The importance of each of these channels depends on the shape of the primordial power spectrum and on the smallest mass halo that can be formed, both of which, in turn, depend on the nature of dark matter particles. For instance, in WIMP-CDM models, the minimum halo mass is in the range $(10^{-12}$–$10^{-6})$ M$_\odot$ [23,24], which is many orders of magnitude below the resolution of current cosmological simulations. Thus, in reality, the amount of smooth accretion measured in a simulation consists of a combination of true unclustered dark matter and unresolved dark matter haloes (e.g., [137,172]). However, estimates based, for example, on the excursion set formalism can be used to extend the results of simulations into the unresolved regime. In the resolved regime, the analytical expectations are in good agreement with simulations [94,137]. These calculations show that the amount of smooth accretion onto present-day CDM Milky Way-size haloes is in the range $\sim (10$–$20)\%$ [94,137]. Thus, in this hierarchical structure formation model, haloes today are mainly composed of the remnants of disrupted smaller haloes. Of these, major mergers (i.e., those with progenitor with mass ratios greater than 1:10) contribute, on average, less than 20% of the final mass [137].

When does a halo become a subhalo? We mentioned earlier that the boundary of a halo is not sharply defined, but rather chosen approximately to separate the region within which the dark matter is in dynamical equilibrium from an outer region where the dark matter is still mostly infalling. In a similar way, the moment at which a halo becomes a subhalo, i.e., when it crosses this transition region for the first time, is somewhat arbitrary. For simplicity, it is common to use the virial radius, e.g., r_{200}, as the boundary of the halo and thus, to define a subhalo as a halo that has crossed the virial radius of a larger halo at some point in the past (see left panel of Figure 8). One could argue for a more physical definition, based for instance on the relevance of the tidal forces exerted by the dominant halo host in the local environment but, for simplicity, and because of its common usage, we will use the former, simple definition of a subhalo. We should remark that it is not uncommon for subhaloes to leave the boundary of the main halo at some point after first crossing [173–175], and thus, the subhalo population today extends to radii a few times the current virial radius ([175]; those systems beyond the virial radius are commonly known as backsplash haloes [173]).

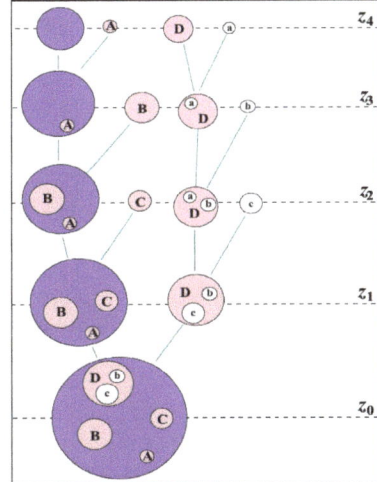 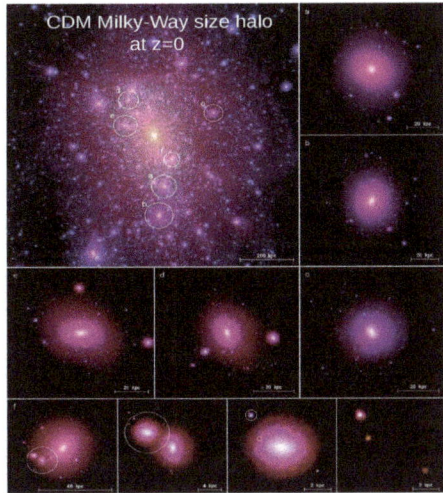

Figure 8. Dark matter subhaloes. *Left:* schematic representation of a dark matter halo *merger tree* (taken from [176][30]) at discrete redshifts. In a hierarchical model, haloes grow by the accretion of smaller neighboring haloes (A,B,C,D), which become subhaloes at the time when they first cross the virial radius of the host halo. The main branch of the tree represents the evolution of the main progenitor (shown in blue). Since this process occurs across the entire hierarchy of structures, there are subhaloes within subhaloes (sub-subhaloes; like a, b, c in system D) and so on. *Right:* a simulated Milky Way-size CDM halo from the Aquarius project (figure taken from [138][31]; this is the same halo illustrated in Figure 6). The circles in the main image mark six subhaloes that are shown enlarged in the surrounding panels, as indicated by the labels. Sub-subhaloes are clearly visible (corresponding to the configuration illustrated in the last step, z_0, in the left panel). The bottom row shows several generations of sub-subhaloes contained within subhalo f.

3.2. Evolution of Subhaloes: Initial Conditions

Halo merger trees and merger rates—N-body simulations have been instrumental in determining the mass assembly history of haloes. A particular powerful tool are halo merger trees (for an overview of different algorithms to construct these trees see [177]). A halo at the redshift of interest is regarded as the trunk of the tree and the merger tree structure consists of a catalogue of progenitors, which constitute the secondary branches that eventually merge onto the main branch of the tree (see left panel of Figure 8). Thus, a merger tree contains information about the accretion times and masses of the haloes that eventually become subhaloes. Both properties, together with the corresponding position and velocity vectors, represent the initial conditions for the subsequent dynamical evolution of the subhalo.

An interesting statistical property that can be extracted from a merger tree is the *merger rate per halo*, $dN_m/d\xi/dz$ [178], which gives the mean number of mergers, dN_m, per mass ratio, $d\xi$ (relative to

---

[30] Reproduced from Carlo Giocoli et al. The substructure hierarchy in dark matter haloes. MNRAS (2010) 404 (1): 502–517, doi: 10.1111/j.1365-2966.2010.16311.x. By permission of Oxford University Press on behalf of the Royal Astronomical Society. For the original article, please visit the following https://academic.oup.com/mnras/article/404/1/502/3101607. This figure is not included under the CC-BY license of this publication. For permissions, please email: journals.permissions@oup.com.

[31] Reproduced from Volker Springel et al. The Aquarius Project: the subhaloes of galactic haloes. MNRAS (2008) 391 (4): 1685–1711, doi: 10.1111/j.1365-2966.2008.14066.x. By permission of Oxford University Press on behalf of the Royal Astronomical Society. For the original article, please visit the following https://academic.oup.com/mnras/article/391/4/1685/1747035. This figure is not included under the CC-BY license of this publication. For permissions, please email: journals.permissions@oup.com.

the main progenitor at the time of accretion), per redshift interval, dz. This quantity has been found to have a functional form that is nearly universal [178,179]:

$$\frac{dN_m}{d\xi dz}(M_0, \xi, z) = A \left(\frac{M_0}{10^{12}\, M_\odot}\right)^\alpha \xi^\beta \exp\left[\left(\frac{\xi}{\tilde{\xi}}\right)^\gamma\right] (1+z)^{\eta_z}, \qquad (12)$$

where M_0 is the mass of the descendant and the fitting parameters (for the Millennium simulation) are given in Table 1 of [179] (see also [180]). In fact, η_z is very small, which implies a very weak redshift dependence; $\alpha > 0$ and $\beta > 0$, implying that the merger rate is higher in more massive haloes and for small mass ratios (the expected outcome in a hierarchical model). We should remark that halo merger trees can also be constructed from Monte-Carlo realizations based on merger rates computed using the extended Press–Schechter formalism, e.g., [181]. This analytical approach, calibrated to simulations [182], is also widely used to model the assembly of dark matter haloes, particularly in the context of semi-analytic models of galaxy formation (e.g., [183]).

Distribution of accretion times and orbital properties—Equation (12) can be used to compute the *average* number of haloes that become subhaloes of a host at a given redshift in a certain mass range. Halo merger trees can be used to compute other statistics of the subhalo population that are directly linked to their subsequent evolution, specifically: (i) the distribution of accretion (or infall) times, and (ii) the distribution of orbital properties at the time of accretion. In a time-independent spherical potential only two variables are needed to specify the orbit of a tracer particle (plus the orientation of the orbit). Although subhaloes orbiting a host halo are far from this idealized case, it is nevertheless useful to describe the initial orbital parameters in this way since this provides a point of comparison with the simple spherical potential. These two parameters can be chosen to be the radial, V_r, and tangential, V_θ, velocities of the subhalo at the time of infall [184], typically expressed in terms of the virial velocity of the host halo, $V_{200} = \sqrt{GM_{200}/r_{200}}$. Another common choice is to characterize the orbital properties at infall in terms of a circular orbit of the same energy, E, and the same magnitude of the angular momentum, j [185]. The initial orbit is then defined by the *circularity*, $\eta = j(E)/j_{\rm circ}(E)$, and the infall radius, $r_{200}/r_{\rm circ}(E)$, at the time of accretion[32].

Figure 9 shows a sample of the orbital parameters of the subhalo population at the time of infall, calculated from high-resolution N-body simulations [186]. The bottom left panel shows the distribution of subhalo infall times for Milky Way-size haloes, $M_{200}(z=0) = 10^{12}\, M_\odot$. This plot only includes subhaloes accreted *after* the *formation redshift* of the halo, $z_{\rm HF}$, defined as the time at which the main progenitor had half the mass of the final halo. Since the merger rate is higher for larger descendant masses (Equation (12)), and since the halo is growing from $z = z_{\rm HF}$ until today, we expect the distribution of infall times to decrease with redshift down to a minimum at $z_{\rm HF}$, independently of the mass ratio. Naturally, recently accreted haloes will be found mostly near the virial radius of the host, while haloes accreted long ago will be mostly found in the central regions (we will return to this point below). For subhaloes in bound orbits, we would expect orbital velocities at infall to be close to the virial velocity of the host halo, V_{200} [187]. The distributions of radial and tangential velocities of infalling satellites for Milky Way-size haloes are shown in the middle panels of Figure 9. Although broad, the distributions of orbital velocities do indeed have median values around V_{200}. In fact, these distributions are not independent since the total velocity of the subhalo orbit, $(V_r^2 + V_\theta^2)^{1/2}$, is determined by the potential of the host halo. Therefore, the ridge line of the bivariate distribution, (V_r, V_θ), shown in the right panel of Figure 9, is close to circular (see also [187]).

---

[32] Another set of parameters that can be used to define the orbit are the apocentre and pericenter. Different parametrizations can be transformed into one another since they are all related to the potential, $\phi(r)$.

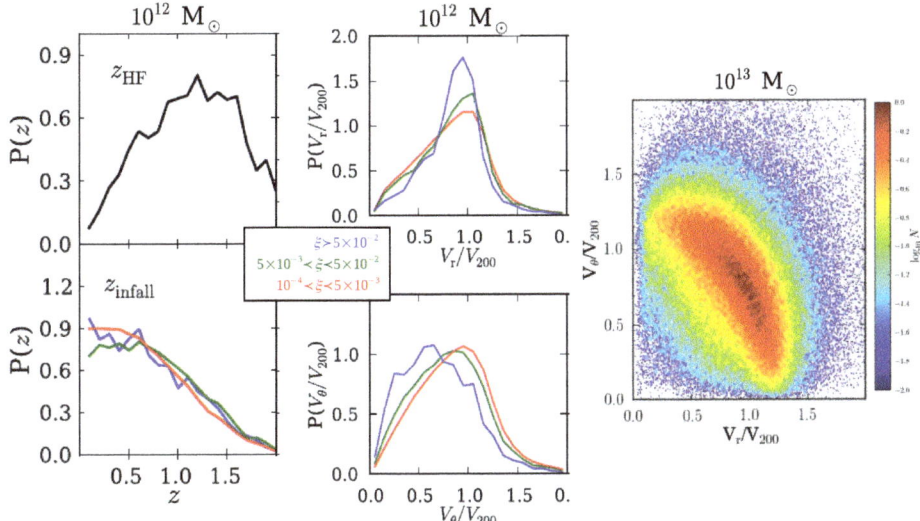

Figure 9. Initial conditions for the orbits of subhaloes infalling into haloes of present-day mass $M_{200} = 10^{12}$ M$_\odot$ (figures taken from [186][33]; see that paper for similar plots for other host masses). *Upper left:* the distribution of *formation redshifts* (defined as the redshift at which the mass of the main progenitor of the halo was half its present value). These and the other histograms in this figure are normalized such that the integral over the distribution is unity. *Lower left:* distribution of infall (accretion) redshifts of subhaloes of different mass ratios, ζ (relative to the host halo at the time of accretion; see legend). *Middle:* distributions of radial (upper panel) and tangential (lower panel) subhalo orbital velocities at infall, relative to the virial velocity of the host, for the same halo mass and subhalo-to-halo mass ratios as in the lower-left panel. *Right:* bivariate distribution of orbital parameters for infalling haloes into hosts of mass $M_{200}(z=0) = 10^{13}$ M$_\odot$.

3.3. Dynamics of Subhaloes

The material content of a halo consists of: (i) a smooth component made up mostly of the debris of disrupted subhaloes but also of material that was accreted in diffuse form; (ii) gravitationally self-bound substructure—the subhaloes. As mentioned in Section 3.1, the contribution of truly smooth accretion to the total mass of a halo at $z = 0$ is subdominant, and even though most dark matter is accreted by (minor) mergers, only a small fraction, ~10%, remains today in bound subhaloes, at least in the CDM model [138][34]. Thus, most of the mass in a halo consists of the remnants of the environmental processes responsible for stripping mass from subhaloes after infall.

---

[33] Reproduced from Lilian Jian et al. Orbital parameters of infalling satellite haloes in the hierarchical ΛCDM model. MNRAS (2015) 448 (2): 1674–1686, doi: 10.1093/mnras/stv053. By permission of Oxford University Press on behalf of the Royal Astronomical Society. For the original article, please visit the following https://academic.oup.com/mnras/article/448/2/1674/1053529. This figure is not included under the CC-BY license of this publication. For permissions, please email: journals.permissions@oup.com.

[34] This estimate is based on an extrapolation over many orders of magnitude of the subhalo mass function determined in simulations down to the free-streaming mass of WIMP-CDM particles. We discuss this in more detail below.

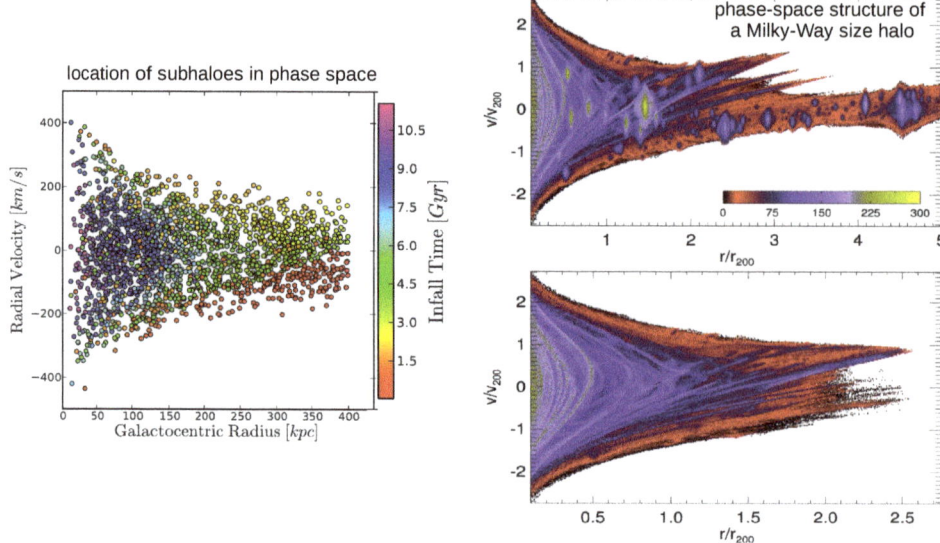

Figure 10. **Left:** Distribution in the 2D radial phase-space plane of subhaloes identified in a Milky Way-size halo simulation (Via Lactea II [188]; figure adapted from [189])[35]. Subhaloes are color-coded according to their infall time (measured from $z = 0$). Subhaloes that are just being accreted are radially infalling, while those that were accreted earlier and have completed many orbits lose energy through dynamical friction and sink towards the halo center. **Right:** the 2D radial phase-space structure of simulation particles in a different Milky Way-size halo simulation (Aquarius [138]; figure adapted from [190])[36]. Each particle is color-coded according to the number of caustics it passes (roughly proportional to the number of orbits executed by a given particle). The top panel includes bound subhaloes, while the bottom one does not. In the latter, tidal streams from disrupted subhaloes are more clearly visible.

Figure 10 provides an illustration of the richness of information contained in the phase-space structure (shown here in its 2D radial projection) of haloes today that is relevant to these environmental processes. In the left panel, the location in phase space (at $z = 0$) of the subhaloes of the Via Lactea II simulation of a Milky Way-size halo [188] is shown color-coded according to the subhalo infall time. This figure reveals the path of a subhalo as it orbits in the host halo: it first falls in radially and then loses energy as it is subjected to dynamical friction and tidal forces in the host halo. The former causes the subhalo to sink towards the center while the latter gradually strip mass from it, creating tidal streams. This picture can be appreciated with clarity in the right-hand panels of the figure where the (2D) phase-space structure of the dark matter particles is shown for a different Milky Way-size halo simulation (Aquarius; [138]). The color in this case encodes the number of caustics that a given particle traverses[37]. Since caustics occur near orbital turning points, the number of caustics is roughly

---

[35] Reproduced from Miguel Rocha et al. Infall times for Milky Way satellites from their present-day kinematics. MNRAS (2012) 425 (1): 231–244, doi: 10.1111/j.1365-2966.2012.21432.x. By permission of Oxford University Press on behalf of the Royal Astronomical Society. For the original article, please visit the following https://academic.oup.com/mnras/article/425/1/231/998181. This figure is not included under the CC-BY license of this publication. For permissions, please email: journals.permissions@oup.com.

[36] Reproduced from Mark Vogelsberger and Simon D. M. White. Streams and caustics: the fine-grained structure of Λ cold dark matter haloes. MNRAS (2011) 413 (2): 1419–1438, doi: 10.1111/j.1365-2966.2011.18224.x. By permission of Oxford University Press on behalf of the Royal Astronomical Society. For the original article, please visit the following https://academic.oup.com/mnras/article/413/2/1419/1070092. This figure is not included under the CC-BY license of this publication. For permissions, please email: journals.permissions@oup.com.

[37] Caustics represent folds in the fine-grained phase-space distribution function, which in CDM evolves according to the collisionless Boltzmann equation (Equation (1)). Before the formation of non-linear structures, CDM particles are distributed nearly uniformly in space with small density and velocity perturbations and very small thermal velocities. CDM particles

proportional to the number of orbits each particle traverses. The caustic count is thus an excellent way to highlight substructure in the 2D phase-space structure seen in the right-hand panel of Figure 10 since particles that are part, or were part, of a subhalo have undergone more particle orbits in their earlier host. This plot thus shows the richness of the substructure present in haloes today. As we mentioned earlier, most of the matter in a halo today has been accreted through mergers and consists of material that was stripped from subhaloes.

Identifying substructure—Several algorithms are in common use to identify subhaloes in *N*- body simulations and define their boundaries and properties. These *subhalo finders* are based on different techniques. Here we will list only the most popular ones; for a comprehensive comparison study see [193]. A common approach consists of finding local density maxima in the parent halo density field and then associating adjacent particles with this peak using a binding energy criterion; a subhalo is thus defined as the collection of particles that is gravitationally self-bound, with the density peak at its center. Examples are: SUBFIND [91]; Bound Density Maxima [194]; VOBOZ [195]; Amiga Halo Finder [196]. A different approach is the "time domain subhalo finder" which follows the time evolution of haloes and tracks them when they become subhalos by identifying their bound particles, as in the Hierachical Bound-Tracing or HBT [197,198]. A similar procedure is used in SURV [176,199], which uses the merger history tree of a halo across all its branches to identify subhalos.

An improvement over this 3D spatial approach can be made by including information on the particle velocities, which has the advantage that subhaloes that are in close spatial proximity with one another can be more easily disentangled. In this case, a density criterion is not enough, but the relative velocity between merging subhaloes is a telltale sign of a merger. Modifications of these algorithms can be used to identify the tidal streams that are the remnants of the tidal stripping process (see below) and which are not localized in real space, but have clear signatures in phase space. Examples of phase-space finders are the Hierarchical Structure Finder [200] and ROCKSTAR [201].

Current subhalo finders can identify subhaloes down to 20–100 particles and different algorithms roughly agree with one another on their location and main properties [193]. Below we review the main physical processes that affect the evolution and inner structure of subhaloes along their orbits, as well as relevant lessons learned from *N*-body simulations.

Tidal stripping—Once a halo reaches the outer boundary of the host halo into which it will merge, tidal forces begin to act, suppressing the accretion of matter into the merging halo and stripping mass from its outer layers in a process known as tidal stripping. Since the enclosed overdensity of a subhalo depends on its position within the host, the virial radius is no longer a meaningful concept. A more relevant scale is the *tidal radius*, r_t, defined as the radius at which the differential tidal force of the host halo is equal to the gravitational force due to the mass of the subhalo, or equivalently, as the radius within which the enclosed mean mass density of the satellite is comparable to the mean mass density of the main halo interior to the distance, R, to the satellite. The expectation is that the matter beyond the tidal radius will be removed from the subhalo, reducing its mass as it orbits around the host. For a circular orbit and assuming that the subhalo mass, $m_{\text{sub}}(< r_t)$, is much smaller than the enclosed mass of the host, $M(< R)$, and that $r_t \ll R$, the tidal radius is given by [202][38]:

$$r_t = R \left[\frac{m_{\text{sub}}(< r_t)}{(2 - \mathrm{d}\ln M/\mathrm{d}\ln r)\, M(< R)} \right]^{1/3}. \tag{13}$$

---

thus occupy a thin, approximately three -dimensional, sheet in 6D phase-space volume. Since CDM particles are collisionless and evolve according to Equation (1), the fine-grained phase-space density is conserved during gravitational evolution (this was discussed earlier in the context of the maximum phase-space density in WDM models in Section 2.4), which implies that the original thin sheet can be stretched and folded but it cannot be broken. Caustics appear where folds occur, and have very large spatial densities, limited only by primordial thermal motions (e.g., [190–192]).

38 This equation ignores the effects of the centrifugal force on the satellite as it orbits around the host. Including this effect (assuming circular orbits) modifies Equation (13) slightly by changing the factor of 2 in the denominator to 3 [203].

For non-circular orbits, the situation is more complex (in fact in the most general cases, the tidal radius can be ill-defined; see e.g., [203]), but the principle behind the concept of tidal radius remains valid: the relevant physical quantity to determine the boundary of the subhalo is the relative strength of the gravitational attraction of the subhalo and the tidal forces of the host. In this way, the tidal radius is commonly used to model tidal stripping in a variety of collisionless systems, not only subhaloes, but also, for example, globular clusters. In particular, for a slowly varying tidal field (i.e., in the adiabatic approximation), the mass loss due to stripping may be modeled as: [204–206]:

$$dm_{\rm sub} = \alpha_t m_{\rm sub}(>r_t) \frac{dt}{t_{\rm orb}(R)}, \tag{14}$$

where $t_{\rm orb}(R)$ is the instantaneous orbital period at the radius of the subhalo, and α_t is a tuning parameter, which encapsulates departures from this simple approximation (e.g., the details of subhalo structure); α_t is typically calibrated from simulations but the values used vary significantly in the literature, which is a major limitation. Equation (14) assumes that the relevant timescale for mass loss is the orbital period of the subhalo, which is justified by noting that the energy scale for the tidally stripped material is given by the change in the potential of the host across the body of the satellite $\sim r_t d\phi_{\rm host}(R)/dR$ [207]. The left panel of Figure 11 shows an example of tidal stripping in this slow (adiabatic) mode for a subhalo in a circular orbit around a static host potential (from [208]). As pointed out by [209], a combination of the ill-defined tidal radius and uncertainty in the parameter α_t, makes the modeling of this adiabatic case quite complicated, with the end result that models based on Equation (14) do not, in general, match simulation results accurately. In any case, it is rare for subhalo orbits to be nearly circular; for realistic orbits, most of the tidal mass loss happens near pericentre.

Tidal shock heating—While tidal stripping (Equation (14)) refers to the gradual loss of loosely bound material from a subhalo due to a slowly varying external potential, a rapid (impulsive) variation in the potential causes a transfer of the satellite's orbital energy to the internal energy of its particles. These tidal shocks are most important at pericentre where the impulsive condition is best satisfied. The redistribution of internal energy produced by the shock alters the inner structure of the subhalo and can unbind some of its particles [204,209,210]. This process is well described by the "impulsive approximation" (see [211,212] for the case of globular clusters) in which tidal forces are assumed to act during a much shorter time than the dynamical timescale of the satellite (see [213,214]). The approximation gives the specific energy change suffered by particles in the subhalo due to a tidal shock as[39]:

$$(\Delta E)_{\rm i,tid} = (\Delta v)_{\rm i,tid}^2 \approx \left| \int_{\rm orbit} \vec{a}_{\rm i,tid}(t) dt \right|^2, \tag{15}$$

where $\vec{a}_{\rm i,tid}$ is the acceleration experienced by a particle in the subhalo and the integral is performed along the orbit of the subhalo. If $(\Delta E)_{\rm i,tid} > E_{i,b}$, where $E_{i,b}$ is the binding energy of the particle, we may assume that the particle will become unbound. The mass fraction of particles that satisfies the inequality is then assumed to be removed *instantaneously* from the subhalo. The impulsive approximation accurately captures the results of simulations for radial orbits. An example is shown in the right-hand panel of Figure 11 (taken from [209]).

Although the energy injection in Equation (15) from tidal shocks might not be enough to unbind particles, it can still affect the inner structure of a subhalo. As a result of the shock, the orbits of dark matter particles in the center expand, reducing the inner density, although this process is not strong enough to create a central core [210,215]. The resulting density profile is, in fact, still well described by an NFW profile in the inner regions, albeit with a higher concentration, while the outer regions are considerably steeper than NFW due to stripping. For instance Reference [215], using idealized

---

[39] It is possible to evaluate Equation (15) for a given fixed potential and a given subhalo orbit (see e.g., Equation (20) of [209] for an NFW halo).

simulations in a static external halo potential, found a profile of the form $\rho \propto r^\gamma \exp(-r/r_b)$, with a central slope, $\gamma \sim 1$, and a cutoff radius due to tidal shocks, r_b (see bottom right panel of Figure 11).

For a general subhalo orbit, a combination of the adiabatic (Equation 14) and impulsive (Equation (15)) approximations provides a good estimate of the amount of stripped mass; the former is valid particularly near the apocentre of nearly circular orbits while the latter is more appropriate near the pericentre of eccentric orbits. The impulsive approximation reproduces the results of simulations quite accurately for radial orbits, but the adiabatic approximation is not very adequate for the reasons discussed above (see also Section 4.3 of [209]).

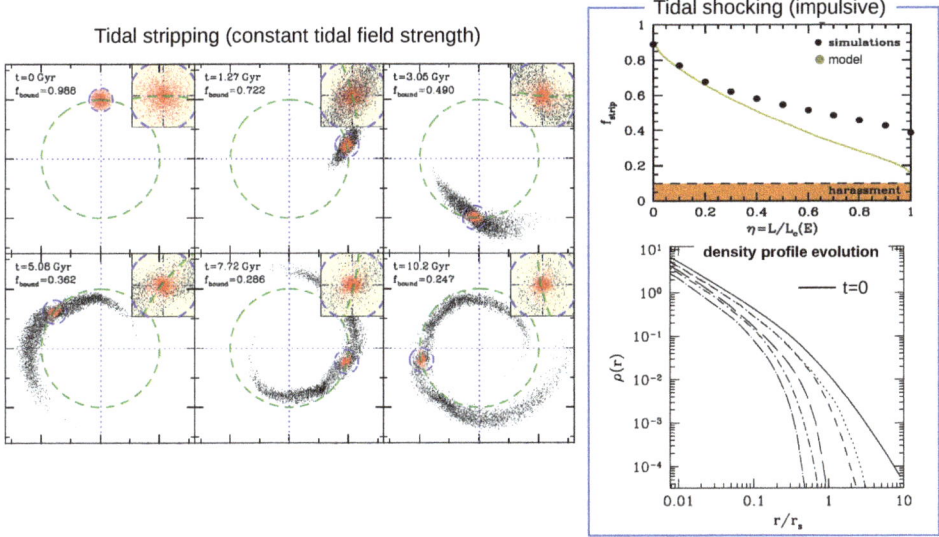

Figure 11. Tidal effects in subhaloes. **Left:** evolution of a subhalo in a circular orbit in a static host halo potential. Since the tidal field strength is constant, the subhalo gradually loses mass (red particles are bound to the subhalo, black particles are unbound) as it orbits in the host halo creating characteristic tidal streams (figure adapted from [208] [40]). **Right:** the effect of tidal shocks. For nearly radial orbits, the variations in the potential near pericentre are rapid (relative to the internal dynamical timescale of the subhalo) and this leads to an impulsive *tidal shock*, which causes a drastic removal of mass (upper right) and a change in the dark matter distribution (bottom right). In the upper panel the fraction of stripped mass as a function of circularity (see Section 3.2), given by the impulsive approximation, is compared with that in a controlled simulation (figure adapted from [209][41]). The model works quite well for radial orbits but it fails for circular orbits (as in the left panel), for which an adiabatic model is more appropriate (Equation (14)). In the lower panel, tidal shocking is seen to reduce the mass in the central regions but preserves the asymptotic NFW behavior, while the outer regions become considerably steeper than NFW (figure adapted from [215][42]).

---

[40] Reproduced from Frank C van den Bosch and Go Ogiya. Dark matter substructure in numerical simulations: a tale of discreteness noise, runaway instabilities, and artificial disruption. MNRAS (2018) 475 (3): 4066–4087, doi: 10.1093/mnras/sty084. By permission of Oxford University Press on behalf of the Royal Astronomical Society. For the original article, please visit the following https://academic.oup.com/mnras/article/475/3/4066/4797185. This figure is not included under the CC-BY license of this publication. For permissions, please email: journals.permissions@oup.com.

[41] Reproduced from Frank C van den Bosch et al. Disruption of dark matter substructure: fact or fiction? MNRAS (2018) 474 (3): 3043–3066, doi: 10.1093/mnras/stx2956. By permission of Oxford University Press on behalf of the Royal Astronomical Society. For the original article, please visit the following https://academic.oup.com/mnras/article/474/3/3043/4638541. This figure is not included under the CC-BY license of this publication. For permissions, please email: journals.permissions@oup.com.

[42] ©AAS. Reproduced with permission. For the original article, please visit the following https://iopscience.iop.org/article/10.1086/420840.

Subhalo-subhalo encounters—Tidal heating can also be caused by impulsive encounters with other subhaloes, which can add up to produce a net effect on the subhalo inner structure and mass loss (a process called *galaxy harassment* in the context of satellite galaxies; [216]). A similar impulsive approach to the one above can be used to estimate the strength of this form of tidal heating, but the calculation is more complicated because, among other things, the distribution of subhaloes in the host and the encounter rate need to be modeled. A recent study finds that tidal shocking from encounters is subdominant (by a factor of several) compared to shocking during pericentric passages [209].

Dynamical friction—When an object of mass, M_s, moves through an ambient medium of collisionless particles of mass $m \ll M_s$, the object experiences a drag force known as dynamical friction. This force may be thought of as the gravitational pull exerted by a local enhancement in the ambient density formed behind the moving object (a trailing wake) as the object gravitationally focuses the surrounding particles (see left panels of Figure 12).

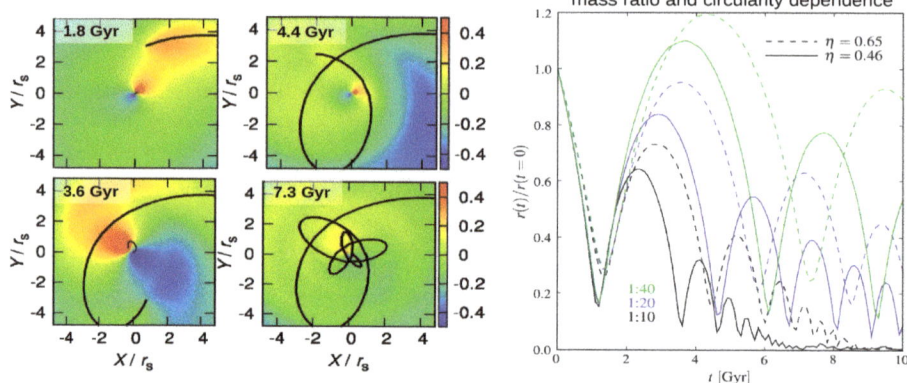

Figure 12. Dynamical friction experienced by subhaloes. **Left:** simulation of a subhalo orbiting a Milky Way-size halo; the initial mass ratio and circularity of the orbit are 0.1 and 0.5, respectively. The images show the projected over- (or under-) density relative to the initial value at $t = 0$, at different times during the evolution. The thick solid line marks the subhalo orbit, which decays over time due to dynamical friction. This gravitational process induces a wake in the host halo trailing behind the satellite (most clearly visible in the top left panel). The dipole feature at the center of the host is caused by the tidal effect of the subhalo, which perturbs the position of the halo potential minimum. This effect diminishes with time as the satellite is stripped of mass (figure adapted from [217][43]). **Right:** evolution of the radial distance of a simulated subhalo orbiting a Milky Way-size halo (figure taken from [218][44]). The orbit decays by dynamical friction on a timescale that strongly depends on the initial mass ratio (different colors) and circularity of the orbit (dashed and solid lines). The timescales are well approximated by the fitting formula (Equation (18)), which is an improvement over the classical Chandrasekhar formula (Equation (17)).

The net result of dynamical friction is a transfer of orbital angular momentum and energy from the moving object into the surrounding medium. The process can be analyzed as a series of uncorrelated sequential encounters between the object of mass, M_s, and velocity, $\vec{v}_s$, and particles randomly taken

---

[43] Reproduced from Go Ogiya and Andreas Burkert. Dynamical friction and scratches of orbiting satellite galaxies on host systems. MNRAS (2016) 457 (2): 2164–2172, doi: 10.1093/mnras/stw091. By permission of Oxford University Press on behalf of the Royal Astronomical Society. For the original article, please visit the following https://academic.oup.com/mnras/article/457/2/2164/970692. This figure is not included under the CC-BY license of this publication. For permissions, please email: journals.permissions@oup.com.

[44] Reproduced from Michael Boylan-Kolchin et al. Dynamical friction and galaxy merging timescales. MNRAS (2008) 383 (1): 93–101, doi: 10.1111/j.1365-2966.2007.12530.x. By permission of Oxford University Press on behalf of the Royal Astronomical Society. For the original article, please visit the following https://academic.oup.com/mnras/article/383/1/93/1067887. This figure is not included under the CC-BY license of this publication. For permissions, please email: journals.permissions@oup.com.

from the ambient medium with velocity distribution, $f_m(\vec{u})$. These interactions occur on a timescale much shorter than the variations in the velocity, $\vec{v}_s$, of the object. If the ambient medium is assumed to have a homogeneous density, ρ_m, then the changes to the velocity of the object perpendicular to its motion average to zero by symmetry, while the velocity changes parallel to the direction of motion are given by the well-known Chandrasekhar dynamical friction formula [219], which, for the drag force, $\vec{F}_{df}$, takes the form:

$$\begin{aligned}\vec{F}_{df} = M_s \frac{d\vec{v}_s}{dt} &= -16\pi^2 M_s^2 m \ln\Lambda \left[\int_0^{|\vec{v}_s|} f_m(|\vec{u}|)u^2 d|\vec{u}|\right] \frac{\vec{v}_s}{|\vec{v}_s|} \\ &= -4\pi \left(\frac{GM_s}{v_s}\right)^2 \ln\Lambda \, \rho(<|\vec{v}_s|) \frac{\vec{v}_s}{|\vec{v}_s|},\end{aligned} \quad (16)$$

where $\rho(<|\vec{v}_s|)$ is the density of ambient particles with speed less than $|\vec{v}_s|$ and $\ln\Lambda \equiv \ln\left[1 + (b_{max}/b_{90})^2\right]$ is the Coulomb logarithm, with $b_{90} = G(M_s + m)/v_\infty^2$, v_∞ the initial relative velocity of an individual encounter, and b_{max} the maximum impact parameter ($b_{max} \gg b_{90}$). As a consequence of dynamical friction, the orbit of the object decays in time sinking towards the center of the host halo. For circular orbits in a spherical singular isothermal host halo (implying a Maxwellian velocity distribution, f_m) of mass, M_h, the timescale for the orbit to decay to zero (i.e., the dynamical friction time) is approximately given by (e.g., [68]):

$$\frac{t_{df}}{t_{dyn}} \approx \frac{1.17}{\ln(M_h/M_s)}\left(\frac{M_h}{M_s}\right), \quad (17)$$

where $t_{dyn} = r_{vir}/V_c$ is the dynamical timescale at the virial radius of the halo, r_h, with V_c the circular velocity of the host halo, which is independent of radius for a singular isothermal sphere. For more general orbits, Equation (17) requires a correction that scales with the circularity, η, as $t_{df} \propto \eta^{\gamma_\eta}$, where $\gamma_\eta \sim 0.53$ [220], implying that more eccentric orbits decay more rapidly (see right panel of Figure 12).

Equation (16) is derived under the following assumptions: (i) both the satellite and the particles that make up the ambient medium can be treated as point masses; (ii) the self-gravity of the ambient medium can be ignored, and (iii) the distribution of ambient medium particles is infinite, homogeneous and isotropic. None of these assumptions is strictly valid in realistic situations. Nevertheless, Chandrasekhar's formula provides a reasonable description of dynamical friction, particularly when modifications are included to account for the density profile of the subhaloes and their orbits; in practice, this can be done by regarding the Coulomb logarithm as a free parameter that depends on these properties. An example of this is provided in [218] with a series of idealized N-body simulations of a subhalo infalling into a host, both described by a Hernquist density profile[45]. This study found the following fitting function to the dynamical friction timescale (a few examples of the orbital evolution in this study are shown in the right panel of Figure 12):

$$\frac{t_{df}}{t_{dyn}} = A \frac{(M_h/M_s)^b}{\ln(1 + M_h/M_s)} \exp[c\eta(E)] \left[\frac{r_c(E)}{r_{200}}\right]^d, \quad (18)$$

where E is the initial orbital energy of the satellite (we recall that $r_c(E)$ is the radius of a circular orbit of the same energy, E), and the fitting parameters are of order one[46]. Equation (18) was found to be valid over a wide range of orbital parameters; the most relevant restriction is $0.025 \leq M_s/M_h \leq 0.3$.

---

[45] The Hernquist halo profile [221] has the same asymptotic behavior at the center as the NFW halo and has the advantage that the velocity distribution function in the isotropic case has an analytic form (see Equation (11)), which makes it particularly simple to set up initial conditions for simulating haloes in dynamical equilibrium.
[46] The values for these parameters reported in [218] are: $A = 0.216$, $b = 1.3$, $c = 1.9$, and $d = 1.0$, but we point out that in this study both the halo and the subhalo were modeled as Hernquist profiles.

For smaller mass ratios, the dynamical friction timescale becomes much larger than the age of the universe, while for larger mass ratios, the relevant timescale is just the dynamical or free-fall time[47].

3.4. The Abundance, Spatial Distribution and Internal Structure of Dark Matter Subhaloes

The abundance, spatial distribution within the host halo and internal structure of subhaloes are determined by the combined effects of the initial conditions at the time of accretion, which depend on cosmology, and the dynamical processes described in the previous section. These properties are best derived in full generality using cosmological N-body simulations but analytic models can provide valuable physical insights [222]. In this section, we present some of the key structural properties of the subhalo population, as revealed by simulations. Naturally, these properties are closely related to those of *isolated* haloes (discussed in Section 2.4) with a few relevant modifications.

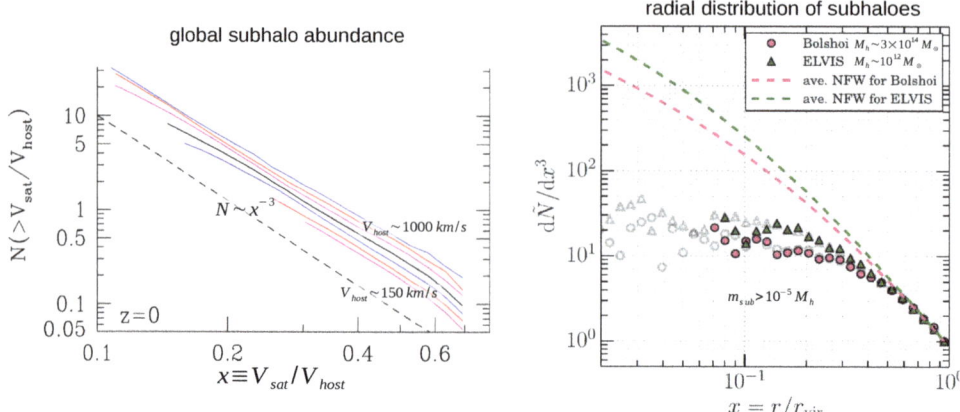

Figure 13. Subhalo abundance. **Left:** the subhalo velocity function at $z = 0$ for haloes of different maximum circular velocity, from ~150 km/s to ~1000 km/s (bottom to top). In terms of the velocity ratio, $x = V_{sub}/V_h$, the velocity function is nearly universal, scaling as x^{-3} (dashed line) with a scale-dependent normalization (see Equation (20); figure adapted from [223][48]). **Right:** the number density of subhaloes as a function of halocentric distance in units of the virial radius for Milky Way-size haloes (triangles) and cluster-size haloes (circles). All subhaloes with $m_{sub}/M_h > 10^{-5}$ have been included. The dashed lines are the average NFW fits to the density profiles of the hosts. These functions have been normalized to unity at the virial radius. Figure adapted from [224].

The subhalo mass function—As in the case of isolated haloes, the total CDM subhalo mass function (measured within the virial radius) is remarkably close to universal and, in fact, has a similar low-mass slope as the halo mass function [138,225–229]: $dn_{sub}/dm_{sub} \propto m_{sub}^\alpha$, where $\alpha \sim -1.9$ (see Equation (7)). This similarity to the halo mass function is partly because most subhaloes identified at a given time were accreted relatively recently and thus tidal effects have not had time to act; see Figure 9. The normalization of the subhalo mass function depends on the mass of the host halo, with more massive haloes having, on average, larger subhalo populations [137,227,230]. This reflects the earlier assembly of low-mass haloes, which allows tidal effects more time to act and disrupt subhaloes. For similar reasons other properties of the host halo can have second-order effects on the amplitude of the subhalo

---

[47] Equation (18) was only explored for values of the circularity in the range $0.3 \leq \eta \leq 1.0$ and for $0.65 \leq r_c/r_{200} \leq 1.0$; the lower limits were imposed to avoid radial orbits that would take the subhalo so close to the center of the halo in the first orbit that the tidal effects of the galaxy cannot be ignored. So far we have not discussed baryonic effects, but it is worth mentioning them here since Equation (18) was not investigated outside this range and might not be valid there even in the absence of a central galaxy.

[48] ©AAS. Reproduced with permission. For the original article, please visit the following https://iopscience.iop.org/article/10.1088/0004-637X/740/2/102.

mass function, e.g., at fixed mass, more concentrated haloes (which assemble earlier) have fewer subhaloes.

When the subhalo mass function is scaled to the host halo mass it becomes nearly universal across halo masses, with a functional form that is well fitted by [230–233]:

$$N(>\mu \equiv m_{\rm sub}/M_h) = \left(\frac{\mu}{\widetilde{\mu}_1}\right)^{1+\alpha} \exp\left[-\left(\frac{\mu}{\mu_{\rm cut}}\right)^b\right], \quad (19)$$

where the exponential cutoff accounts for the increasing rarity of subhaloes of mass close to that of the host halo mass[49]. The parameter, $\widetilde{\mu}_1$, is the typical mass fraction of the most massive subhalo (relative to the host halo mass), which, for a Milky Way-size halo, is of order 0.01, but with a large spread [232]. The universality of the subhalo mass function is, however, not perfect; the remaining dependence on host halo mass can be captured by allowing a relatively weak scaling of the normalization parameter, $\widetilde{\mu}_1$. This dependence is amplified, and perhaps better expressed, if the subhalo *velocity function* is used instead, i.e., if the abundance is given in terms of the maximum circular velocity instead of the mass (see left panel of Figure 13). In this case, an accurate approximation is given by [223]:

$$N(>x \equiv V_{\rm sub}/V_h) \propto V_h^{1/2} x^{-3}, \quad x < 0.7, \quad (20)$$

which implies, for instance, that a cluster-sized halo ($V_h \sim 1000$ km/s) has ~ 2.2 times more substructure of a given velocity ratio than a Milky Way-size halo ($V_h \sim 200$ km/s). Notice that this difference is considerably weaker if the mass ratio is used since, in this case, the abundance scales as a power law of exponent ~ -0.9 rather than ~ -3.

The fact that the power-law exponent of the subhalo mass function at low masses, α, is greater than -2 is important; a steeper slope would imply that the total mass in substructures diverges when extrapolated to arbitrarily low masses. For a given particle dark matter model we know, of course, that this extrapolation cannot be continued beyond the truncation mass below which the properties of the dark matter particle prevents the formation of smaller structures (due to the suppression mechanisms mentioned in Section 2.1). In the case of CDM-WIMPs, the extrapolation of the subhalo mass function down to the Earth's mass (10^{-6} $M_\odot$) implies that the fraction of mass contained in *unresolved* subhaloes is $\sim 4.5\%$, in contrast to the $\sim 13\%$ mass fraction found in the highest resolution simulation (with a particle mass of 2×10^4 $M_\odot$) to date of a Milky Way-size halo [138]. As mentioned earlier, most of the mass in haloes is not, in fact, in the form of self-bound subhaloes, but in the remnants of the tidal stripping process accumulated over the entire history of the halo.

The radial distribution of subhaloes—The spatial distribution of a subhalo population reflects the balance between accretion of new subhaloes and tidal disruption of older ones. This distribution has been studied extensively in N-body simulations [138,225,231,234–237] and the picture that emerges is that the radial distribution of subhaloes is significantly less centrally concentrated than the dark matter distribution (i.e., the smooth halo), and is relatively independent of the host halo mass (see right panel of Figure 13). Most remarkably, when subhaloes are selected according to mass (rather than maximum circular velocity) and the distribution is normalized to the mean number density of subhaloes of a given mass within the virial radius, there appears to be no trend in the shape of the number density profile with subhalo mass [138,175,231]. A recent analysis using the HBT finder, however, has shown that the most massive subhaloes are actually more concentrated in the central regions than lower-mass subhaloes [198], which seemingly reflects the resilience of very massive subhaloes to tidal stripping despite suffering from substantial dynamical friction. The near universality of the radial distribution of subhaloes is then the result of a convolution of the distribution of subhaloes before infall (sometimes

---

[49] The fitting parameters in Equation (19) in the case of the Millennium simulations may be found in [230], where a redshift dependence is also provided.

called the *unevolved* radial distribution of subhaloes), which is nearly scale free, with a tidal stripping process that is also nearly scale free, except at the massive end (see [222] for an analytical model of the subhalo distribution). It is interesting that the radial distribution of subhaloes with maximum circular velocity $>V_{\text{sub}}$ is steeper than that of subhaloes with mass $>m_{\text{sub}}$ (see e.g., [238]), since the latter is more heavily influenced by tidal stripping.

The ratio between the average, mass-selected subhalo radial distribution and the average NFW mass density profile of their host haloes (both normalized to the virial radius as defined in [224] and shown in the right-hand panel of Figure 13), is approximated quite accurately by the following functional form [224]:

$$\phi(x \equiv r/r_{\text{vir}}) = \frac{d\widetilde{N}/dx^3|_{\text{sub}}}{d\widetilde{N}/dx^3|_{\text{NFW}}} = 4\frac{x^4}{(1+x)^2} \tag{21}$$

The inner structure of dark matter subhaloes—Cosmological N-body simulations have shown that the density profiles of subhaloes retain the near universal properties of isolated field haloes but with modifications that reflect the tidal effects discussed in Section 3.3. These modifications are consistent with expectations of analytical estimates and controlled simulations. In particular, for CDM, the subhalo radial density profile exhibits the same central cusp as an isolated halo in equilibrium (left panel of Figure 14), while the outer regions show a steep truncation at a radius approximately equal to the tidal radius given in Equation (13) (see Figure 15 of [138]). We should remark that, as has been found for field haloes [121], a better fit to the density profile of subhaloes is given by the 3-parameter Einasto profile [239][50]:

$$\rho_E(x_E \equiv r/r_{-2}) = \rho_{-2} e^{-2(x_E^{\alpha_E}-1)/\alpha_E}, \tag{22}$$

where α_E is a shape parameter and ρ_{-2} and r_{-2} are the density and radius at which the logarithmic slope of the density profile is equal to -2. The Einasto and NFW profiles are quite similar, and both give good fits to the subhalo profiles in the range $0.01 < x_E < 100$ if $\alpha_E \sim 0.22$ [241,242]. Although for isolated haloes the parameters α_E and r_{-2} can be related to the virial mass of the halo, M_{200} [135,240], in a similar way as the halo (NFW) concentration is connected to the virial mass, the situation is less clear for subhaloes [241], and the spread of the parameters across subhalo masses is large. Thus, for its simplicity, the NFW profile remains a reasonable approximation to the structure of both haloes and subhaloes.

Since for subhaloes the virial radius no longer has a proper meaning as the "boundary" of the object, the concentration parameter, defined as $c = r_{200}/r_s$ commonly used to characterize NFW haloes, is no longer appropriate. Instead, it is convenient to define the concentration of a subhalo in a way that is independent of its size. One such measure of concentration is the characteristic overdensity, δ_V, defined as the mean density within the radius, r_{max}, where the circular velocity peaks, at a value of V_{max}, relative to the critical density [138,226]:

$$\delta_V = \frac{\overline{\rho}(<r_{\text{max}})}{\rho_{\text{crit}}} = 2\left(\frac{V_{\text{max}}}{Hr_{\text{max}}}\right)^2, \tag{23}$$

where H is the Hubble constant. Equation (23) can be related to the standard scale density of the NFW profile (δ_c in Equation (9)), and thus to the NFW concentration, in a straightforward way [226]:

$$\delta_V = \left(\frac{c}{2.163}\right)^3 \frac{K_c(2.163)}{K_c(c)} \Delta, \tag{24}$$

where K_c was defined just after Equation (9). We note that for the NFW profile, $r_{\text{max}}/r_s = 2.163$.

---

[50] Although the introduction of a third parameter will obviously improve the quality of the fit, the Einasto profile is, in fact, a slightly better fit to simulations than the 2-parameter NFW profile even after one of the parameters (α_E) is fixed to an appropriate value. For instance, fixing $\alpha_E \sim 0.16$ gives a better fit than NFW to haloes across a range of halo masses [240].

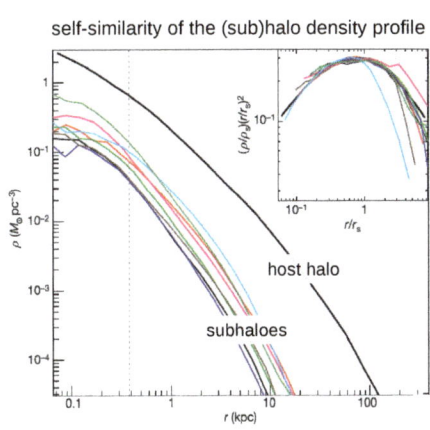
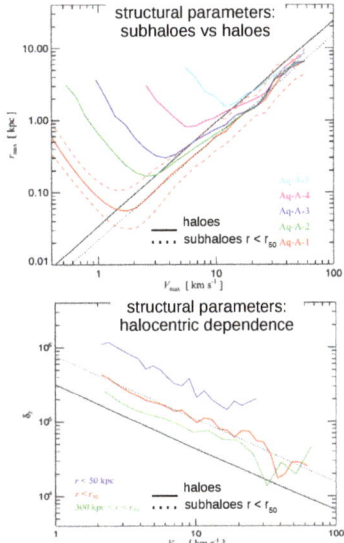

Figure 14. The inner structure of subhaloes. *Left*: spherically averaged density profile of subhaloes (which is remarkably similar to that of isolated haloes). The plot shows the density profile of a Milky Way-size halo (solid black line) and eight of its largest subhaloes (color lines). The vertical dotted line marks the radius beyond which the simulation results are converged. The self-similarity in the central region is better appreciated in the inset where the density and radius are scaled to their values at the scale radius, r_s. The figure is for the Via Lactea II simulation and is adapted from [188]. *Upper right*: mean relation between the maximum circular velocity, V_{max}, and the radius at which it is achieved, r_{max}, for subhaloes within r_{50} (the radius within which the mean enclosed density is 50 times the critical density) of one the Milky Way-size halo simulations in the Aquarius project, at different resolution levels (color lines). The red dashed lines show the scatter (68% of the distribution) for the highest resolution level. The dotted line is a fit to the mean relation for subhaloes and lies systematically below the equivalent line for isolated haloes (solid line). *Lower right*: a measure of concentration for subhaloes (see Equation (24)) within different radial ranges, as given in the legend. The solid line corresponds to isolated haloes. Figures adapted from [138][51].

Since the concentration of haloes (and subhaloes) is tightly correlated with their mass (see Section 2.4), Equation (23) implies a tight correlation between V_{max} and r_{max}, which indeed has been found and characterized in simulations (see right panel of Figure 14). For instance in the case of the Aquarius-A Milky Way-size halo, the following fitting functions (to the mean relations in subhaloes) provide a direct connection in terms of the subhalo mass[52]:

$$V_{max} = 10 \text{ km/s} \left(\frac{m_{sub}}{3.37 \times 10^7 M_\odot}\right)^{0.29}$$

$$\frac{\delta_V(z=0)}{2} = \left(\frac{V_{max}}{H_0 r_{max}}\right)^2 = 2.9 \times 10^4 \left(\frac{m_{sub}}{10^8 M_\odot}\right)^{-0.18}. \quad (25)$$

---

[51] Reproduced from Volker Springel et al. The Aquarius Project: the subhaloes of galactic haloes. MNRAS (2008) 391 (4): 1685–1711, doi: 10.1111/j.1365-2966.2008.14066.x. By permission of Oxford University Press on behalf of the Royal Astronomical Society. For the original article, please visit the following https://academic.oup.com/mnras/article/391/4/1685/1747035. This figure is not included under the CC-BY license of this publication. For permissions, please email: journals.permissions@oup.com.

[52] We note that there is a typo in the caption of Figure 28 in [138], which gives the fitting function for δ_V and m_{sub} (5.8 × 10^8 M$_\odot$ → 5.8 × 10^4 M$_\odot$).

The fitting function for the $r_{max} - V_{max}$ relation implied by Equation (25) is shown as a dotted line in the upper right panel of Figure 14, while the corresponding relation for isolated haloes in the Millennium I simulation is shown as a solid line [129][53]. The most relevant result when comparing haloes and subhaloes in the $r_{max} - V_{max}$ plane is that both share the same relation, but subhaloes have systematically higher concentrations at a given V_{max} [126]: in Figure 14 the dotted line is a factor of 0.62 lower than the solid line, i.e., subhaloes have on average r_{max} values that are smaller than haloes of the same V_{max} by this factor. Equivalently, the characteristic overdensity, δ_V, in subhaloes is roughly a factor of $(1/0.62)^2 \sim 2.6$ larger for subhaloes than for haloes of the same V_{max} (lower right panel of Figure 14), which roughly corresponds to a 30% increase in the NFW concentration. This relative increase in concentration is larger for subhaloes nearer the center of the host, as expected from the nature of the tidal forces experienced by the subhaloes as described in Section 3.3: while tidal stripping naturally reduces V_{max}, it reduces r_{max} even further [210][54], making the subhalo effectively more concentrated; the stronger the mass loss, the stronger the effect, and hence the trend with halocentric distance.

It is thus possible to model the inner density profile of the subhalo population by assuming a model for the concentration-mass relation of field haloes and making a simple correction to the subhalo concentration depending on the location of the subhalo. More exhaustive studies of subhalo concentration exist that provide fitting functions across a wide range of subhalo masses, host halo masses, and distance to the halo center (e.g., [244] for the case of Milky Way-size haloes).

The shapes and internal kinematics of subhaloes—The impact of tidal forces in the structure of subhaloes is reflected also in their shapes. Although tides tend to elongate objects, these distortions are short-lived features accentuated during pericentric passages. Once the tidal streams cease to be bound to the subhalo, simulations have shown that the bound material remains in an equilibrium configuration that is, in fact, more spherical than it was at the time of infall; the stronger the tidal effects, the more spherical the subhalo becomes [245]. Although these differences are significant for the fraction of the subhalo population whose orbits are strongly influenced by the tides of the host, the subhalo population as a whole is only slightly affected and exhibits a small systematic shift towards less aspherical shapes compared to field haloes [246]. This is because the global subhalo population is dominated in number by subhaloes near the virial radius of the host, which have only recently fallen in.

When tidal effects are strong, the internal kinematics of subhaloes are also substantially altered. In particular, the velocity anisotropy of the dark matter particles becomes increasingly tangential ($\beta < 0$) from the subhalo center outwards [246], in contrast to field haloes that are radially anisotropic at larger radii. This is the result of the preferential stripping by tides (when the subhalo is near pericentre) of subhalo particles with radial orbits. On the other hand, the pseudo-phase-space density, Q_{sub}, of subhaloes in equilibrium retains the universal power-law behavior of CDM field haloes, but with a slightly shallower slope, $Q_{sub} \sim r^{-1.6}$ [246], compared to $\sim r^{-1.9}$ for field haloes.

3.5. The Impact of the Nature of the Dark Matter

Subhalo abundance—By far the main difference in the subhalo populations predicted in models with different kinds of dark matter is the abundance of low-mass subhaloes. In particular, as we discussed in Section 2.4, models in which the primordial power spectrum of density perturbations has a cutoff at relatively low k (such as WDM and interacting dark matter) have a corresponding cutoff in the mass function of haloes and subhaloes. These models predict far fewer haloes and subhaloes than

---

[53] Obtained by taking the power-law concentration-mass relation in [129] (their Equation (4)), and using Equations (23) and (24).
[54] For a clear illustration of the evolutionary track of subhaloes in the $r_{max} - V_{max}$ plane due to tidal stripping, see Figure 8 of [243].

CDM, and this offers the best prospect for distinguishing between them and perhaps constraining the properties of the particles themselves (such as the WDM particle mass).

A cutoff in the mass function breaks the universal behavior of the halo and subhalo mass functions at low masses in a way that also depends on the nature of the dark matter particle. For example, the self-similarity of the abundance of CDM subhaloes as a function of relative mass, exhibited in Equations (19) and (20), is broken [247] because the cutoff scale expressed in terms of the ratio, $\mu = m_{\text{sub}}/M_h$, occurs at larger values of μ for smaller values of M_h. The radial distribution of subhaloes in WDM models is quite similar to that in the CDM case, with only minor differences explained by the enhanced tidal stripping of low-mass WDM haloes resulting from their lower concentrations [222,247].

In many SIDM models the subhalo mass function remains largely unchanged as long as the interaction cross-section, $\sigma_T/m_\chi < 10$ cm^2/g [57,58,248]. For higher values, collisions between dark matter particles within subhaloes and in the host are frequent enough to unbind material from the halo. This form of *subhalo evaporation* is energetically efficient because the energy transfer is determined by the relative velocity of the colliding particles, which is of the order of the orbital velocity. In this case, the mass loss in subhaloes is enhanced and the subhalo abundance is depleted relative to the CDM case, particularly in the central regions [57].

Inner structure of subhaloes—The inner structure of WDM and SIDM subhaloes is rather similar to that of field haloes (see Figure 7), and the outer structure is altered by tidal effects in a very similar way as in CDM. The main difference is an enhancement in the concentration of subhaloes relative to their field counterparts in WDM models [247] due to the increased efficiency of tidal stripping of WDM subhaloes, which are less concentrated than in CDM at the time of infall. Tidal stripping also plays a greater role in SIDM subhaloes [248]. Most importantly, it can trigger a gravothermal catastrophe and this can give rise to segregation according to particular orbits, with cuspy profiles for subhaloes which have experienced substantial tidal mass loss and central cores for those where tidal effects have been minimal [249].

4. Outlook

It is fair to say that the evolution of the phase-space distribution of classical, non-relativistic, collisionless dark matter (CDM) down to galactic-scale haloes and subhaloes is now essentially a solved problem, largely through the application of N-body simulations over the past 40 years[55]. This strong statement carries a couple of major caveats, which define today's frontier in N-body simulations of cosmological structure formation.

First, the statement above can still hold if, instead of CDM, most of the dark matter consists of other types of particles, such as WDM and SIDM[56], for which N-body simulations with appropriate modifications have been applied at a similar level of detail as in CDM; in this review we have discussed the most important changes in the dark matter phase-space structure that occur in these alternative models. Nevertheless, there are still dark matter models that remain unexplored, or only partially explored with N-body simulations, e.g., hidden dark sector models with DAOs [21,114] and inelastic SIDM[57]. Secondly, and crucially, the statement above does not take account of the interplay between baryons and dark matter, which are dynamically coupled through gravity. Several mechanisms that can radically modify the dark-matter-only predictions of N-body simulations and which are, of course,

---

[55] By galactic-scale haloes and subhaloes, we mean self-bound dark matter structures that can potentially host a galaxy, i.e., is haloes of mass above $\sim 10^8$ M$_\odot$, in which gas can cool by atomic processes (e.g., [68,250]).

[56] This is true only for elastic SIDM, and for cross sections that do not exceed the gravothermal collapse threshold, $\sigma_T/m_\chi \sim 10$ cm^2/g, for dwarf-size haloes (see the last paragraphs of Section 2.4). Although the regime of gravothermal collapse has been known for a couple of decades [162,163], a comprehensive analysis of this regime has yet to be carried out (see [32,249,251,252] for recent developments in this interesting regime).

[57] There is a class of inelastic SIDM models in which the dark matter can have ground and excited states (e.g., [253]), and in which scattering between the excited and ground states can result in energy injection at the center of dark matter haloes thus altering their structure. Only until very recently have these models began to be explored with simulations [166,254].

crucial for a complete theory of structure formation and its connection to reality, have been studied extensively for several decades. We briefly summarize these in Section 4.1 below.

Finally, we make two remarks concerning the limited resolution of current *N*-simulations: (i) there have been recent claims that subhaloes can be artificially disrupted in cosmological simulations due to discreteness effects and inadequate force resolution [208,209]; if correct, these effects could alter some of the current results on the abundance and structure of subhaloes, particularly at low masses; (ii) as we have seen, the best current *N*-body simulations only resolve haloes of mass greater that $\sim 10^5$ M$_\odot$, many orders of magnitude larger than the cutoff mass in the linear density power spectrum for CDM. Yet, if the dark matter is made of Majorana particles, these, so far unresolved, haloes could be crucial for predicting the properties of their annihilation radiation and thus for elucidating the nature of dark matter. The first attempts at understanding the properties of haloes down to the cutoff in the CDM primordial power spectrum have been made [122–124] but new techniques will be required to tackle this problem in full generality.

4.1. The Impact of Baryonic Physics on Dark Matter Structure

In the linear regime, the (gravitational) impact of baryons (and electrons and photons) in the dark matter distribution, of which baryonic acoustic oscillations is perhaps the best-known outcome (e.g., [255]), is fairly well understood. In the non-linear regime, on the other hand, the complexity of baryonic physics is much greater, and the list of relevant processes is extensive: gas dynamics, radiative processes, star formation and evolution, supermassive black hole formation and evolution, etc. Here we focus on some of the most important mechanisms that modify the predictions for the abundance and structure of CDM haloes from *N*-body simulations.

Condensation of baryons into haloes: adiabatic gas cooling and mergers.- In the classical theory of galaxy formation, gas initially follows dark matter; as haloes collapse and virialize, the associated gas heats up by shocks and adiabatic compression to the virial temperature of the halo [87,256][58]. Subsequently, the gas can radiatively cool and condense towards the center of the halo if the cooling time is shorter than the free-fall time. The halo mass threshold for effective cooling depends on the density, temperature and metallicity of the gas; cooling is quite efficient in low-mass haloes down to the atomic cooling limit (virial temperatures $\sim 10^4$ K, corresponding to halo masses $\sim 10^8$ M$_\odot$) below which cooling becomes highly inefficient. At higher masses ($\sim 10^{13}$ M$_\odot$ for gas with solar metallicity, e.g., [87]), cooling is also suppressed because the cooling time exceeds the free-fall time, limiting the condensation of baryons, a process that can be exacerbated when the gas is heated by energy input from Active Galactic Nuclei (AGN) [84,258]. A hot, quasi-hydrostatic corona forms from which gas can subsequently cool at the center. Additional gas may be brought in by galaxy mergers. Regardless of the condensation mode, the assembly of the central galaxy ultimately results in an enhancement of the central gravitational potential, compared to the situation where the galaxy is absent. The dark matter distribution reacts dynamically, becoming more concentrated, a process first modeled assuming an adiabatic response leading to the contraction of the halo [259,260]. Even though the assembly of baryonic matter by mergers is not, in general, adiabatic, the simple adiabatic model remains a reasonable approximation [261]. In the absence of heating processes, the general expectation is thus that haloes should be cuspier than the NFW profile in the central regions, as indeed is seen in cosmological hydrodynamics simulations (e.g., [262,263]).

Energy injection into haloes: UV background photoheating.- The hydrogen emerging from recombination is, of course, neutral. However, the UV radiation produced by stars in the first generations of galaxies reionizes this gas and heats it up, suppressing gas cooling into low-mass haloes

---

[58] Large relative velocities between gas and dark matter inherited from the photon-baryon coupling before recombination can impede the growth of gravitational perturbations and stop gas from accreting into the first haloes [257]. This process, however, is only thought to be relevant for the formation of the first stars.

and subsequent star formation [264,265]. This heating mechanism moves the minimum scale for galaxy formation from the atomic cooling limit to larger halo masses of order 10^9 M$_\odot$ today [59] [250,267–274]. This baryonic process is important also because, in conjunction with the expulsion of gas from haloes by supernova feedback (see below) at high redshift, it reduces the overall baryonic content, and thus, the total mass content of low-mass haloes; this reduces the growth rate and final masses of these haloes compared to their counterparts in simulations without baryons [250,275].

Energy injection into haloes: supernova and AGN feedback.- When massive stars explode as supernovae in the final stages of their evolution they release vast amounts of energy, a fraction of which may couple effectively to the surrounding interstellar medium (ISM), heating it and pushing it in a violent blowout. The combined impulsive removal of baryonic outflows from several supernovae creates a collective effect in the host galaxy known as supernova feedback, which has a fundamental role in regulating star formation [87]. Supernova feedback affects the evolution of the galaxy population at all galactic masses, but is particularly important in low-mass haloes which have shallow potential wells; supernova-driven galactic winds affect both the abundance [256,276,277] and inner structure of low-luminosity galaxies. Acting in conjunction with reionization, such winds strongly suppress galaxy formation in small haloes, reducing the abundance of luminous low-mass galaxies [269,271,272].

Energy injection from supernova can potentially alter the inner structure of dark matter haloes: if gas becomes gravitationally dominant in the center and most of it is removed suddenly, as could happen in a starburst, energy can be transferred from the gas to the dark matter and this can cause the center to expand, turning the original NFW cusp into a core. This mechanism, first proposed in the 1990s [278], became fashionable again several years later [279–286] when tentative observational evidence for cores, particularly in dwarf galaxies, began to emerge [287]. This evidence, however, is controversial [288,289]. While the proof of concept in [278] was based on a single explosive event, recent simulations have shown that repeated outflows can create rapid fluctuations in the gravitational potential which efficiently transfer energy to the dark matter [282]. This core-formation mechanism depends on the details of the baryon physics implemented in the simulation [290] and not all cosmological simulations produce cores in dwarf galaxies [291]. On scales larger than dwarfs, energy injection by AGN has been invoked as a mechanism for core formation; however, the conditions required to alter the deep potential wells of massive galaxies appear quite extreme [292–296].

Energy injection into subhaloes: tidal effects from baryonic structures. In Section 3.3 we described the tidal effects that the host halo induces on the dynamics and structure of subhaloes. The presence of a central galaxy enhances these effects both in subhaloes and in the satellite galaxies within them, particularly when their orbits cross the region where the central galaxy dominates the tidal field. Tidal shocking by a galactic disc can result in the total disruption of subhaloes around the central regions of the host [297,298] and other structural changes. Current hydrodynamical simulations of Milky-Way-like galaxies and their environment seem to agree that the overall effect is a substantial reduction in the number of subhaloes near the center [299–301].

There has been great progress in the past decade in incorporating baryonic physics into full cosmological simulations; today galaxy formation and evolution can be modeled in unprecedented detail [302–306]. In this way the effect on the dark matter phase-space distribution of the complex interplay between the cooling and heating mechanisms of baryons described above can be studied in their full cosmological setting. Despite this undeniable progress, many aspects of baryonic physics remain poorly understood and, when they involve processes on scales below the resolution of the simulation, they need to be included as a *subgrid model*. There are different approaches to this problem which are often difficult to validate, and this translates into substantial uncertainties in some of the predictions of the simulations (see [304] for a discussion of the limitations of gas dynamics simulations).

---

[59] This mass threshold is smaller at higher redshifts, see e.g., Figure 3 of [266].

4.2. Astrophysical Tests of the Nature of the Dark Matter

Laboratory searches for dark matter have so far proved unsuccessful. This, and the failure to find evidence for SUPERSYMMETRY, has generated gloom among proponents of the lightest stable supersymmetric particle as the dark matter (even though the mass of the Higgs boson suggests that the supersymmetry scale is likely to be larger than a few TeV, beyond the reach of the LHC). There have been, however, claims that both CDM-WIMPs, and WDM particles in the form of sterile neutrinos of mass 7 keV, have been discovered, the former through γ-ray annihilation radiation from the Galactic Center [307], the latter through a 3.5 keV decay line in the X-ray spectra of galaxies and clusters [308,309] [60]. These claims are highly controversial but, since cosmogonic models based on such particles have strong predictive power, they are disprovable with appropriate astrophysical observations.

The standard CDM model has naturally come under the closest scrutiny. Perhaps the two most important predictions of this model (derived from *N*-body simulations) are: (i) the existence of a vast population of haloes and subhaloes which, below a mass of order 10^9 $M_\odot$, are dark; (ii) the presence (in the absence of the baryon effects discussed in the preceding section) of a steep cusp ($\rho \propto r^{-1}$) in the density profile of dark matter haloes of all masses. These two predictions are related to three of the much publicized four problems of the CDM model on subgalactic scales (often referred to as the "small-scale crisis" of CDM): the *(i)* missing satellites; *(ii)* too-big-to-fail and *(iii)* core-cusp problems. The fourth is the so-called *(iv)* planes of satellites problem. Indeed, some of alternative dark matter particle models, such as SIDM, have been proposed specifically to solve some or all these perceived astrophysical problems.

The missing satellites problem is the discrepancy between the relatively small number of satellites observed around the Milky Way and M31 and the many orders of magnitude larger number of halo substructures predicted by CDM *N*-body simulations [312,313]. The "too-big-to-fail" problem is the existence in CDM *N*-body simulations of massive, dense galactic subhaloes (maximum circular velocities, $V_{\max} > 30$ km/s) whose kinematics appear inconsistent with those of the brightest Milky Way satellites [314]. The core-cusp problem is the discrepancy between the cuspy universal NFW density profiles predicted for pure CDM/WDM haloes and the inference of central cores in some galaxies, particularly dwarfs (e.g., [315]). The "planes of satellites" problem is the arrangement of the bright satellites of the Milky Way, M31 (and a few others) on a thin plane in which the satellites seem to be coherently rotating and which have been claimed to be incompatible with CDM [316–318].

The first three of the four perceived problems can be solved once the effects of baryons discussed in Section 4.1 are taken into account. Perhaps paradoxically, the solution to what later became known as the "missing satellites" problem was understood long before it came to be regarded as a problem for CDM. The strong suppression of galaxy formation in haloes below a mass of $\sim 10^{10} M_\odot$ was originally calculated using semi-analytic techniques [264,265,267], as were the implications for the abundance of galactic satellites in the CDM model [269,271,272]. This solution has been repeatedly confirmed by modern gas dynamic simulations (e.g., [250,266,319,320]). Similarly, the "too-big-to-fail" problem disappears when baryons are taken into account, in this case through the more subtle effect of the reduced growth of subhaloes arising from the early loss of baryons mentioned above [250]. The "core-cusp problem", if it exists at all, can also be solved by the type of explosive baryonic effects discussed in Section 4.1, which can transform NFW cusps into cores[61]. The existence of "planes of satellites" in the Milky Way and M31 turns out not to be as unlikely as has been claimed (e.g., [317,323]), once the statistics are calculated rigorously, taking into account the "look elsewhere effect" [324][62]. The origin of these planes is almost certainly the anisotropic nature of the accretion of satellites along

---

[60] Contrary to some claims, XMM data for Draco, and Hitomi data for Perseus, are consistent with a 7 keV neutrino [310,311].
[61] Other baryon effects that can transform cusps into cores have been proposed (e.g., [321,322]) but have been less studied.
[62] See [325] for an opposed view.

filaments of the cosmic web [326,327] although the exact mechanism is still unclear as is the expected frequency of these structures.

While it is now generally agreed among practitioners of the field that CDM is not afflicted by a "missing satellite" or a "too-big-to-fail" problem, the data on the satellites of the Milky Way can be used to constrain alternative dark matter models, particularly those with a cutoff in the primordial power spectrum. In WDM, the cutoff length scale varies roughly inversely with the mass of particle. Thus, if this mass is too small, then too few small-mass haloes would form and their abundance could be too low to account for the observed number of satellites of the Milky Way. The expected subhalo abundance increases roughly in proportion to the mass of the parent halo [328] so, in reality, the observed abundance of satellites constrains both the particle mass and the host halo mass simultaneously. For instance, using a semi-analytic model of galaxy formation, the thermal WDM model was found to conflict with the data if the Milky Way halo mass is smaller than 1.1×10^{12} $M_\odot$ [329]. Using a similar approach, [330] have ruled out a significant fraction of the parameter space of sterile neutrinos and conclude that the models that are in best agreement with the observed 3.5 keV line require the Milky Way halo to have a mass no smaller than 1.5×10^{12} $M_\odot$, a value that may already be in conflict with the most recent determinations of the Milky Way halo mass [331]. We should note that since the central densities of WDM haloes are lower than those of their CDM counterparts, the "too-big-to-fail problem" is easily avoided in WDM [153,332].

Although the strongest constraints on the SIDM cross-section come from the shapes and dynamics of massive haloes (particularly of galaxy clusters, see e.g., Table 1 of [18]), the Milky Way satellites are perhaps the best testbed for SIDM, since it is in these systems that the model shows its greatest promise as an alternative to CDM. A few years ago it was suggested that the interesting range of cross-sections for the SIDM model to alleviate the "too-big-to-fail" problem (without taking into account the baryonic processes just mentioned) is $0.1 \lesssim \sigma_T/m_\chi \lesssim 10$ cm^2/g [168]. Since then, several studies have taken a closer look at the properties of the Milky Way satellites within the context of SIDM and the picture that is emerging points to promising tests for the near future which will either strengthen SIDM as an alternative to CDM or narrow the range of allowed cross-sections. For instance, the diversity of dark matter densities on subkiloparsec scales in the Milky Way satellites is difficult to accommodate for SIDM cross sections $\sigma/m_\chi \sim 1$ cm^2/g [32]. The inferred high dark matter densities in the ultra-faint satellites (albeit uncertain due to possible systematic effects) are at first sight difficult to explain within SIDM, which naturally predicts cores, particularly in low-mass dark matter-dominated haloes. However, a gravothermal collapse phase in SIDM haloes has recently been proposed [32,249,251,252] as a mechanism to create a diverse population of dwarf-size haloes, some of which would be cuspy (those that collapse), and others that would have cores. If cores are indeed shown to be present in (some) dwarf galaxies, then dark matter self-interactions and the explosive baryon effects in CDM mentioned above provide alternative explanations that need to be contrasted. A promising way to achieve this, recently put forward [333], is to search for distinct signatures in the detailed kinematics of the stellar population as they respond differently to these two core-formation mechanisms, one impulsive (supernova feedback) and the other adiabatic (SIDM).

Since, as we have seen, the simplicity of the predictions of N-body simulations can be easily obscured by the complexity of baryon effects, testing dark matter models with astronomical observations might, at first sight, seem a hopeless task. In fact, this is not the case: most haloes in CDM (and in many alternative dark matter models) are dark, i.e., unaffected or almost unaffected by baryons. It is the existence of a vast population of such small-mass haloes ($m \lesssim 5 \times 10^9$ $M_\odot$) that is the hallmark of the CDM model that distinguishes it from, for example, the WDM model. Fortunately, nature has provided us with several tools to detect dark objects in the universe. One of these takes advantage of a side effect of cosmic reionization which allows haloes in a small mass window ($10^8 \lesssim 5 \times 10^9$) $M_\odot$ to retain neutral hydrogen in hydrostatic equilibrium with the dark matter potential and in thermal equilibrium with the ionizing UV background, gas which is, however, too diffuse to make stars [334]. These objects called RELHICs (REionization-Limited HI Clouds, [335]) may be detectable in forthcoming blind HI surveys and provide, in principle, a critical test of CDM and related models in a regime that has not been proved before.

An interesting idea that has been proposed to infer the existence of small dark subhalos orbiting in the Milky Way halo is the disturbance they cause when they cross a tidal stellar stream [336]. When a subhalo crosses a stream, it induces velocity changes along and across the stream that can give rise to a visible gap, particularly in cold streams such as those stripped from globular clusters. The cross-section for gap creation is dominated by the smallest subhalos so gaps can, in principle, constrain the identity of the dark matter. The creation of gaps has been investigated with analytical treatments or idealized numerical studies and it has been suggested that perturbers of mass $\sim 10^7$ $M_\odot$ could be detected in the GD-1 and Pal 5 globular cluster stellar streams [337]. A complication of this method is that perturbations on the streams can be induced not only by dark subhaloes but also by giant molecular clouds and the bar at the center of the Milky Way [338]. Recent deep imaging around the Pal 5 stellar stream does indeed reveal significant disturbances, in particular two gaps which have been attributed to the impact of subhalos of mass in the range 10^6–10^7 $M_\odot$ and 10^7–10^8 $M_\odot$ respectively (although the smaller gap could also be due to the impact of a giant molecular cloud) [339,340].

However, perhaps the most direct method to search for the ubiquitous small-mass dark haloes is gravitational lensing. There are two specific instances where strong gravitational lensing could provide the means to do this. The first are the "flux-ratio anomalies" seen in some multiple-lensed quasars; the second are small distortions of Einstein rings and large arcs.

In a multiple-lensed image, the magnifications are determined by high order derivatives of the lensing potential and are therefore particularly sensitive to small changes in the potential such as those produced by intervening small-mass structures. If the mass distribution of the lens is smooth, the ratios of the fluxes of close images (formed when the sources are close to a fold or a cusp of the caustic) follow a certain asymptotic relation [341,342]. These smooth-lens relations are violated if there are intervening structures or substructures in the lens giving rise to flux-ratio anomalies, which probe the total amount of mass in structures along the line of sight to the lens [341,343,344]. Flux-ratio anomalies have been observed in several quadruple-lensed quasars but dark substructures alone are insufficient to explain the observed anomalies [345], implying that other effects such as inadequate lens modeling may be at work. With better modeling of the lens (including stellar discs and luminous satellites), it has been possible to set a lower limit to the mass of a thermal WDM particle (see [346] and Harvey et al. in preparation), similar to the limits from satellite counts discussed above and to those derived from the observed inhomogeneity of the gas distribution at high redshift probed by the Lyman-α forest [347].

A more direct strategy for detecting dark structures and substructures is to search for distortions in strongly lensed images. When the source (a background galaxy), the lens (a massive halo) and the observer are perfectly aligned, a circular feature near the center of the lens, an Einstein ring, is formed; if the alignment is not perfect, then giant arcs are formed. If the lens is a halo of mass larger than $\sim 10^{13}$ $M_\odot$, the radius of the Einstein ring is generally larger than the image of the central galaxy and can thus be studied in detail. If a halo or subhalo happens to be projected onto the Einstein ring, it too will gravitationally lens the light from the source producing a small distortion in the image of the Einstein ring or giant arc [348]. This strategy has already yielded a halo of $\sim 10^8$ $M_\odot$ [349] [63] and could detect haloes as small as $\sim 10^7$ $M_\odot$ [353,354].

Detecting the small signal generated by individual projected haloes or subhaloes requires accurate modeling of the source and the lens (the "macro" model; (e.g., [348,355])) and sophisticated statistical techniques to analyze the image residuals. Dark haloes imprint other observable features onto strong arcs. For example, distortions to the lensing potential caused by the cumulative contribution of many

---

[63] This halo mass was estimated assuming a truncated pseudo-Jaffe profile (see e.g., Equation (42) in [350]). The inferred mass is likely to be larger if an NFW profile is assumed instead. For instance, a similar dark matter substructure detected with lensing was reported by [351] with a mass of $\sim 3.5 \times 10^9$ $M_\odot$ assuming a truncated pseudo-Jaffe profile, while assuming an NFW profile this substructure is estimated to have a mass of $\sim 10^{10}$ $M_\odot$ [352].

hundreds of projected structures produce unique correlated residuals in the lensed image, the nature of which is dependent on the abundance and mass distribution of the halo population and, therefore, on the nature of the dark matter [356,357]. The mass function of dark haloes may also be detectable through the N-point functions of the projected density field or the substructure convergence power spectrum [357].

A very attractive feature of strong lensing as a means to detect small-mass objects is that, for lens configurations of interest, the dominant source of strong arc distortions are field haloes along the line of sight, rather than subhaloes resident in the lens [358,359][64]. This makes this test uniquely powerful because, as we have seen, the haloes of interest, of mass less than $\sim 10^8$ M$_\odot$, are completely dark: they have never been modified in any way by baryons. Thus, the test depends mostly on the abundance of pristine "field" dark matter haloes which we know very well how to calculate rigorously and precisely with N-body simulations for cosmological models of interest.

Approximately a few hundred high quality strong lens systems would suffice to rule out either the 7 keV sterile neutrino model or CDM itself [360]. Very high-resolution imaging is the primary requirement, either in the optical or UV, or using interferometry at submillimeter and longer wavelengths [354]. At least several tens of systems with high quality data are already available and future imaging facilities such as LSST and Euclid will increase the number of suitable strong lenses by orders of magnitude. By bypassing the complications introduced by baryons, which have spoiled all previous efforts to test the CDM model unambiguously and distinguish it from alternative models, be they on small or large scales, gravitational lensing offers a unique opportunity for a breakthrough in this quest from astrophysics evidence alone.

Funding: This research received funding from the Icelandic Research Fund (Rannís), Grant of Excellence number 173929, the European Research Council (ERC) Advanced Investigator grant DMIDAS (GA 786910) and the Science and Technology Facilities Council (STFC) [grant number ST/F001166/1, ST/I00162X/1, ST/P000541/1].

Acknowledgments: We are very grateful to Alejandro Benitez-Llambay, Sownak Bose, Jiaxin Han, Mark Lovell and Simon White for their valuable comments. Special thanks to Simon for his careful reading of the manuscript, insightful comments and suggestions. All of these have significantly improved our paper. We also thank Sebastian Bohr for his help in creating the right panel of Figure 2. JZ acknowledges support by a Grant of Excellence from the Icelandic Research fund (grant number 173929-052). CSF acknowledges support from the European Research Council (ERC) Advanced Investigator grant DMIDAS (GA 786910) and the Science and Technology Facilities Council (STFC) [grant number ST/F001166/1, ST/I00162X/1, ST/P000541/1]. Some of the simulations reported in this paper were carried out on the DiRAC Data Centric system at Durham University, operated by the ICC on behalf of the STFC DiRAC HPC Facility (www.dirac.ac.uk). This equipment was funded by BIS National E-infrastructure capital grant ST/K00042X/1, STFC capital grant ST/H008519/1, and STFC DiRAC Operations grant ST/K003267/1 and Durham University. DiRAC is part of the National E-Infrastructure.

Conflicts of Interest: The authors declare no conflict of interest.

References

1. Planck Collaboration. Planck 2018 results. VI. Cosmological parameters. *arXiv* **2018**, arXiv:1807.06209.
2. Walker, M.G.; Mateo, M.; Olszewski, E.W.; Peñarrubia, J.; Evans, N.W.; Gilmore, G. A Universal Mass Profile for Dwarf Spheroidal Galaxies? *Astrophys. J.* **2009**, *704*, 1274–1287. [CrossRef]
3. Łokas, E.L.; Mamon, G.A. Dark matter distribution in the Coma cluster from galaxy kinematics: Breaking the mass-anisotropy degeneracy. *Mon. Not. R. Astron. Soc.* **2003**, *343*, 401–412.
4. Springel, V.; White, S.D.M.; Jenkins, A.; Frenk, C.S.; Yoshida, N.; Gao, L.; Navarro, J.; Thacker, R.; Croton, D.; Helly, J.; et al. Simulations of the formation, evolution and clustering of galaxies and quasars. *Nature* **2005**, *435*, 629–636. [CrossRef] [PubMed]
5. Guth, A.H. Inflationary universe: A possible solution to the horizon and flatness problems. *Phys. Rev. D* **1981**, *23*, 347–356. [CrossRef]

---

[64] This has been demonstrated explicitly for the case of Einstein ring distortions but it may hold true for other tests as well.

6. Linde, A.D. A new inflationary universe scenario: A possible solution of the horizon, flatness, homogeneity, isotropy and primordial monopole problems. *Phys. Lett. B* **1982**, *108*, 389–393. [CrossRef]
7. Kahlhoefer, F. Review of LHC dark matter searches. *Int. J. Mod. Phys. A* **2017**, *32*, 1730006. [CrossRef]
8. Xenon Collaboration. Dark Matter Search Results from a One Ton-Year Exposure of XENON1T. *Phys. Rev. Lett.* **2018**, *121*, 111302, [CrossRef]
9. ADMX Collaboration. Search for Invisible Axion Dark Matter with the Axion Dark Matter Experiment. *Phys. Rev. Lett.* **2018**, *120*, 151301. [CrossRef]
10. Fermi-LAT Collaboration; DES Collaboration. Searching for Dark Matter Annihilation in Recently Discovered Milky Way Satellites with Fermi-Lat. *Astrophys. J.* **2017**, *834*, 110. [CrossRef]
11. Horiuchi, S.; Humphrey, P.J.; Oñorbe, J.; Abazajian, K.N.; Kaplinghat, M.; Garrison-Kimmel, S. Sterile neutrino dark matter bounds from galaxies of the Local Group. *Phys. Rev. D* **2014**, *89*, 025017. [CrossRef]
12. Gondolo, P.; Gelmini, G. Cosmic abundances of stable particles: Improved analysis. *Nucl. Phys. B* **1991**, *360*, 145–179. [CrossRef]
13. Jungman, G.; Kamionkowski, M.; Griest, K. Supersymmetric dark matter. *Phys. Rep.* **1996**, *267*, 195–373. [CrossRef]
14. Davis, M.; Efstathiou, G.; Frenk, C.S.; White, S.D.M. The evolution of large-scale structure in a universe dominated by cold dark matter. *Astrophys. J.* **1985**, *292*, 371–394. [CrossRef]
15. Preskill, J.; Wise, M.B.; Wilczek, F. Cosmology of the invisible axion. *Phys. Lett. B* **1983**, *120*, 127–132. [CrossRef]
16. Boyarsky, A.; Drewes, M.; Lasserre, T.; Mertens, S.; Ruchayskiy, O. Sterile Neutrino Dark Matter. *arXiv* **2018**, arXiv:1807.07938.
17. Hui, L.; Ostriker, J.P.; Tremaine, S.; Witten, E. Ultralight scalars as cosmological dark matter. *Phys. Rev. D* **2017**, *95*, 043541. [CrossRef]
18. Tulin, S.; Yu, H.B. Dark matter self-interactions and small scale structure. *Phys. Rep.* **2018**, *730*, 1–57. [CrossRef]
19. Zurek, K.M. Asymmetric Dark Matter: Theories, signatures, and constraints. *Phys. Rep.* **2014**, *537*, 91–121. [CrossRef]
20. Buckley, M.R.; Zavala, J.; Cyr-Racine, F.Y.; Sigurdson, K.; Vogelsberger, M. Scattering, damping, and acoustic oscillations: Simulating the structure of dark matter halos with relativistic force carriers. *Phys. Rev. D* **2014**, *90*, 043524. [CrossRef]
21. Cyr-Racine, F.Y.; Sigurdson, K.; Zavala, J.; Bringmann, T.; Vogelsberger, M.; Pfrommer, C. ETHOS: An effective theory of structure formation: From dark particle physics to the matter distribution of the Universe. *Phys. Rev. D* **2016**, *93*, 123527. [CrossRef]
22. Meszaros, P. The behaviour of point masses in an expanding cosmological substratum. *Astron. Astrophys.* **1974**, *37*, 225–228.
23. Green, A.M.; Hofmann, S.; Schwarz, D.J. The first WIMPy halos. *J. Cosm. Part. Phys.* **2005**, *8*, 3. [CrossRef]
24. Bringmann, T. Particle models and the small-scale structure of dark matter. *New J. Phys.* **2009**, *11*, 105027. [CrossRef]
25. Bœhm, C.; Riazuelo, A.; Hansen, S.H.; Schaeffer, R. Interacting dark matter disguised as warm dark matter. *Phys. Rev. D* **2002**, *66*, 083505. [CrossRef]
26. Bœhm, C.; Schewtschenko, J.A.; Wilkinson, R.J.; Baugh, C.M.; Pascoli, S. Using the Milky Way satellites to study interactions between cold dark matter and radiation. *Mon. Not. R. Astron. Soc.* **2014**, *445*, L31–L35. [CrossRef]
27. Loeb, A.; Zaldarriaga, M. Small-scale power spectrum of cold dark matter. *Phys. Rev. D* **2005**, *71*, 103520. [CrossRef]
28. Viel, M.; Becker, G.D.; Bolton, J.S.; Haehnelt, M.G. Warm dark matter as a solution to the small scale crisis: New constraints from high redshift Lyman-α forest data. *Phys. Rev. D* **2013**, *88*, 043502. [CrossRef]
29. Schive, H.Y.; Chiueh, T.; Broadhurst, T.; Huang, K.W. Contrasting Galaxy Formation from Quantum Wave Dark Matter, ψDM, with ΛCDM, using Planck and Hubble Data. *Astrophys. J.* **2016**, *818*, 89. [CrossRef]
30. Cole, S.; Percival, W.J.; Peacock, J.A.; Norberg, P.; Baugh, C.M.; Frenk, C.S.; Baldry, I.; Bland-Hawthorn, J.; Bridges, T.; Cannon, R.; et al. The 2dF Galaxy Redshift Survey: Power-spectrum analysis of the final data set and cosmological implications. *Mon. Not. R. Astron. Soc.* **2005**, *362*, 505–534. [CrossRef]

31. Percival, W.J.; Nichol, R.C.; Eisenstein, D.J.; Frieman, J.A.; Fukugita, M.; Loveday, J.; Pope, A.C.; Schneider, D.P.; Szalay, A.S.; Tegmark, M.; et al. The Shape of the Sloan Digital Sky Survey Data Release 5 Galaxy Power Spectrum. *Astrophys. J.* **2007**, *657*, 645–663. [CrossRef]
32. Zavala, J.; Lovell, M.R.; Vogelsberger, M.; Burger, J.D. The diverse dark matter density at sub-kiloparsec scales in Milky Way satellites:implications for the nature of dark matter. *arXiv* **2019**, arXiv:1904.09998.
33. Carroll, S.M.; Press, W.H.; Turner, E.L. The cosmological constant. *Annu. Rev. Astron. Astrophys.* **1992**, *30*, 499–542. [CrossRef]
34. Baumann, D.; Nicolis, A.; Senatore, L.; Zaldarriaga, M. Cosmological non-linearities as an effective fluid. *J. Cosmol. Astropart. Phys.* **2012**, *7*, 51. [CrossRef]
35. Carrasco, J.J.M.; Hertzberg, M.P.; Senatore, L. The effective field theory of cosmological large scale structures. *J. High Energy Phys.* **2012**, *9*, 82. [CrossRef]
36. Gunn, J.E.; Gott, J.R, III. On the Infall of Matter Into Clusters of Galaxies and Some Effects on Their Evolution. *Astrophys. J.* **1972**, *176*, 1. [CrossRef]
37. Sheth, R.K.; Mo, H.J.; Tormen, G. Ellipsoidal collapse and an improved model for the number and spatial distribution of dark matter haloes. *Mon. Not. R. Astron. Soc.* **2001**, *323*, 1–12. [CrossRef]
38. Cooray, A.; Sheth, R. Halo models of large scale structure. *Phys. Rep.* **2002**, *372*, 1–129. [CrossRef]
39. Davis, M.; Peebles, P.J.E. On the integration of the BBGKY equations for the development of strongly nonlinear clustering in an expanding universe. *Astrophys. J. Suppl.* **1977**, *34*, 425–450. [CrossRef]
40. Smith, R.E.; Peacock, J.A.; Jenkins, A.; White, S.D.M.; Frenk, C.S.; Pearce, F.R.; Thomas, P.A.; Efstathiou, G.; Couchman, H.M.P. Stable clustering, the halo model and non-linear cosmological power spectra. *Mon. Not. R. Astron. Soc.* **2003**, *341*, 1311–1332. [CrossRef]
41. Afshordi, N.; Mohayaee, R.; Bertschinger, E. Hierarchy in the phase space and dark matter astronomy. *Phys. Rev. D* **2010**, *81*, 101301. [CrossRef]
42. Zavala, J.; Afshordi, N. Clustering in the phase space of dark matter haloes—II. Stable clustering and dark matter annihilation. *Mon. Not. R. Astron. Soc.* **2014**, *441*, 1329–1339. [CrossRef]
43. Zavala, J.; Afshordi, N. Universal clustering of dark matter in phase space. *Mon. Not. R. Astron. Soc.* **2016**, *457*, 986–992. [CrossRef]
44. Hahn, O.; Abel, T.; Kaehler, R. A new approach to simulating collisionless dark matter fluids. *Mon. Not. R. Astron. Soc.* **2013**, *434*, 1171–1191. [CrossRef]
45. Angulo, R.E.; Hahn, O.; Abel, T. The warm dark matter halo mass function below the cut-off scale. *Mon. Not. R. Astron. Soc.* **2013**, *434*, 3337–3347. [CrossRef]
46. Dehnen, W.; Read, J.I. N-body simulations of gravitational dynamics. *Eur. Phys. J. Plus* **2011**, *126*, 55. [CrossRef]
47. Dehnen, W. Towards optimal softening in three-dimensional N-body codes—I. Minimizing the force error. *Mon. Not. R. Astron. Soc.* **2001**, *324*, 273–291. [CrossRef]
48. Klypin, A.A.; Shandarin, S.F. Three-dimensional numerical model of the formation of large-scale structure in the Universe. *Mon. Not. R. Astron. Soc.* **1983**, *204*, 891–907. [CrossRef]
49. Melott, A.L. Massive neutrinos in large-scale gravitational clustering. *Astrophys. J.* **1983**, *264*, 59–78. [CrossRef]
50. Frenk, C.S.; White, S.D.M.; Davis, M. Nonlinear evolution of large-scale structure in the universe. *Astrophys. J.* **1983**, *271*, 417–430. [CrossRef]
51. Hockney, R.W.; Eastwood, J.W. *Computer Simulation Using Particles*; CRC Press: Boca Raton, FL, USA, 1988.
52. Efstathiou, G.; Eastwood, J.W. On the clustering of particles in an expanding universe. *Mon. Not. R. Astron. Soc.* **1981**, *194*, 503–525. [CrossRef]
53. Barnes, J.; Hut, P. A hierarchical O(N log N) force-calculation algorithm. *Nature* **1986**, *324*, 446–449. [CrossRef]
54. Springel, V. The cosmological simulation code GADGET-2. *Mon. Not. R. Astron. Soc.* **2005**, *364*, 1105–1134. [CrossRef]
55. Kochanek, C.S.; White, M. A Quantitative Study of Interacting Dark Matter in Halos. *Astrophys. J.* **2000**, *543*, 514–520. [CrossRef]
56. Davé, R.; Spergel, D.N.; Steinhardt, P.J.; Wandelt, B.D. Halo Properties in Cosmological Simulations of Self-interacting Cold Dark Matter. *Astrophys. J.* **2001**, *547*, 574–589. [CrossRef]

57. Vogelsberger, M.; Zavala, J.; Loeb, A. Subhaloes in self-interacting galactic dark matter haloes. *Mon. Not. R. Astron. Soc.* **2012**, *423*, 3740–3752. [CrossRef]
58. Rocha, M.; Peter, A.H.G.; Bullock, J.S.; Kaplinghat, M.; Garrison-Kimmel, S.; Oñorbe, J.; Moustakas, L.A. Cosmological simulations with self-interacting dark matter—I. Constant-density cores and substructure. *Mon. Not. R. Astron. Soc.* **2013**, *430*, 81–104. [CrossRef]
59. Robertson, A.; Massey, R.; Eke, V. Cosmic particle colliders: Simulations of self-interacting dark matter with anisotropic scattering. *Mon. Not. R. Astron. Soc.* **2017**, *467*, 4719–4730. [CrossRef]
60. Lynden-Bell, D.; Eggleton, P.P. On the consequences of the gravothermal catastrophe. *Mon. Not. R. Astron. Soc.* **1980**, *191*, 483–498. [CrossRef]
61. Koda, J.; Shapiro, P.R. Gravothermal collapse of isolated self-interacting dark matter haloes: N-body simulation versus the fluid model. *Mon. Not. R. Astron. Soc.* **2011**, *415*, 1125–1137. [CrossRef]
62. Schive, H.Y.; Chiueh, T.; Broadhurst, T. Cosmic structure as the quantum interference of a coherent dark wave. *Nat. Phys.* **2014**, *10*, 496–499. [CrossRef]
63. Mocz, P.; Vogelsberger, M.; Robles, V.H.; Zavala, J.; Boylan-Kolchin, M.; Fialkov, A.; Hernquist, L. Galaxy formation with BECDM—I. Turbulence and relaxation of idealized haloes. *Mon. Not. R. Astron. Soc.* **2017**, *471*, 4559–4570. [CrossRef] [PubMed]
64. Efstathiou, G.; Davis, M.; White, S.D.M.; Frenk, C.S. Numerical techniques for large cosmological N-body simulations. *Astrophys. J. Suppl.* **1985**, *57*, 241–260. [CrossRef]
65. Hahn, O.; Abel, T. Multi-scale initial conditions for cosmological simulations. *Mon. Not. R. Astron. Soc.* **2011**, *415*, 2101–2121. [CrossRef]
66. Jenkins, A. A new way of setting the phases for cosmological multiscale Gaussian initial conditions. *Mon. Not. R. Astron. Soc.* **2013**, *434*, 2094–2120. [CrossRef]
67. Sirko, E. Initial Conditions to Cosmological N-Body Simulations, or, How to Run an Ensemble of Simulations. *Astrophys. J.* **2005**, *634*, 728–743. [CrossRef]
68. Mo, H.; van den Bosch, F.C.; White, S. *Galaxy Formation and Evolution*; Cambridge University Press: Cambridge, UK, 2010.
69. White, S.D.M. Formation and Evolution of Galaxies. In *Cosmology and Large Scale Structure*; Schaeffer, R.; Silk, J.; Spiro, M.; Zinn-Justin, J., Eds.; Cambridge University Press: Cambridge, UK, 1996; p. 349.
70. Zel'dovich, Y.B. Gravitational instability: An approximate theory for large density perturbations. *Astron. Astrophys.* **1970**, *5*, 84–89.
71. Jenkins, A. Second-order Lagrangian perturbation theory initial conditions for resimulations. *Mon. Not. R. Astron. Soc.* **2010**, *403*, 1859–1872. [CrossRef]
72. Pontzen, A.; Roškar, R.; Stinson, G.S.; Woods, R.; Reed, D.M.; Coles, J.; Quinn, T.R. Pynbody: Astrophysics Simulation Analysis for Python. In *Astrophysics Source Code Library*; ascl:1305.002; 2013. Available online: https://pynbody.github.io/pynbody/#acknowledging-pynbody-in-scientific-publications (accessed on 24 September 2019).
73. Smith, R.E.; Markovic, K. Testing the warm dark matter paradigm with large-scale structures. *Phys. Rev. D* **2011**, *84*, 063507. [CrossRef]
74. Wang, J.; White, S.D.M. Discreteness effects in simulations of hot/warm dark matter. *Mon. Not. R. Astron. Soc.* **2007**, *380*, 93–103. [CrossRef]
75. Lovell, M.R.; Frenk, C.S.; Eke, V.R.; Jenkins, A.; Gao, L.; Theuns, T. The properties of warm dark matter haloes. *Mon. Not. R. Astron. Soc.* **2014**, *439*, 300–317. [CrossRef]
76. Hobbs, A.; Read, J.I.; Agertz, O.; Iannuzzi, F.; Power, C. NOVel Adaptive softening for collisionless N-body simulations: Eliminating spurious haloes. *Mon. Not. R. Astron. Soc.* **2016**, *458*, 468–479. [CrossRef]
77. Boylan-Kolchin, M.; Springel, V.; White, S.D.M.; Jenkins, A.; Lemson, G. Resolving cosmic structure formation with the Millennium-II Simulation. *Mon. Not. R. Astron. Soc.* **2009**, *398*, 1150–1164. [CrossRef]
78. Frenk, C.S.; White, S.D.M.; Davis, M.; Efstathiou, G. The formation of dark halos in a universe dominated by cold dark matter. *Astrophys. J.* **1988**, *327*, 507–525. [CrossRef]
79. Kuhlen, M.; Vogelsberger, M.; Angulo, R. Numerical simulations of the dark universe: State of the art and the next decade. *Phys. Dark Univ.* **2012**, *1*, 50–93. [CrossRef]
80. Potter, D.; Stadel, J.; Teyssier, R. PKDGRAV3: beyond trillion particle cosmological simulations for the next era of galaxy surveys. *Comput. Astrophys. Cosmol.* **2017**, *4*, 2. [CrossRef]
81. Geller, M.J.; Huchra, J.P. Mapping the universe. *Science* **1989**, *246*, 897–903. [CrossRef]

82. Gott, J.R., III; Jurić, M.; Schlegel, D.; Hoyle, F.; Vogeley, M.; Tegmark, M.; Bahcall, N.; Brinkmann, J. A Map of the Universe. *Astrophys. J.* **2005**, *624*, 463–484. [CrossRef]
83. Colless, M.; Dalton, G.; Maddox, S.; Sutherland, W.; Norberg, P.; Cole, S.; Bland-Hawthorn, J.; Bridges, T.; Cannon, R.; Collins, C.; et al. The 2dF Galaxy Redshift Survey: Spectra and redshifts. *Mon. Not. R. Astron. Soc.* **2001**, *328*, 1039–1063. [CrossRef]
84. Croton, D.J.; Springel, V.; White, S.D.M.; De Lucia, G.; Frenk, C.S.; Gao, L.; Jenkins, A.; Kauffmann, G.; Navarro, J.F.; Yoshida, N. The many lives of active galactic nuclei: Cooling flows, black holes and the luminosities and colours of galaxies. *Mon. Not. R. Astron. Soc.* **2006**, *365*, 11–28. [CrossRef]
85. Springel, V.; Frenk, C.S.; White, S.D.M. The large-scale structure of the Universe. *Nature* **2006**, *440*, 1137–1144. [CrossRef] [PubMed]
86. Bardeen, J.M.; Bond, J.R.; Kaiser, N.; Szalay, A.S. The statistics of peaks of Gaussian random fields. *Astrophys. J.* **1986**, *304*, 15–61. [CrossRef]
87. White, S.D.M.; Frenk, C.S. Galaxy formation through hierarchical clustering. *Astrophys. J.* **1991**, *379*, 52–79. [CrossRef]
88. Kauffmann, G.; Nusser, A.; Steinmetz, M. Galaxy formation and large-scale bias. *Mon. Not. R. Astron. Soc.* **1997**, *286*, 795–811. [CrossRef]
89. Kauffmann, G.; Colberg, J.M.; Diaferio, A.; White, S.D.M. Clustering of galaxies in a hierarchical universe—I. Methods and results at z = 0. *Mon. Not. R. Astron. Soc.* **1999**, *303*, 188–206. [CrossRef]
90. Benson, A.J.; Cole, S.; Frenk, C.S.; Baugh, C.M.; Lacey, C.G. The nature of galaxy bias and clustering. *Mon. Not. R. Astron. Soc.* **2000**, *311*, 793–808. [CrossRef]
91. Springel, V.; White, S.D.M.; Tormen, G.; Kauffmann, G. Populating a cluster of galaxies—I. Results at [formmu2]z = 0. *Mon. Not. R. Astron. Soc.* **2001**, *328*, 726–750. [CrossRef]
92. Lacey, C.G.; Baugh, C.M.; Frenk, C.S.; Benson, A.J.; Bower, R.G.; Cole, S.; Gonzalez-Perez, V.; Helly, J.C.; Lagos, C.D.P.; Mitchell, P.D. A unified multiwavelength model of galaxy formation. *Mon. Not. R. Astron. Soc.* **2016**, *462*, 3854–3911. [CrossRef]
93. White, S.D.M.; Frenk, C.S.; Davis, M. Clustering in a neutrino-dominated universe. *Astrophys. J. Lett.* **1983**, *274*, L1–L5. [CrossRef]
94. Angulo, R.E.; White, S.D.M. The birth and growth of neutralino haloes. *Mon. Not. R. Astron. Soc.* **2010**, *401*, 1796–1803. [CrossRef]
95. Stücker, J.; Busch, P.; White, S.D.M. The median density of the Universe. *Mon. Not. R. Astron. Soc.* **2018**, *477*, 3230–3246. [CrossRef]
96. White, M. The mass of a halo. *Astron. Astrophys.* **2001**, *367*, 27–32. [CrossRef]
97. Cuesta, A.J.; Prada, F.; Klypin, A.; Moles, M. The virialized mass of dark matter haloes. *Mon. Not. R. Astron. Soc.* **2008**, *389*, 385–397. [CrossRef]
98. Cole, S.; Lacey, C. The structure of dark matter haloes in hierarchical clustering models. *Mon. Not. R. Astron. Soc.* **1996**, *281*, 716. [CrossRef]
99. Eke, V.R.; Cole, S.; Frenk, C.S. Cluster evolution as a diagnostic for Omega. *Mon. Not. R. Astron. Soc.* **1996**, *282*, 263–280. [CrossRef]
100. Bryan, G.L.; Norman, M.L. Statistical Properties of X-Ray Clusters: Analytic and Numerical Comparisons. *Astrophys. J.* **1998**, *495*, 80–99. [CrossRef]
101. Jenkins, A.; Frenk, C.S.; White, S.D.M.; Colberg, J.M.; Cole, S.; Evrard, A.E.; Couchman, H.M.P.; Yoshida, N. The mass function of dark matter haloes. *Mon. Not. R. Astron. Soc.* **2001**, *321*, 372–384. [CrossRef]
102. Warren, M.S.; Abazajian, K.; Holz, D.E.; Teodoro, L. Precision Determination of the Mass Function of Dark Matter Halos. *Astrophys. J.* **2006**, *646*, 881–885. [CrossRef]
103. Lukić, Z.; Heitmann, K.; Habib, S.; Bashinsky, S.; Ricker, P.M. The Halo Mass Function: High-Redshift Evolution and Universality. *Astrophys. J.* **2007**, *671*, 1160–1181. [CrossRef]
104. Tinker, J.; Kravtsov, A.V.; Klypin, A.; Abazajian, K.; Warren, M.; Yepes, G.; Gottlöber, S.; Holz, D.E. Toward a Halo Mass Function for Precision Cosmology: The Limits of Universality. *Astrophys. J.* **2008**, *688*, 709–728. [CrossRef]
105. Trujillo-Gomez, S.; Klypin, A.; Primack, J.; Romanowsky, A.J. Galaxies in ΛCDM with Halo Abundance Matching: Luminosity-Velocity Relation, Baryonic Mass-Velocity Relation, Velocity Function, and Clustering. *Astrophys. J.* **2011**, *742*, 16. [CrossRef]

106. Hellwing, W.A.; Frenk, C.S.; Cautun, M.; Bose, S.; Helly, J.; Jenkins, A.; Sawala, T.; Cytowski, M. The Copernicus Complexio: a high-resolution view of the small-scale Universe. *Mon. Not. R. Astron. Soc.* **2016**, *457*, 3492–3509. [CrossRef]
107. Crain, R.A.; Theuns, T.; Dalla Vecchia, C.; Eke, V.R.; Frenk, C.S.; Jenkins, A.; Kay, S.T.; Peacock, J.A.; Pearce, F.R.; Schaye, J.; et al. Galaxies-intergalactic medium interaction calculation—I. Galaxy formation as a function of large-scale environment. *Mon. Not. R. Astron. Soc.* **2009**, *399*, 1773–1794. [CrossRef]
108. Press, W.H.; Schechter, P. Formation of Galaxies and Clusters of Galaxies by Self-Similar Gravitational Condensation. *Astrophys. J.* **1974**, *187*, 425–438. [CrossRef]
109. Bond, J.R.; Cole, S.; Efstathiou, G.; Kaiser, N. Excursion set mass functions for hierarchical Gaussian fluctuations. *Astrophys. J.* **1991**, *379*, 440–460. [CrossRef]
110. Bower, R.G. The evolution of groups of galaxies in the Press-Schechter formalism. *Mon. Not. R. Astron. Soc.* **1991**, *248*, 332–352. [CrossRef]
111. Schneider, A.; Smith, R.E.; Reed, D. Halo mass function and the free streaming scale. *Mon. Not. R. Astron. Soc.* **2013**, *433*, 1573–1587. [CrossRef]
112. Bose, S.; Hellwing, W.A.; Frenk, C.S.; Jenkins, A.; Lovell, M.R.; Helly, J.C.; Li, B. The Copernicus Complexio: Statistical properties of warm dark matter haloes. *Mon. Not. R. Astron. Soc.* **2016**, *455*, 318–333. [CrossRef]
113. Schewtschenko, J.A.; Wilkinson, R.J.; Baugh, C.M.; Bœhm, C.; Pascoli, S. Dark matter-radiation interactions: The impact on dark matter haloes. *Mon. Not. R. Astron. Soc.* **2015**, *449*, 3587–3596. [CrossRef]
114. Vogelsberger, M.; Zavala, J.; Cyr-Racine, F.Y.; Pfrommer, C.; Bringmann, T.; Sigurdson, K. ETHOS—An effective theory of structure formation: Dark matter physics as a possible explanation of the small-scale CDM problems. *Mon. Not. R. Astron. Soc.* **2016**, *460*, 1399–1416. [CrossRef]
115. Benson, A.J.; Farahi, A.; Cole, S.; Moustakas, L.A.; Jenkins, A.; Lovell, M.; Kennedy, R.; Helly, J.; Frenk, C. Dark matter halo merger histories beyond cold dark matter—I. Methods and application to warm dark matter. *Mon. Not. R. Astron. Soc.* **2013**, *428*, 1774–1789. [CrossRef]
116. Leo, M.; Baugh, C.M.; Li, B.; Pascoli, S. A new smooth-k space filter approach to calculate halo abundances. *J. Cosmol. Astropart. Phys.* **2018**, *4*, 010, [CrossRef]
117. Sameie, O.; Benson, A.J.; Sales, L.V.; Yu, H.B.; Moustakas, L.A.; Creasey, P. The effect of dark matter-dark radiation interactions on halo abundance—A Press-Schechter approach. *arXiv* **2018**, arXiv:1810.11040 [CrossRef]
118. Cyr-Racine, F.Y.; Sigurdson, K. Cosmology of atomic dark matter. *Phys. Rev. D* **2013**, *87*, 103515. [CrossRef]
119. Navarro, J.F.; Frenk, C.S.; White, S.D.M. The Structure of Cold Dark Matter Halos. *Astrophys. J.* **1996**, *462*, 563. [CrossRef]
120. Navarro, J.F.; Frenk, C.S.; White, S.D.M. A Universal Density Profile from Hierarchical Clustering. *Astrophys. J.* **1997**, *490*, 493–508. [CrossRef]
121. Navarro, J.F.; Ludlow, A.; Springel, V.; Wang, J.; Vogelsberger, M.; White, S.D.M.; Jenkins, A.; Frenk, C.S.; Helmi, A. The diversity and similarity of simulated cold dark matter haloes. *Mon. Not. R. Astron. Soc.* **2010**, *402*, 21–34. [CrossRef]
122. Anderhalden, D.; Diemand, J. Density profiles of CDM microhalos and their implications for annihilation boost factors. *J. Cosmol. Astropart. Phys.* **2013**, *4*, 009. [CrossRef]
123. Ishiyama, T. Hierarchical Formation of Dark Matter Halos and the Free Streaming Scale. *Astrophys. J.* **2014**, *788*, 27. [CrossRef]
124. Angulo, R.E.; Hahn, O.; Ludlow, A.D.; Bonoli, S. Earth-mass haloes and the emergence of NFW density profiles. *Mon. Not. R. Astron. Soc.* **2017**, *471*, 4687–4701. [CrossRef]
125. Delos, M.S.; Erickcek, A.L.; Bailey, A.P.; Alvarez, M.A. Density profiles of ultracompact minihalos: Implications for constraining the primordial power spectrum. *Phys. Rev. D* **2018**, *98*, 063527. [CrossRef]
126. Bullock, J.S.; Kolatt, T.S.; Sigad, Y.; Somerville, R.S.; Kravtsov, A.V.; Klypin, A.A.; Primack, J.R.; Dekel, A. Profiles of dark haloes: Evolution, scatter and environment. *Mon. Not. R. Astron. Soc.* **2001**, *321*, 559–575. [CrossRef]
127. Eke, V.R.; Navarro, J.F.; Steinmetz, M. The Power Spectrum Dependence of Dark Matter Halo Concentrations. *Astrophys. J.* **2001**, *554*, 114–125. [CrossRef]
128. Wechsler, R.H.; Bullock, J.S.; Primack, J.R.; Kravtsov, A.V.; Dekel, A. Concentrations of Dark Halos from Their Assembly Histories. *Astrophys. J.* **2002**, *568*, 52–70. [CrossRef]

129. Neto, A.F.; Gao, L.; Bett, P.; Cole, S.; Navarro, J.F.; Frenk, C.S.; White, S.D.M.; Springel, V.; Jenkins, A. The statistics of Λ CDM halo concentrations. *Mon. Not. R. Astron. Soc.* **2007**, *381*, 1450–1462. [CrossRef]
130. Zhao, D.H.; Jing, Y.P.; Mo, H.J.; Börner, G. Accurate Universal Models for the Mass Accretion Histories and Concentrations of Dark Matter Halos. *Astrophys. J.* **2009**, *707*, 354–369. [CrossRef]
131. Prada, F.; Klypin, A.A.; Cuesta, A.J.; Betancort-Rijo, J.E.; Primack, J. Halo concentrations in the standard Λ cold dark matter cosmology. *Mon. Not. R. Astron. Soc.* **2012**, *423*, 3018–3030. [CrossRef]
132. Ludlow, A.D.; Navarro, J.F.; Angulo, R.E.; Boylan-Kolchin, M.; Springel, V.; Frenk, C.; White, S.D.M. The mass-concentration-redshift relation of cold dark matter haloes. *Mon. Not. R. Astron. Soc.* **2014**, *441*, 378–388. [CrossRef]
133. Sánchez-Conde, M.A.; Prada, F. The flattening of the concentration-mass relation towards low halo masses and its implications for the annihilation signal boost. *Mon. Not. R. Astron. Soc.* **2014**, *442*, 2271–2277. [CrossRef]
134. Diemer, B.; Kravtsov, A.V. A Universal Model for Halo Concentrations. *Astrophys. J.* **2015**, *799*, 108. [CrossRef]
135. Klypin, A.; Yepes, G.; Gottlöber, S.; Prada, F.; Heß, S. MultiDark simulations: The story of dark matter halo concentrations and density profiles. *Mon. Not. R. Astron. Soc.* **2016**, *457*, 4340–4359. [CrossRef]
136. Pilipenko, S.V.; Sánchez-Conde, M.A.; Prada, F.; Yepes, G. Pushing down the low-mass halo concentration frontier with the Lomonosov cosmological simulations. *Mon. Not. R. Astron. Soc.* **2017**, *472*, 4918–4927. [CrossRef]
137. Wang, J.; Navarro, J.F.; Frenk, C.S.; White, S.D.M.; Springel, V.; Jenkins, A.; Helmi, A.; Ludlow, A.; Vogelsberger, M. Assembly history and structure of galactic cold dark matter haloes. *Mon. Not. R. Astron. Soc.* **2011**, *413*, 1373–1382. [CrossRef]
138. Springel, V.; Wang, J.; Vogelsberger, M.; Ludlow, A.; Jenkins, A.; Helmi, A.; Navarro, J.F.; Frenk, C.S.; White, S.D.M. The Aquarius Project: The subhaloes of galactic haloes. *Mon. Not. R. Astron. Soc.* **2008**, *391*, 1685–1711. [CrossRef]
139. Vera-Ciro, C.A.; Sales, L.V.; Helmi, A.; Frenk, C.S.; Navarro, J.F.; Springel, V.; Vogelsberger, M.; White, S.D.M. The shape of dark matter haloes in the Aquarius simulations: Evolution and memory. *Mon. Not. R. Astron. Soc.* **2011**, *416*, 1377–1391. [CrossRef]
140. Vogelsberger, M.; Helmi, A.; Springel, V.; White, S.D.M.; Wang, J.; Frenk, C.S.; Jenkins, A.; Ludlow, A.; Navarro, J.F. Phase-space structure in the local dark matter distribution and its signature in direct detection experiments. *Mon. Not. R. Astron. Soc.* **2009**, *395*, 797–811. [CrossRef]
141. Binney, J.; Tremaine, S. *Galactic Dynamics: Second Edition*; Princeton University Press: Princeton, NJ, USA, 2008.
142. Ludlow, A.D.; Navarro, J.F.; White, S.D.M.; Boylan-Kolchin, M.; Springel, V.; Jenkins, A.; Frenk, C.S. The density and pseudo-phase-space density profiles of cold dark matter haloes. *Mon. Not. R. Astron. Soc.* **2011**, *415*, 3895–3902. [CrossRef]
143. Hansen, S.H.; Moore, B. A universal density slope Velocity anisotropy relation for relaxed structures. *New Astron.* **2006**, *11*, 333–338. [CrossRef]
144. Taylor, J.E.; Navarro, J.F. The Phase-Space Density Profiles of Cold Dark Matter Halos. *Astrophys. J.* **2001**, *563*, 483–488. [CrossRef]
145. Bertschinger, E. Self-similar secondary infall and accretion in an Einstein-de Sitter universe. *Astrophys. J. Suppl.* **1985**, *58*, 39–65. [CrossRef]
146. Eddington, A.S. The distribution of stars in globular clusters. *Mon. Not. R. Astron. Soc.* **1916**, *76*, 572–585. [CrossRef]
147. Jing, Y.P.; Suto, Y. Triaxial Modeling of Halo Density Profiles with High-Resolution N-Body Simulations. *Astrophys. J.* **2002**, *574*, 538–553. [CrossRef]
148. Hayashi, E.; Navarro, J.F.; Springel, V. The shape of the gravitational potential in cold dark matter haloes. *Mon. Not. R. Astron. Soc.* **2007**, *377*, 50–62. [CrossRef]
149. Ganeshaiah Veena, P.; Cautun, M.; van de Weygaert, R.; Tempel, E.; Jones, B.J.T.; Rieder, S.; Frenk, C.S. The Cosmic Ballet: Spin and shape alignments of haloes in the cosmic web. *Mon. Not. R. Astron. Soc.* **2018**, *481*, 414–438. [CrossRef]

150. Bonamigo, M.; Despali, G.; Limousin, M.; Angulo, R.; Giocoli, C.; Soucail, G. Universality of dark matter haloes shape over six decades in mass: Insights from the Millennium XXL and SBARBINE simulations. *Mon. Not. R. Astron. Soc.* **2015**, *449*, 3171–3182. [CrossRef]
151. Vega-Ferrero, J.; Yepes, G.; Gottlöber, S. On the shape of dark matter haloes from MultiDark Planck simulations. *Mon. Not. R. Astron. Soc.* **2017**, *467*, 3226–3238. [CrossRef]
152. Despali, G.; Giocoli, C.; Tormen, G. Some like it triaxial: The universality of dark matter halo shapes and their evolution along the cosmic time. *Mon. Not. R. Astron. Soc.* **2014**, *443*, 3208–3217. [CrossRef]
153. Lovell, M.R.; Eke, V.; Frenk, C.S.; Gao, L.; Jenkins, A.; Theuns, T.; Wang, J.; White, S.D.M.; Boyarsky, A.; Ruchayskiy, O. The haloes of bright satellite galaxies in a warm dark matter universe. *Mon. Not. R. Astron. Soc.* **2012**, *420*, 2318–2324. [CrossRef]
154. Colín, P.; Avila-Reese, V.; Valenzuela, O. Substructure and Halo Density Profiles in a Warm Dark Matter Cosmology. *Astrophys. J.* **2000**, *542*, 622–630. [CrossRef]
155. Avila-Reese, V.; Colín, P.; Valenzuela, O.; D'Onghia, E.; Firmani, C. Formation and Structure of Halos in a Warm Dark Matter Cosmology. *Astrophys. J.* **2001**, *559*, 516–530. [CrossRef]
156. Colín, P.; Valenzuela, O.; Avila-Reese, V. On the Structure of Dark Matter Halos at the Damping Scale of the Power Spectrum with and without Relict Velocities. *Astrophys. J.* **2008**, *673*, 203–214. [CrossRef]
157. Schneider, A.; Smith, R.E.; Macciò, A.V.; Moore, B. Non-linear evolution of cosmological structures in warm dark matter models. *Mon. Not. R. Astron. Soc.* **2012**, *424*, 684–698. [CrossRef]
158. Ludlow, A.D.; Bose, S.; Angulo, R.E.; Wang, L.; Hellwing, W.A.; Navarro, J.F.; Cole, S.; Frenk, C.S. The mass-concentration-redshift relation of cold and warm dark matter haloes. *Mon. Not. R. Astron. Soc.* **2016**, *460*, 1214–1232. [CrossRef]
159. Dalcanton, J.J.; Hogan, C.J. Halo Cores and Phase-Space Densities: Observational Constraints on Dark Matter Physics and Structure Formation. *Astrophys. J.* **2001**, *561*, 35–45. [CrossRef]
160. Macciò, A.V.; Paduroiu, S.; Anderhalden, D.; Schneider, A.; Moore, B. Cores in warm dark matter haloes: A Catch 22 problem. *Mon. Not. R. Astron. Soc.* **2012**, *424*, 1105–1112. [CrossRef]
161. Shao, S.; Gao, L.; Theuns, T.; Frenk, C.S. The phase-space density of fermionic dark matter haloes. *Mon. Not. R. Astron. Soc.* **2013**, *430*, 2346–2357. [CrossRef]
162. Colín, P.; Avila-Reese, V.; Valenzuela, O.; Firmani, C. Structure and Subhalo Population of Halos in a Self-interacting Dark Matter Cosmology. *Astrophys. J.* **2002**, *581*, 777–793. [CrossRef]
163. Balberg, S.; Shapiro, S.L.; Inagaki, S. Self-Interacting Dark Matter Halos and the Gravothermal Catastrophe. *Astrophys. J.* **2002**, *568*, 475–487. [CrossRef]
164. Lynden-Bell, D.; Wood, R. The gravo-thermal catastrophe in isothermal spheres and the onset of red-giant structure for stellar systems. *Mon. Not. R. Astron. Soc.* **1968**, *138*, 495. [CrossRef]
165. Pollack, J.; Spergel, D.N.; Steinhardt, P.J. Supermassive Black Holes from Ultra-strongly Self-interacting Dark Matter. *Astrophys. J.* **2015**, *804*, 131. [CrossRef]
166. Vogelsberger, M.; Zavala, J.; Schutz, K.; Slatyer, T.R. Evaporating the Milky Way halo and its satellites with inelastic self-interacting dark matter. *Mon. Not. R. Astron. Soc.* **2019**, *484*, 5437–5452. [CrossRef]
167. Yoshida, N.; Springel, V.; White, S.D.M.; Tormen, G. Weakly Self-interacting Dark Matter and the Structure of Dark Halos. *Astrophys. J. Lett.* **2000**, *544*, L87–L90. [CrossRef]
168. Zavala, J.; Vogelsberger, M.; Walker, M.G. Constraining self-interacting dark matter with the Milky Way's dwarf spheroidals. *Mon. Not. R. Astron. Soc.* **2013**, *431*, L20–L24. [CrossRef]
169. Brinckmann, T.; Zavala, J.; Rapetti, D.; Hansen, S.H.; Vogelsberger, M. The structure and assembly history of cluster-sized haloes in self-interacting dark matter. *Mon. Not. R. Astron. Soc.* **2018**, *474*, 746–759. [CrossRef]
170. Peter, A.H.G.; Rocha, M.; Bullock, J.S.; Kaplinghat, M. Cosmological simulations with self-interacting dark matter—II. Halo shapes versus observations. *Mon. Not. R. Astron. Soc.* **2013**, *430*, 105–120. [CrossRef]
171. Vogelsberger, M.; Zavala, J. Direct detection of self-interacting dark matter. *Mon. Not. R. Astron. Soc.* **2013**, *430*, 1722–1735. [CrossRef]
172. Genel, S.; Bouché, N.; Naab, T.; Sternberg, A.; Genzel, R. The Growth of Dark Matter Halos: Evidence for Significant Smooth Accretion. *Astrophys. J.* **2010**, *719*, 229–239. [CrossRef]
173. Gill, S.P.D.; Knebe, A.; Gibson, B.K. The evolution of substructure—III. The outskirts of clusters. *Mon. Not. R. Astron. Soc.* **2005**, *356*, 1327–1332. [CrossRef]
174. Sales, L.V.; Navarro, J.F.; Abadi, M.G.; Steinmetz, M. Cosmic ménage à trois: The origin of satellite galaxies on extreme orbits. *Mon. Not. R. Astron. Soc.* **2007**, *379*, 1475–1483. [CrossRef]

175. Ludlow, A.D.; Navarro, J.F.; Springel, V.; Jenkins, A.; Frenk, C.S.; Helmi, A. The Unorthodox Orbits of Substructure Halos. *Astrophys. J.* **2009**, *692*, 931–941. [CrossRef]
176. Giocoli, C.; Tormen, G.; Sheth, R.K.; van den Bosch, F.C. The substructure hierarchy in dark matter haloes. *Mon. Not. R. Astron. Soc.* **2010**, *404*, 502–517. [CrossRef]
177. Srisawat, C.; Knebe, A.; Pearce, F.R.; Schneider, A.; Thomas, P.A.; Behroozi, P.; Dolag, K.; Elahi, P.J.; Han, J.; Helly, J.; et al. Sussing Merger Trees: The Merger Trees Comparison Project. *Mon. Not. R. Astron. Soc.* **2013**, *436*, 150–162. [CrossRef]
178. Fakhouri, O.; Ma, C.P. The nearly universal merger rate of dark matter haloes in ΛCDM cosmology. *Mon. Not. R. Astron. Soc.* **2008**, *386*, 577–592. [CrossRef]
179. Fakhouri, O.; Ma, C.P.; Boylan-Kolchin, M. The merger rates and mass assembly histories of dark matter haloes in the two Millennium simulations. *Mon. Not. R. Astron. Soc.* **2010**, *406*, 2267–2278. [CrossRef]
180. Poole, G.B.; Mutch, S.J.; Croton, D.J.; Wyithe, S. Convergence properties of halo merger trees; halo and substructure merger rates across cosmic history. *Mon. Not. R. Astron. Soc.* **2017**, *472*, 3659–3682. [CrossRef]
181. Lacey, C.; Cole, S. Merger rates in hierarchical models of galaxy formation. *Mon. Not. R. Astron. Soc.* **1993**, *262*, 627–649. [CrossRef]
182. Parkinson, H.; Cole, S.; Helly, J. Generating dark matter halo merger trees. *Mon. Not. R. Astron. Soc.* **2008**, *383*, 557–564. [CrossRef]
183. Cole, S.; Lacey, C.G.; Baugh, C.M.; Frenk, C.S. Hierarchical galaxy formation. *Mon. Not. R. Astron. Soc.* **2000**, *319*, 168–204. [CrossRef]
184. Benson, A.J. Orbital parameters of infalling dark matter substructures. *Mon. Not. R. Astron. Soc.* **2005**, *358*, 551–562. [CrossRef]
185. Tormen, G. The rise and fall of satellites in galaxy clusters. *Mon. Not. R. Astron. Soc.* **1997**, *290*, 411–421. [CrossRef]
186. Jiang, L.; Cole, S.; Sawala, T.; Frenk, C.S. Orbital parameters of infalling satellite haloes in the hierarchical ΛCDM model. *Mon. Not. R. Astron. Soc.* **2015**, *448*, 1674–1686. [CrossRef]
187. Wetzel, A.R. On the orbits of infalling satellite haloes. *Mon. Not. R. Astron. Soc.* **2011**, *412*, 49–58. [CrossRef]
188. Diemand, J.; Kuhlen, M.; Madau, P.; Zemp, M.; Moore, B.; Potter, D.; Stadel, J. Clumps and streams in the local dark matter distribution. *Nature* **2008**, *454*, 735–738. [CrossRef] [PubMed]
189. Rocha, M.; Peter, A.H.G.; Bullock, J. Infall times for Milky Way satellites from their present-day kinematics. *Mon. Not. R. Astron. Soc.* **2012**, *425*, 231–244. [CrossRef]
190. Vogelsberger, M.; White, S.D.M. Streams and caustics: The fine-grained structure of Λ cold dark matter haloes. *Mon. Not. R. Astron. Soc.* **2011**, *413*, 1419–1438. [CrossRef]
191. Natarajan, A.; Sikivie, P. Inner caustics of cold dark matter halos. *Phys. Rev. D* **2006**, *73*, 023510. [CrossRef]
192. Vogelsberger, M.; White, S.D.M.; Mohayaee, R.; Springel, V. Caustics in growing cold dark matter haloes. *Mon. Not. R. Astron. Soc.* **2009**, *400*, 2174–2184. [CrossRef]
193. Onions, J.; Knebe, A.; Pearce, F.R.; Muldrew, S.I.; Lux, H.; Knollmann, S.R.; Ascasibar, Y.; Behroozi, P.; Elahi, P.; Han, J.; et al. Subhaloes going Notts: The subhalo-finder comparison project. *Mon. Not. R. Astron. Soc.* **2012**, *423*, 1200–1214. [CrossRef]
194. Klypin, A.; Gottlöber, S.; Kravtsov, A.V.; Khokhlov, A.M. Galaxies in N-Body Simulations: Overcoming the Overmerging Problem. *Astrophys. J.* **1999**, *516*, 530–551. [CrossRef]
195. Neyrinck, M.C.; Gnedin, N.Y.; Hamilton, A.J.S. VOBOZ: An almost-parameter-free halo-finding algorithm. *Mon. Not. R. Astron. Soc.* **2005**, *356*, 1222–1232. [CrossRef]
196. Knollmann, S.R.; Knebe, A. AHF: Amiga's Halo Finder. *Astrophys. J. Suppl.* **2009**, *182*, 608–624. [CrossRef]
197. Han, J.; Jing, Y.P.; Wang, H.; Wang, W. Resolving subhaloes' lives with the Hierarchical Bound-Tracing algorithm. *Mon. Not. R. Astron. Soc.* **2012**, *427*, 2437–2449. [CrossRef]
198. Han, J.; Cole, S.; Frenk, C.S.; Benitez-Llambay, A.; Helly, J. HBT+: An improved code for finding subhaloes and building merger trees in cosmological simulations. *Mon. Not. R. Astron. Soc.* **2018**, *474*, 604–617. [CrossRef]
199. Tormen, G.; Moscardini, L.; Yoshida, N. Properties of cluster satellites in hydrodynamical simulations. *Mon. Not. R. Astron. Soc.* **2004**, *350*, 1397–1408. [CrossRef]
200. Maciejewski, M.; Colombi, S.; Springel, V.; Alard, C.; Bouchet, F.R. Phase-space structures—II. Hierarchical Structure Finder. *Mon. Not. R. Astron. Soc.* **2009**, *396*, 1329–1348. [CrossRef]

201. Behroozi, P.S.; Wechsler, R.H.; Wu, H.Y. The ROCKSTAR Phase-space Temporal Halo Finder and the Velocity Offsets of Cluster Cores. *Astrophys. J.* **2013**, *762*, 109. [CrossRef]
202. Tormen, G.; Diaferio, A.; Syer, D. Survival of substructure within dark matter haloes. *Mon. Not. R. Astron. Soc.* **1998**, *299*, 728–742. [CrossRef]
203. Tollet, É.; Cattaneo, A.; Mamon, G.A.; Moutard, T.; van den Bosch, F.C. On stellar mass loss from galaxies in groups and clusters. *Mon. Not. R. Astron. Soc.* **2017**, *471*, 4170–4193. [CrossRef]
204. Taylor, J.E.; Babul, A. The Dynamics of Sinking Satellites around Disk Galaxies: A Poor Man's Alternative to High-Resolution Numerical Simulations. *Astrophys. J.* **2001**, *559*, 716–735. [CrossRef]
205. Zentner, A.R.; Bullock, J.S. Halo Substructure and the Power Spectrum. *Astrophys. J.* **2003**, *598*, 49–72. [CrossRef]
206. Zentner, A.R.; Berlind, A.A.; Bullock, J.S.; Kravtsov, A.V.; Wechsler, R.H. The Physics of Galaxy Clustering. I. A Model for Subhalo Populations. *Astrophys. J.* **2005**, *624*, 505–525. [CrossRef]
207. Johnston, K.V. A Prescription for Building the Milky Way's Halo from Disrupted Satellites. *Astrophys. J.* **1998**, *495*, 297–308. [CrossRef]
208. van den Bosch, F.C.; Ogiya, G. Dark matter substructure in numerical simulations: A tale of discreteness noise, runaway instabilities, and artificial disruption. *Mon. Not. R. Astron. Soc.* **2018**, *475*, 4066–4087. [CrossRef]
209. van den Bosch, F.C.; Ogiya, G.; Hahn, O.; Burkert, A. Disruption of dark matter substructure: Fact or fiction? *Mon. Not. R. Astron. Soc.* **2018**, *474*, 3043–3066. [CrossRef]
210. Hayashi, E.; Navarro, J.F.; Taylor, J.E.; Stadel, J.; Quinn, T. The Structural Evolution of Substructure. *Astrophys. J.* **2003**, *584*, 541–558. [CrossRef]
211. Spitzer, Jr., L. Disruption of Galactic Clusters. *Astrophys. J.* **1958**, *127*, 17. [CrossRef]
212. Gnedin, O.Y.; Hernquist, L.; Ostriker, J.P. Tidal Shocking by Extended Mass Distributions. *Astrophys. J.* **1999**, *514*, 109–118. [CrossRef]
213. Aguilar, L.A.; White, S.D.M. Tidal interactions between spherical galaxies. *Astrophys. J.* **1985**, *295*, 374. [CrossRef]
214. Aguilar, L.A.; White, S.D.M. The Density Profiles of Tidally Stripped Galaxies. *Astrophys. J.* **1986**, *307*, 97. [CrossRef]
215. Kazantzidis, S.; Mayer, L.; Mastropietro, C.; Diemand, J.; Stadel, J.; Moore, B. Density Profiles of Cold Dark Matter Substructure: Implications for the Missing-Satellites Problem. *Astrophys. J.* **2004**, *608*, 663–679. [CrossRef]
216. Moore, B.; Katz, N.; Lake, G.; Dressler, A.; Oemler, A. Galaxy harassment and the evolution of clusters of galaxies. *Nature* **1996**, *379*, 613–616. [CrossRef]
217. Ogiya, G.; Burkert, A. Dynamical friction and scratches of orbiting satellite galaxies on host systems. *Mon. Not. R. Astron. Soc.* **2016**, *457*, 2164–2172. [CrossRef]
218. Boylan-Kolchin, M.; Ma, C.P.; Quataert, E. Dynamical friction and galaxy merging time-scales. *Mon. Not. R. Astron. Soc.* **2008**, *383*, 93–101. [CrossRef]
219. Chandrasekhar, S. Dynamical Friction. I. General Considerations: The Coefficient of Dynamical Friction. *Astrophys. J.* **1943**, *97*, 255. [CrossRef]
220. van den Bosch, F.C.; Lewis, G.F.; Lake, G.; Stadel, J. Substructure in Dark Halos: Orbital Eccentricities and Dynamical Friction. *Astrophys. J.* **1999**, *515*, 50–68. [CrossRef]
221. Hernquist, L. An analytical model for spherical galaxies and bulges. *Astrophys. J.* **1990**, *356*, 359–364. [CrossRef]
222. Han, J.; Cole, S.; Frenk, C.S.; Jing, Y. A unified model for the spatial and mass distribution of subhaloes. *Mon. Not. R. Astron. Soc.* **2016**, *457*, 1208–1223, [CrossRef]
223. Klypin, A.A.; Trujillo-Gomez, S.; Primack, J. Dark Matter Halos in the Standard Cosmological Model: Results from the Bolshoi Simulation. *Astrophys. J.* **2011**, *740*, 102. [CrossRef]
224. Jiang, F.; van den Bosch, F.C. Statistics of dark matter substructure—III. Halo-to-halo variance. *Mon. Not. R. Astron. Soc.* **2017**, *472*, 657–674. [CrossRef]
225. Gao, L.; White, S.D.M.; Jenkins, A.; Stoehr, F.; Springel, V. The subhalo populations of ΛCDM dark haloes. *Mon. Not. R. Astron. Soc.* **2004**, *355*, 819–834. [CrossRef]
226. Diemand, J.; Kuhlen, M.; Madau, P. Formation and Evolution of Galaxy Dark Matter Halos and Their Substructure. *Astrophys. J.* **2007**, *667*, 859–877. [CrossRef]

227. Gao, L.; Navarro, J.F.; Frenk, C.S.; Jenkins, A.; Springel, V.; White, S.D.M. The Phoenix Project: The dark side of rich Galaxy clusters. *Mon. Not. R. Astron. Soc.* **2012**, *425*, 2169–2186. [CrossRef]
228. Garrison-Kimmel, S.; Boylan-Kolchin, M.; Bullock, J.S.; Lee, K. ELVIS: Exploring the Local Volume in Simulations. *Mon. Not. R. Astron. Soc.* **2014**, *438*, 2578–2596. [CrossRef]
229. Griffen, B.F.; Ji, A.P.; Dooley, G.A.; Gómez, F.A.; Vogelsberger, M.; O'Shea, B.W.; Frebel, A. The Caterpillar Project: A Large Suite of Milky Way Sized Halos. *Astrophys. J.* **2016**, *818*, 10. [CrossRef]
230. Gao, L.; Frenk, C.S.; Boylan-Kolchin, M.; Jenkins, A.; Springel, V.; White, S.D.M. The statistics of the subhalo abundance of dark matter haloes. *Mon. Not. R. Astron. Soc.* **2011**, *410*, 2309–2314. [CrossRef]
231. Angulo, R.E.; Lacey, C.G.; Baugh, C.M.; Frenk, C.S. The fate of substructures in cold dark matter haloes. *Mon. Not. R. Astron. Soc.* **2009**, *399*, 983–995. [CrossRef]
232. Boylan-Kolchin, M.; Springel, V.; White, S.D.M.; Jenkins, A. There's no place like home? Statistics of Milky Way-mass dark matter haloes. *Mon. Not. R. Astron. Soc.* **2010**, *406*, 896–912. [CrossRef]
233. Rodríguez-Puebla, A.; Behroozi, P.; Primack, J.; Klypin, A.; Lee, C.; Hellinger, D. Halo and subhalo demographics with Planck cosmological parameters: Bolshoi-Planck and MultiDark-Planck simulations. *Mon. Not. R. Astron. Soc.* **2016**, *462*, 893–916. [CrossRef]
234. Ghigna, S.; Moore, B.; Governato, F.; Lake, G.; Quinn, T.; Stadel, J. Density Profiles and Substructure of Dark Matter Halos: Converging Results at Ultra-High Numerical Resolution. *Astrophys. J.* **2000**, *544*, 616–628. [CrossRef]
235. Diemand, J.; Moore, B.; Stadel, J. Velocity and spatial biases in cold dark matter subhalo distributions. *Mon. Not. R. Astron. Soc.* **2004**, *352*, 535–546. [CrossRef]
236. Nagai, D.; Kravtsov, A.V. The Radial Distribution of Galaxies in Λ Cold Dark Matter Clusters. *Astrophys. J.* **2005**, *618*, 557–568. [CrossRef]
237. Diemand, J.; Kuhlen, M.; Madau, P. Dark Matter Substructure and Gamma-Ray Annihilation in the Milky Way Halo. *Astrophys. J.* **2007**, *657*, 262–270. [CrossRef]
238. Gao, L.; De Lucia, G.; White, S.D.M.; Jenkins, A. Galaxies and subhaloes in ΛCDM galaxy clusters. *Mon. Not. R. Astron. Soc.* **2004**, *352*, L1–L5. [CrossRef]
239. Navarro, J.F.; Hayashi, E.; Power, C.; Jenkins, A.R.; Frenk, C.S.; White, S.D.M.; Springel, V.; Stadel, J.; Quinn, T.R. The inner structure of ΛCDM haloes—III. Universality and asymptotic slopes. *Mon. Not. R. Astron. Soc.* **2004**, *349*, 1039–1051. [CrossRef]
240. Gao, L.; Navarro, J.F.; Cole, S.; Frenk, C.S.; White, S.D.M.; Springel, V.; Jenkins, A.; Neto, A.F. The redshift dependence of the structure of massive Λ cold dark matter haloes. *Mon. Not. R. Astron. Soc.* **2008**, *387*, 536–544. [CrossRef]
241. Vera-Ciro, C.A.; Helmi, A.; Starkenburg, E.; Breddels, M.A. Not too big, not too small: The dark haloes of the dwarf spheroidals in the Milky Way. *Mon. Not. R. Astron. Soc.* **2013**, *428*, 1696–1703. [CrossRef]
242. Dutton, A.A.; Macciò, A.V. Cold dark matter haloes in the Planck era: Evolution of structural parameters for Einasto and NFW profiles. *Mon. Not. R. Astron. Soc.* **2014**, *441*, 3359–3374. [CrossRef]
243. Peñarrubia, J.; Navarro, J.F.; McConnachie, A.W. The Tidal Evolution of Local Group Dwarf Spheroidals. *Astrophys. J.* **2008**, *673*, 226–240. [CrossRef]
244. Moliné, Á.; Sánchez-Conde, M.A.; Palomares-Ruiz, S.; Prada, F. Characterization of subhalo structural properties and implications for dark matter annihilation signals. *Mon. Not. R. Astron. Soc.* **2017**, *466*, 4974–4990. [CrossRef]
245. Barber, C.; Starkenburg, E.; Navarro, J.F.; McConnachie, A.W. Galactic tides and the shape and orientation of dwarf galaxy satellites. *Mon. Not. R. Astron. Soc.* **2015**, *447*, 1112–1125. [CrossRef]
246. Vera-Ciro, C.A.; Sales, L.V.; Helmi, A.; Navarro, J.F. The shape of dark matter subhaloes in the Aquarius simulations. *Mon. Not. R. Astron. Soc.* **2014**, *439*, 2863–2872. [CrossRef]
247. Bose, S.; Hellwing, W.A.; Frenk, C.S.; Jenkins, A.; Lovell, M.R.; Helly, J.C.; Li, B.; Gonzalez-Perez, V.; Gao, L. Substructure and galaxy formation in the Copernicus Complexio warm dark matter simulations. *Mon. Not. R. Astron. Soc.* **2017**, *464*, 4520–4533. [CrossRef]
248. Dooley, G.A.; Peter, A.H.G.; Vogelsberger, M.; Zavala, J.; Frebel, A. Enhanced tidal stripping of satellites in the galactic halo from dark matter self-interactions. *Mon. Not. R. Astron. Soc.* **2016**, *461*, 710–727. [CrossRef]
249. Nishikawa, H.; Boddy, K.K.; Kaplinghat, M. Accelerated core collapse in tidally stripped self-interacting dark matter halos. *arXiv* **2019**, arXiv:1901.00499.

250. Sawala, T.; Frenk, C.S.; Fattahi, A.; Navarro, J.F.; Theuns, T.; Bower, R.G.; Crain, R.A.; Furlong, M.; Jenkins, A.; Schaller, M.; et al. The chosen few: The low-mass haloes that host faint galaxies. *Mon. Not. R. Astron. Soc.* **2016**, *456*, 85–97. [CrossRef]
251. Sameie, O.; Yu, H.B.; Sales, L.V.; Vogelsberger, M.; Zavala, J. Self-Interacting Dark Matter Subhalos in the Milky Way's Tides. *arXiv* **2019**, arXiv:1904.07872.
252. Kahlhoefer, F.; Kaplinghat, M.; Slatyer, T.R.; Wu, C.L. Diversity in density profiles of self-interacting dark matter satellite halos. *arXiv* **2019**, arXiv:1904.10539.
253. Arkani-Hamed, N.; Finkbeiner, D.P.; Slatyer, T.R.; Weiner, N. A theory of dark matter. *Phys. Rev. D* **2009**, *79*, 015014. [CrossRef]
254. Todoroki, K.; Medvedev, M.V. Dark matter haloes in the multicomponent model—I. Substructure. *Mon. Not. R. Astron. Soc.* **2019**, *483*, 3983–4003. [CrossRef]
255. Eisenstein, D.J.; Hu, W. Baryonic Features in the Matter Transfer Function. *Astrophys. J.* **1998**, *496*, 605–614. [CrossRef]
256. White, S.D.M.; Rees, M.J. Core condensation in heavy halos — A two-stage theory for galaxy formation and clustering. *Mon. Not. R. Astron. Soc.* **1978**, *183*, 341–358. [CrossRef]
257. Tseliakhovich, D.; Hirata, C. Relative velocity of dark matter and baryonic fluids and the formation of the first structures. *Phys. Rev. D* **2010**, *82*, 083520. [CrossRef]
258. Bower, R.G.; Benson, A.J.; Malbon, R.; Helly, J.C.; Frenk, C.S.; Baugh, C.M.; Cole, S.; Lacey, C.G. Breaking the hierarchy of galaxy formation. *Mon. Not. R. Astron. Soc.* **2006**, *370*, 645–655. [CrossRef]
259. Blumenthal, G.R.; Faber, S.M.; Flores, R.; Primack, J.R. Contraction of dark matter galactic halos due to baryonic infall. *Astrophys. J.* **1986**, *301*, 27–34. [CrossRef]
260. Mo, H.J.; Mao, S.; White, S.D.M. The formation of galactic discs. *Mon. Not. R. Astron. Soc.* **1998**, *295*, 319–336. [CrossRef]
261. Gnedin, O.Y.; Kravtsov, A.V.; Klypin, A.A.; Nagai, D. Response of Dark Matter Halos to Condensation of Baryons: Cosmological Simulations and Improved Adiabatic Contraction Model. *Astrophys. J.* **2004**, *616*, 16–26. [CrossRef]
262. Schaller, M.; Frenk, C.S.; Bower, R.G.; Theuns, T.; Jenkins, A.; Schaye, J.; Crain, R.A.; Furlong, M.; Dalla Vecchia, C.; McCarthy, I.G. Baryon effects on the internal structure of ΛCDM haloes in the EAGLE simulations. *Mon. Not. R. Astron. Soc.* **2015**, *451*, 1247–1267. [CrossRef]
263. Lovell, M.R.; Pillepich, A.; Genel, S.; Nelson, D.; Springel, V.; Pakmor, R.; Marinacci, F.; Weinberger, R.; Torrey, P.; Vogelsberger, M.; et al. The fraction of dark matter within galaxies from the IllustrisTNG simulations. *Mon. Not. R. Astron. Soc.* **2018**, *481*, 1950–1975. [CrossRef]
264. Efstathiou, G. Suppressing the formation of dwarf galaxies via photoionization. *Mon. Not. R. Astron. Soc.* **1992**, *256*, 43P–47P. [CrossRef]
265. Babul, A.; Rees, M.J. On dwarf elliptical galaxies and the faint blue counts. *Mon. Not. R. Astron. Soc.* **1992**, *255*, 346–350. [CrossRef]
266. Okamoto, T.; Gao, L.; Theuns, T. Mass loss of galaxies due to an ultraviolet background. *Mon. Not. R. Astron. Soc.* **2008**, *390*, 920–928. [CrossRef]
267. Thoul, A.A.; Weinberg, D.H. Hydrodynamic Simulations of Galaxy Formation. II. Photoionization and the Formation of Low-Mass Galaxies. *Astrophys. J.* **1996**, *465*, 608. [CrossRef]
268. Barkana, R.; Loeb, A. The Photoevaporation of Dwarf Galaxies during Reionization. *Astrophys. J.* **1999**, *523*, 54–65. [CrossRef]
269. Bullock, J.S.; Kravtsov, A.V.; Weinberg, D.H. Reionization and the Abundance of Galactic Satellites. *Astrophys. J.* **2000**, *539*, 517–521. [CrossRef]
270. Gnedin, N.Y. Effect of Reionization on Structure Formation in the Universe. *Astrophys. J.* **2000**, *542*, 535–541. [CrossRef]
271. Benson, A.J.; Lacey, C.G.; Baugh, C.M.; Cole, S.; Frenk, C.S. The effects of photoionization on galaxy formation—I. Model and results at z = 0. *Mon. Not. R. Astron. Soc.* **2002**, *333*, 156–176. [CrossRef]
272. Somerville, R.S. Can Photoionization Squelching Resolve the Substructure Crisis? *Astrophys. J. Lett.* **2002**, *572*, L23–L26. [CrossRef]
273. Hoeft, M.; Yepes, G.; Gottlöber, S.; Springel, V. Dwarf galaxies in voids: Suppressing star formation with photoheating. *Mon. Not. R. Astron. Soc.* **2006**, *371*, 401–414. [CrossRef]

274. Ocvirk, P.; Gillet, N.; Shapiro, P.R.; Aubert, D.; Iliev, I.T.; Teyssier, R.; Yepes, G.; Choi, J.H.; Sullivan, D.; Knebe, A.; et al. Cosmic Dawn (CoDa): The First Radiation-Hydrodynamics Simulation of Reionization and Galaxy Formation in the Local Universe. *Mon. Not. R. Astron. Soc.* **2016**, *463*, 1462–1485. [CrossRef]
275. Sawala, T.; Frenk, C.S.; Crain, R.A.; Jenkins, A.; Schaye, J.; Theuns, T.; Zavala, J. The abundance of (not just) dark matter haloes. *Mon. Not. R. Astron. Soc.* **2013**, *431*, 1366–1382. [CrossRef]
276. Larson, R.B. Effects of supernovae on the early evolution of galaxies. *Mon. Not. R. Astron. Soc.* **1974**, *169*, 229–246. [CrossRef]
277. Dekel, A.; Silk, J. The origin of dwarf galaxies, cold dark matter, and biased galaxy formation. *Astrophys. J.* **1986**, *303*, 39–55. [CrossRef]
278. Navarro, J.F.; Eke, V.R.; Frenk, C.S. The cores of dwarf galaxy haloes. *Mon. Not. R. Astron. Soc.* **1996**, *283*, L72–L78. [CrossRef]
279. Read, J.I.; Gilmore, G. Mass loss from dwarf spheroidal galaxies: The origins of shallow dark matter cores and exponential surface brightness profiles. *Mon. Not. R. Astron. Soc.* **2005**, *356*, 107–124. [CrossRef]
280. Gnedin, O.Y.; Zhao, H. Maximum feedback and dark matter profiles of dwarf galaxies. *Mon. Not. R. Astron. Soc.* **2002**, *333*, 299–306. [CrossRef]
281. Governato, F.; Brook, C.; Mayer, L.; Brooks, A.; Rhee, G.; Wadsley, J.; Jonsson, P.; Willman, B.; Stinson, G.; Quinn, T.; et al. Bulgeless dwarf galaxies and dark matter cores from supernova-driven outflows. *Nature* **2010**, *463*, 203–206. [CrossRef] [PubMed]
282. Pontzen, A.; Governato, F. How supernova feedback turns dark matter cusps into cores. *Mon. Not. R. Astron. Soc.* **2012**, *421*, 3464–3471. [CrossRef]
283. Di Cintio, A.; Brook, C.B.; Macciò, A.V.; Stinson, G.S.; Knebe, A.; Dutton, A.A.; Wadsley, J. The dependence of dark matter profiles on the stellar-to-halo mass ratio: A prediction for cusps versus cores. *Mon. Not. R. Astron. Soc.* **2014**, *437*, 415–423. [CrossRef]
284. Chan, T.K.; Kereš, D.; Oñorbe, J.; Hopkins, P.F.; Muratov, A.L.; Faucher-Giguère, C.A.; Quataert, E. The impact of baryonic physics on the structure of dark matter haloes: The view from the FIRE cosmological simulations. *Mon. Not. R. Astron. Soc.* **2015**, *454*, 2981–3001. [CrossRef]
285. Tollet, E.; Macciò, A.V.; Dutton, A.A.; Stinson, G.S.; Wang, L.; Penzo, C.; Gutcke, T.A.; Buck, T.; Kang, X.; Brook, C.; et al. NIHAO - IV: Core creation and destruction in dark matter density profiles across cosmic time. *Mon. Not. R. Astron. Soc.* **2016**, *456*, 3542–3552. [CrossRef]
286. Read, J.I.; Agertz, O.; Collins, M.L.M. Dark matter cores all the way down. *Mon. Not. R. Astron. Soc.* **2016**, *459*, 2573–2590. [CrossRef]
287. Moore, B. Evidence against dissipation-less dark matter from observations of galaxy haloes. *Nature* **1994**, *370*, 629–631. [CrossRef]
288. Oman, K.A.; Navarro, J.F.; Fattahi, A.; Frenk, C.S.; Sawala, T.; White, S.D.M.; Bower, R.; Crain, R.A.; Furlong, M.; Schaller, M.; et al. The unexpected diversity of dwarf galaxy rotation curves. *Mon. Not. R. Astron. Soc.* **2015**, *452*, 3650–3665. [CrossRef]
289. Oman, K.A.; Marasco, A.; Navarro, J.F.; Frenk, C.S.; Schaye, J.; Benítez-Llambay, A. Non-circular motions and the diversity of dwarf galaxy rotation curves. *Mon. Not. R. Astron. Soc.* **2019**, *482*, 821–847. [CrossRef]
290. Benitez-Llambay, A.; Frenk, C.S.; Ludlow, A.D.; Navarro, J.F. Baryon-induced dark matter cores in the EAGLE simulations. *arXiv* **2018**, arXiv:1810.04186
291. Bose, S.; Frenk, C.S.; Jenkins, A.; Fattahi, A.; Gómez, F.A.; Grand, R.J.J.; Marinacci, F.; Navarro, J.F.; Oman, K.A.; Pakmor, R.; et al. No cores in dark matter-dominated dwarf galaxies with bursty star formation histories. *Mon. Not. R. Astron. Soc.* **2019**, *486*, 4790–4804. [CrossRef]
292. Peirani, S.; Kay, S.; Silk, J. Active galactic nuclei and massive galaxy cores. *Astron. Astrophys.* **2008**, *479*, 123–129. [CrossRef]
293. Duffy, A.R.; Schaye, J.; Kay, S.T.; Dalla Vecchia, C.; Battye, R.A.; Booth, C.M. Impact of baryon physics on dark matter structures: A detailed simulation study of halo density profiles. *Mon. Not. R. Astron. Soc.* **2010**, *405*, 2161–2178. [CrossRef]
294. Teyssier, R.; Moore, B.; Martizzi, D.; Dubois, Y.; Mayer, L. Mass distribution in galaxy clusters: The role of Active Galactic Nuclei feedback. *Mon. Not. R. Astron. Soc.* **2011**, *414*, 195–208. [CrossRef]
295. Martizzi, D.; Teyssier, R.; Moore, B. Cusp-core transformations induced by AGN feedback in the progenitors of cluster galaxies. *Mon. Not. R. Astron. Soc.* **2013**, *432*, 1947–1954. [CrossRef]

296. Peirani, S.; Dubois, Y.; Volonteri, M.; Devriendt, J.; Bundy, K.; Silk, J.; Pichon, C.; Kaviraj, S.; Gavazzi, R.; Habouzit, M. Density profile of dark matter haloes and galaxies in the HORIZON-AGN simulation: The impact of AGN feedback. *Mon. Not. R. Astron. Soc.* **2017**, *472*, 2153–2169. [CrossRef]
297. D'Onghia, E.; Springel, V.; Hernquist, L.; Keres, D. Substructure Depletion in the Milky Way Halo by the Disk. *Astrophys. J.* **2010**, *709*, 1138–1147. [CrossRef]
298. Kazantzidis, S.; Łokas, E.L.; Callegari, S.; Mayer, L.; Moustakas, L.A. On the Efficiency of the Tidal Stirring Mechanism for the Origin of Dwarf Spheroidals: Dependence on the Orbital and Structural Parameters of the Progenitor Disky Dwarfs. *Astrophys. J.* **2011**, *726*, 98. [CrossRef]
299. Zolotov, A.; Brooks, A.M.; Willman, B.; Governato, F.; Pontzen, A.; Christensen, C.; Dekel, A.; Quinn, T.; Shen, S.; Wadsley, J. Baryons Matter: Why Luminous Satellite Galaxies have Reduced Central Masses. *Astrophys. J.* **2012**, *761*, 71. [CrossRef]
300. Sawala, T.; Pihajoki, P.; Johansson, P.H.; Frenk, C.S.; Navarro, J.F.; Oman, K.A.; White, S.D.M. Shaken and stirred: The Milky Way's dark substructures. *Mon. Not. R. Astron. Soc.* **2017**, *467*, 4383–4400. [CrossRef]
301. Garrison-Kimmel, S.; Hopkins, P.F.; Wetzel, A.; Bullock, J.S.; Boylan-Kolchin, M.; Keres, D.; Faucher-Giguere, C.A.; El-Badry, K.; Lamberts, A.; Quataert, E.; et al. The Local Group on FIRE: Dwarf galaxy populations across a suite of hydrodynamic simulations. *arXiv* **2018**, arXiv:1806.04143.
302. Dubois, Y.; Pichon, C.; Welker, C.; Le Borgne, D.; Devriendt, J.; Laigle, C.; Codis, S.; Pogosyan, D.; Arnouts, S.; Benabed, K.; et al. Dancing in the dark: Galactic properties trace spin swings along the cosmic web. *Mon. Not. R. Astron. Soc.* **2014**, *444*, 1453–1468. [CrossRef]
303. Vogelsberger, M.; Genel, S.; Springel, V.; Torrey, P.; Sijacki, D.; Xu, D.; Snyder, G.; Nelson, D.; Hernquist, L. Introducing the Illustris Project: Simulating the coevolution of dark and visible matter in the Universe. *Mon. Not. R. Astron. Soc.* **2014**, *444*, 1518–1547. [CrossRef]
304. Schaye, J.; Crain, R.A.; Bower, R.G.; Furlong, M.; Schaller, M.; Theuns, T.; Dalla Vecchia, C.; Frenk, C.S.; McCarthy, I.G.; Helly, J.C.; et al. The EAGLE project: Simulating the evolution and assembly of galaxies and their environments. *Mon. Not. R. Astron. Soc.* **2015**, *446*, 521–554. [CrossRef]
305. Khandai, N.; Di Matteo, T.; Croft, R.; Wilkins, S.; Feng, Y.; Tucker, E.; DeGraf, C.; Liu, M.S. The MassiveBlack-II simulation: The evolution of haloes and galaxies to z 0. *Mon. Not. R. Astron. Soc.* **2015**, *450*, 1349–1374. [CrossRef]
306. Pillepich, A.; Springel, V.; Nelson, D.; Genel, S.; Naiman, J.; Pakmor, R.; Hernquist, L.; Torrey, P.; Vogelsberger, M.; Weinberger, R.; et al. Simulating galaxy formation with the IllustrisTNG model. *Mon. Not. R. Astron. Soc.* **2018**, *473*, 4077–4106. [CrossRef]
307. Goodenough, L.; Hooper, D. Possible Evidence For Dark Matter Annihilation In The Inner Milky Way From The Fermi Gamma Ray Space Telescope. *arXiv* **2009**, arXiv:0910.2998.
308. Bulbul, E.; Markevitch, M.; Foster, A.; Smith, R.K.; Loewenstein, M.; Randall, S.W. Detection of an Unidentified Emission Line in the Stacked X-Ray Spectrum of Galaxy Clusters. *Astrophys. J.* **2014**, *789*, 13. [CrossRef]
309. Boyarsky, A.; Ruchayskiy, O.; Iakubovskyi, D.; Franse, J. Unidentified Line in X-Ray Spectra of the Andromeda Galaxy and Perseus Galaxy Cluster. *Phys. Rev. Lett.* **2014**, *113*, 251301. [CrossRef] [PubMed]
310. Ruchayskiy, O.; Boyarsky, A.; Iakubovskyi, D.; Bulbul, E.; Eckert, D.; Franse, J.; Malyshev, D.; Markevitch, M.; Neronov, A. Searching for decaying dark matter in deep XMM-Newton observation of the Draco dwarf spheroidal. *Mon. Not. R. Astron. Soc.* **2016**, *460*, 1390–1398. [CrossRef]
311. Aharonian, F.A.; Akamatsu, H.; Akimoto, F.; Allen, S.W.; Angelini, L.; Arnaud, K.A.; Audard, M.; Awaki, H.; Axelsson, M.; Bamba, A.; et al. Hitomi Constraints on the 3.5 keV Line in the Perseus Galaxy Cluster. *Astrophys. J. Lett.* **2017**, *837*, L15. [CrossRef]
312. Klypin, A.; Kravtsov, A.V.; Valenzuela, O.; Prada, F. Where Are the Missing Galactic Satellites? *Astrophys. J.* **1999**, *522*, 82–92. [CrossRef]
313. Moore, B.; Ghigna, S.; Governato, F.; Lake, G.; Quinn, T.; Stadel, J.; Tozzi, P. Dark Matter Substructure within Galactic Halos. *Astrophys. J.* **1999**, *524*, L19–L22. [CrossRef]
314. Boylan-Kolchin, M.; Bullock, J.S.; Kaplinghat, M. Too big to fail? The puzzling darkness of massive Milky Way subhaloes. *Mon. Not. R. Astron. Soc.* **2011**, *415*, L40–L44. [CrossRef]
315. Walker, M.G.; Peñarrubia, J. A Method for Measuring (Slopes of) the Mass Profiles of Dwarf Spheroidal Galaxies. *Astrophys. J.* **2011**, *742*, 20. [CrossRef]

316. Kroupa, P.; Theis, C.; Boily, C.M. The great disk of Milky-Way satellites and cosmological sub-structures. *Astron. Astrophys.* **2005**, *431*, 517–521. [CrossRef]
317. Ibata, R.A.; Ibata, N.G.; Lewis, G.F.; Martin, N.F.; Conn, A.; Elahi, P.; Arias, V.; Fernando, N. A Thousand Shadows of Andromeda: Rotating Planes of Satellites in the Millennium-II Cosmological Simulation. *Astrophys. J. Lett.* **2014**, *784*, L6. [CrossRef]
318. Pawlowski, M.S.; Famaey, B.; Jerjen, H.; Merritt, D.; Kroupa, P.; Dabringhausen, J.; Lüghausen, F.; Forbes, D.A.; Hensler, G.; Hammer, F.; et al. Co-orbiting satellite galaxy structures are still in conflict with the distribution of primordial dwarf galaxies. *Mon. Not. R. Astron. Soc.* **2014**, *442*, 2362–2380, [CrossRef]
319. Wadepuhl, M.; Springel, V. Satellite galaxies in hydrodynamical simulations of Milky Way sized galaxies. *Mon. Not. R. Astron. Soc.* **2011**, *410*, 1975–1992. [CrossRef]
320. Simpson, C.M.; Grand, R.J.J.; Gómez, F.A.; Marinacci, F.; Pakmor, R.; Springel, V.; Campbell, D.J.R.; Frenk, C.S. Quenching and ram pressure stripping of simulated Milky Way satellite galaxies. *Mon. Not. R. Astron. Soc.* **2018**, *478*, 548–567. [CrossRef]
321. Mashchenko, S.; Couchman, H.M.P.; Wadsley, J. The removal of cusps from galaxy centres by stellar feedback in the early Universe. *Nature* **2006**, *442*, 539–542. [CrossRef] [PubMed]
322. Weinberg, M.D.; Katz, N. Bar-driven Dark Halo Evolution: A Resolution of the Cusp-Core Controversy. *Astrophys. J.* **2002**, *580*, 627–633. [CrossRef]
323. Müller, O.; Jerjen, H.; Pawlowski, M.S.; Binggeli, B. Testing the two planes of satellites in the Centaurus group. *Astron. Astrophys.* **2016**, *595*, A119, [CrossRef]
324. Cautun, M.; Bose, S.; Frenk, C.S.; Guo, Q.; Han, J.; Hellwing, W.A.; Sawala, T.; Wang, W. Planes of satellite galaxies: When exceptions are the rule. *Mon. Not. R. Astron. Soc.* **2015**, *452*, 3838–3852. [CrossRef]
325. Müller, O.; Pawlowski, M.S.; Jerjen, H.; Lelli, F. A whirling plane of satellite galaxies around Centaurus A challenges cold dark matter cosmology. *Science* **2018**, *359*, 534–537. [CrossRef]
326. Libeskind, N.I.; Frenk, C.S.; Cole, S.; Helly, J.C.; Jenkins, A.; Navarro, J.F.; Power, C. The distribution of satellite galaxies: The great pancake. *Mon. Not. R. Astron. Soc.* **2005**, *363*, 146–152. [CrossRef]
327. Shao, S.; Cautun, M.; Frenk, C.S. Evolution of galactic planes of satellites in the EAGLE simulation. *arXiv* **2019**, arXiv:1904.02719.
328. Wang, J.; Frenk, C.S.; Navarro, J.F.; Gao, L.; Sawala, T. The missing massive satellites of the Milky Way. *Mon. Not. R. Astron. Soc.* **2012**, *424*, 2715–2721. [CrossRef]
329. Kennedy, R.; Frenk, C.; Cole, S.; Benson, A. Constraining the warm dark matter particle mass with Milky Way satellites. *Mon. Not. R. Astron. Soc.* **2014**, *442*, 2487–2495. [CrossRef]
330. Lovell, M.R.; Bose, S.; Boyarsky, A.; Cole, S.; Frenk, C.S.; Gonzalez-Perez, V.; Kennedy, R.; Ruchayskiy, O.; Smith, A. Satellite galaxies in semi-analytic models of galaxy formation with sterile neutrino dark matter. *Mon. Not. R. Astron. Soc.* **2016**, *461*, 60–72. [CrossRef]
331. Callingham, T.M.; Cautun, M.; Deason, A.J.; Frenk, C.S.; Wang, W.; Gómez, F.A.; Grand, R.J.J.; Marinacci, F.; Pakmor, R. The mass of the Milky Way from satellite dynamics. *Mon. Not. R. Astron. Soc.* **2019**, *484*, 5453–5467. [CrossRef]
332. Lovell, M.R.; Gonzalez-Perez, V.; Bose, S.; Boyarsky, A.; Cole, S.; Frenk, C.S.; Ruchayskiy, O. Addressing the too big to fail problem with baryon physics and sterile neutrino dark matter. *Mon. Not. R. Astron. Soc.* **2017**, *468*, 2836–2849. [CrossRef]
333. Burger, J.D.; Zavala, J. The nature of core formation in dark matter haloes: Adiabatic or impulsive? *Mon. Not. R. Astron. Soc.* **2019**, *485*, 1008–1028. [CrossRef]
334. Rees, M.J. Lyman absorption lines in quasar spectra—Evidence for gravitationally-confined gas in dark minihaloes. *Mon. Not. R. Astron. Soc.* **1986**, *218*, 25P–30P. [CrossRef]
335. Benítez-Llambay, A.; Navarro, J.F.; Frenk, C.S.; Sawala, T.; Oman, K.; Fattahi, A.; Schaller, M.; Schaye, J.; Crain, R.A.; Theuns, T. The properties of 'dark' ΛCDM haloes in the Local Group. *Mon. Not. R. Astron. Soc.* **2017**, *465*, 3913–3926. [CrossRef]
336. Carlberg, R.G. Dark Matter Sub-halo Counts via Star Stream Crossings. *Astrophys. J.* **2012**, *748*, 20. [CrossRef]
337. Erkal, D.; Belokurov, V.; Bovy, J.; Sanders, J.L. The number and size of subhalo-induced gaps in stellar streams. *Mon. Not. R. Astron. Soc.* **2016**, *463*, 102–119. [CrossRef]
338. Amorisco, N.C.; Gómez, F.A.; Vegetti, S.; White, S.D.M. Gaps in globular cluster streams: Giant molecular clouds can cause them too. *Mon. Not. R. Astron. Soc.* **2016**, *463*, L17–L21. [CrossRef]

339. Erkal, D.; Koposov, S.E.; Belokurov, V. A sharper view of Pal 5's tails: Discovery of stream perturbations with a novel non-parametric technique. *Mon. Not. R. Astron. Soc.* **2017**, *470*, 60–84. [CrossRef]
340. Bovy, J.; Erkal, D.; Sanders, J.L. Linear perturbation theory for tidal streams and the small-scale CDM power spectrum. *Mon. Not. R. Astron. Soc.* **2017**, *466*, 628–668. [CrossRef]
341. Mao, S.; Schneider, P. Evidence for substructure in lens galaxies? *Mon. Not. R. Astron. Soc.* **1998**, *295*, 587. [CrossRef]
342. Schneider, P.; Weiss, A. The gravitational lens equation near cusps. *Astron. Astrophys.* **1992**, *260*, 1–13.
343. Metcalf, R.B.; Madau, P. Compound Gravitational Lensing as a Probe of Dark Matter Substructure within Galaxy Halos. *Astrophys. J.* **2001**, *563*, 9–20. [CrossRef]
344. Dalal, N.; Kochanek, C.S. Direct Detection of Cold Dark Matter Substructure. *Astrophys. J.* **2002**, *572*, 25–33. [CrossRef]
345. Xu, D.; Sluse, D.; Gao, L.; Wang, J.; Frenk, C.; Mao, S.; Schneider, P.; Springel, V. How well can cold dark matter substructures account for the observed radio flux-ratio anomalies. *Mon. Not. R. Astron. Soc.* **2015**, *447*, 3189–3206. [CrossRef]
346. Hsueh, J.W.; Enzi, W.; Vegetti, S.; Auger, M.; Fassnacht, C.D.; Despali, G.; Koopmans, L.V.E.; McKean, J.P. SHARP – VII. New constraints on warm dark matter free-streaming properties and substructure abundance from flux-ratio anomalous lensed quasars. *arXiv* **2019**, arXiv:1905.04182.
347. Iršič, V.; Viel, M.; Haehnelt, M.G.; Bolton, J.S.; Cristiani, S.; Becker, G.D.; D'Odorico, V.; Cupani, G.; Kim, T.S.; Berg, T.A.M.; et al. New constraints on the free-streaming of warm dark matter from intermediate and small scale Lyman-α forest data. *Phys. Rev. D* **2017**, *96*, 023522. [CrossRef]
348. Vegetti, S.; Koopmans, L.V.E. Bayesian strong gravitational lens modelling on adaptive grids: Objective detection of mass substructure in Galaxies. *Mon. Not. R. Astron. Soc.* **2009**, *392*, 945–963. [CrossRef]
349. Vegetti, S.; Lagattuta, D.J.; McKean, J.P.; Auger, M.W.; Fassnacht, C.D.; Koopmans, L.V.E. Gravitational detection of a low-mass dark satellite galaxy at cosmological distance. *Nature* **2012**, *481*, 341–343. [CrossRef] [PubMed]
350. Keeton, C.R. A Catalog of Mass Models for Gravitational Lensing. *arXiv* **2001**, arXiv:astro-ph/0102341.
351. Vegetti, S.; Koopmans, L.V.E.; Bolton, A.; Treu, T.; Gavazzi, R. Detection of a dark substructure through gravitational imaging. *Mon. Not. R. Astron. Soc.* **2010**, *408*, 1969–1981. [CrossRef]
352. Vegetti, S.; Despali, G.; Lovell, M.R.; Enzi, W. Constraining sterile neutrino cosmologies with strong gravitational lensing observations at redshift z = 0.2. *Mon. Not. R. Astron. Soc.* **2018**, *481*, 3661–3669. [CrossRef]
353. Vegetti, S.; Koopmans, L.V.E.; Auger, M.W.; Treu, T.; Bolton, A.S. Inference of the cold dark matter substructure mass function at z = 0.2 using strong gravitational lenses. *Mon. Not. R. Astron. Soc.* **2014**, *442*, 2017–2035. [CrossRef]
354. Hezaveh, Y.D.; Dalal, N.; Marrone, D.P.; Mao, Y.Y.; Morningstar, W.; Wen, D.; Blandford, R.D.; Carlstrom, J.E.; Fassnacht, C.D.; Holder, G.P.; et al. Detection of Lensing Substructure Using ALMA Observations of the Dusty Galaxy SDP.81. *Astrophys. J.* **2016**, *823*, 37. [CrossRef]
355. Nightingale, J.; Dye, S.; Massey, R. AutoLens: Automated Modeling of a Strong Lens's Light, Mass and Source. *arXiv* **2017**. [CrossRef]
356. Brewer, B.J.; Huijser, D.; Lewis, G.F. Trans-dimensional Bayesian inference for gravitational lens substructures. *Mon. Not. R. Astron. Soc.* **2016**, *455*, 1819–1829. [CrossRef]
357. Diaz Rivero, A.; Cyr-Racine, F.Y.; Dvorkin, C. On the Power Spectrum of Dark Matter Substructure in Strong Gravitational Lenses. *arXiv* **2017**, arXiv:1707.04590.
358. Li, R.; Frenk, C.S.; Cole, S.; Wang, Q.; Gao, L. Projection effects in the strong lensing study of subhaloes. *Mon. Not. R. Astron. Soc.* **2017**, *468*, 1426–1432. [CrossRef]
359. Despali, G.; Vegetti, S.; White, S.D.M.; Giocoli, C.; van den Bosch, F.C. Modelling the line-of-sight contribution in substructure lensing. *Mon. Not. R. Astron. Soc.* **2018**, *475*, 5424–5442. [CrossRef]
360. Li, R.; Frenk, C.S.; Cole, S.; Gao, L.; Bose, S.; Hellwing, W.A. Constraints on the identity of the dark matter from strong gravitational lenses. *Mon. Not. R. Astron. Soc.* **2016**, *460*, 363–372. [CrossRef]

© 2019 by the authors. Licensee MDPI, Basel, Switzerland. This article is an open access article distributed under the terms and conditions of the Creative Commons Attribution (CC BY) license (http://creativecommons.org/licenses/by/4.0/).

Review

Halo Substructure Boosts to the Signatures of Dark Matter Annihilation

Shin'ichiro Ando [1,2,*], Tomoaki Ishiyama [3] and Nagisa Hiroshima [4,5,6]

1. GRAPPA Institute, Institute of Physics, University of Amsterdam, 1098 XH Amsterdam, The Netherlands
2. Kavli Institute for the Physics and Mathematics of the Universe (WPI), University of Tokyo, Kashiwa 277-8583, Japan
3. Institute of Management and Information Technologies, Chiba University, Chiba 263-8522, Japan
4. Institute for Cosmic Ray Research, University of Tokyo, Kashiwa 277-8582, Japan
5. Institute of Particle and Nuclear Studies, High Energy Accelerator Research Organization (KEK), Tsukuba 305-0801, Japan
6. RIKEN Interdisciplinary Theoretical and Mathematical Sciences (iTHEMS), Wako 351-0198, Japan
* Correspondence: s.ando@uva.nl

Received: 27 March 2019; Accepted: 26 June 2019; Published: 1 July 2019

Abstract: The presence of dark matter substructure will boost the signatures of dark matter annihilation. We review recent progress on estimates of this subhalo boost factor—a ratio of the luminosity from annihilation in the subhalos to that originating the smooth component—based on both numerical N-body simulations and semi-analytic modelings. Since subhalos of all the scales, ranging from the Earth mass (as expected, e.g., the supersymmetric neutralino, a prime candidate for cold dark matter) to galaxies or larger, give substantial contribution to the annihilation rate, it is essential to understand subhalo properties over a large dynamic range of more than twenty orders of magnitude in masses. Even though numerical simulations give the most accurate assessment in resolved regimes, extrapolating the subhalo properties down in sub-grid scales comes with great uncertainties—a straightforward extrapolation yields a very large amount of the subhalo boost factor of $\gtrsim$100 for galaxy-size halos. Physically motivated theoretical models based on analytic prescriptions such as the extended Press-Schechter formalism and tidal stripping modeling, which are well tested against the simulation results, predict a more modest boost of order unity for the galaxy-size halos. Giving an accurate assessment of the boost factor is essential for indirect dark matter searches and thus, having models calibrated at large ranges of host masses and redshifts, is strongly urged upon.

Keywords: halo substructure; dark matter annihilation; indirect dark matter searches; subhalo boost

1. Introduction

One of the most popular candidates for dark matter is weakly interacting massive particles (WIMPs) [1,2]. They are motivated by beyond-the-standard-model physics such as supersymmetry [3] or universal extra-dimensions [4], although the non-discovery of new physics at the TeV scale with the Large Hadron Collider puts these models to serious test [5]. In addition, WIMPs can naturally explain the relic dark matter density with thermal freezeout mechanisms, where the WIMPs following the weak-scale physics were in chemical equilibrium until freezeout—when the expansion of the Universe became faster than the annihilation rate [6]. Since dark matter is often the lightest particle in an extended sector, it can self-annihilate only into the standard-model particles, which end up producing gamma rays, charged cosmic rays, and neutrinos. Indirect detection of dark matter annihilation is therefore a direct test of the thermal freezeout of WIMPs.

WIMPs are also a subcategory of cold dark matter (CDM), where they were nonrelativistic when structure formation started. In the CDM framework, it is known that the structures form hierarchically, from smaller to larger ones. These virialized structures are referred to as halos and they are nearly spherically symmetric. Typical size of the smallest structure is highly model dependent. In the case of the supersymmetric neutralino that is one of the most popular WIMP candidates, the smallest halos tend to be of the Earth mass, $10^{-6} M_\odot$ but with very large range of possible values of $\sim 10^{-12}$–$10^{-3} M_\odot$ [7–14]. Smaller halos collapse at higher redshifts when the Universe was denser, and hence they are of higher density. A larger dark matter halo today contains lots of substructures (or subhalos) of all mass scales, which can go down to the Earth masses or even smaller and hence denser.

Since the annihilation rate depends on the dark matter density squared (and $\langle \rho^2 \rangle \geq \langle \rho \rangle^2$), the presence of the subhalos will *boost* the gamma-ray signatures from dark matter annihilation. This subhalo boost of dark matter annihilation, in relation with the smallest-scale subhalos, has been a topic of interest for very many years [15–50]. The main difficulty is the fact that subhalos of all the scales ranging from the Earth mass (or even smaller) to larger masses (a significant fraction of their host's mass) give a substantial contribution to the annihilation rate. Covering this very large dynamic range is challenging even with the state-of-the-art numerical simulations. In simulations of Milky-Way-size halos ($10^{12} M_\odot$) [37,51,52], one can resolve only down to 10^4–$10^5 M_\odot$, and there still remains more than ten orders of magnitude to reach.

We will review recent progress on the subhalo contribution to dark matter annihilation. (See also Reference [53] for a review on generic processes that subhalos undergo.) We first discuss approaches using the numerical N-body simulations and estimate of the annihilation boost factor by adopting the results and extrapolation down to very-small-mass ranges. To complement the approach based on simulations, we then review an analytical approach. In the CDM framework, fraction of halos that collapse is described with the Press-Schechter formalism [54] based on spherical or ellipsoidal collapse models. This has been further extended to accommodate collapsed regions within larger halos (excursion set or extended Press-Schechter formalism [55]), which can be applied to address statistics of halo substructure. More recent literature suggests that the annihilation boost factor, defined as the luminosity due to subhalos divided by the host luminosity, is modest, ranging from order of unity to a few tens for galaxy-size halos [35,46–50]. This relatively mild amount of the annihilation boost makes the prospect of indirect dark matter searches less promising compared with earlier more optimistic predictions [36,40,41,56]. We note that our focus is mainly on subhalo boost factors in extragalactic halos. For the subhalo boosts in the Galactic halo, on the other hand, we need to assess the spatial distribution of the subhalos too. The N-body simulations described in Section 3 can address this issue but again are subject to resolution issues as well as the baryonic effect. See, for example, Reference [47] for an alternative approach adopting analytical prescription.

This review is organized as follows. In Section 2, we introduce basic concepts of density profiles, mass functions, and the annihilation boost factors of the subhalos, starting with simple formulations. Here we make some simplifying assumptions, which are to be addressed in later sections. In Section 3, we summarize the progress from the numerical simulations for the subhalos and the annihilation boost factors. Section 4 presents more recent approaches based on realistic formulation than Section 2. In Section 4.1, we first show new analytic models that predict the subhalo mass functions well in agreement with the results from the numerical simulations for various ranges of the host masses and redshifts, and that the annihilation boost factors are on the order of unity even for cluster-size halos. Then, we summarize other semi-analytic approaches for computing the annihilation boost factors, based on self-similarity (Section 4.2) and universal phase-space clustering (Section 4.3) of the subhalos. We conclude the review in Section 5. Finally, for convenience, we summarize fitting functions for the subhalo mass functions, and annihilation boost factors that can be applicable to nearly arbitrary masses and redshifts in Appendix A.

2. Formulation

In this section, we introduce several important quantities such as density profiles, subhalo mass function, and the annihilation boost factors. This section is based on a simplified analytic model, which in several aspects are unrealistic but sets the basis for the latter discussions according to numerical simulations (Section 3) and more sophisticated semi-analytical models (Section 4).

2.1. Subhalo Boost Factor

The rate of dark matter annihilation is proportional to dark matter density squared, ρ_χ^2, where χ represents the dark matter particle. In the presence of substructure, ρ_χ is divided into two terms:

$$\rho_\chi(x) = \rho_{\chi,\text{sm}}(x) + \rho_{\chi,\text{sh}}(x), \tag{1}$$

where ρ_{sm} and ρ_{sh} represent smooth and subhalo components, respectively. (In the following, we omit the subscript χ.) The volume average of the density squared in a host halo characterized by its virial mass M and redshift z, which is the relevant quantity for the indirect dark matter researches, is therefore written as

$$\langle \rho^2(x) \rangle_{M,z} = \langle \rho_{\text{sm}}^2(x) \rangle_{M,z} + \langle \rho_{\text{sh}}^2(x) \rangle_{M,z} + 2\langle \rho_{\text{sm}}(x)\rho_{\text{sh}}(x) \rangle_{M,z}. \tag{2}$$

We assume that the smooth component ρ_{sm} is characterized by the following Navarro-Frenk-White (NFW) profile [57,58]:

$$\rho_{\text{NFW}}(r) = \frac{\rho_s}{(r/r_s)(1+r/r_s)^2}, \tag{3}$$

where ρ_s is a characteristic density and r_s is a scale radius. These parameters, ρ_s and r_s, are evaluated such that the volume integral of ρ_{NFW} yields the total halo mass M, and thus we have $\rho_{\text{sm}}(r) = (1 - f_{\text{sh}})\rho_{\text{NFW}}(r)$, where f_{sh} is defined as the mass fraction in the subhalos. The first term is then simply

$$\langle \rho_{\text{sm}}^2(x) \rangle_{M,z} = \frac{1}{V}\int d^3x \rho_{\text{sm}}^2(r) = \frac{4\pi(1-f_{\text{sh}})^2}{3V}\rho_s^2 r_s^3 \left[1 - \frac{1}{(1+c_{\text{vir}})^3}\right], \tag{4}$$

where $V = 4\pi r_{\text{vir}}^3/3$ is the volume of the host out to its virial radius r_{vir}, $c_{\text{vir}} \equiv r_{\text{vir}}/r_s$ is the concentration parameter. The parameters characterizing the host profile—ρ_s, r_s, and c_{vir}—are all functions of M and z.[1]

Next, we evaluate the second term of Equation (2), $\langle \rho_{\text{sh}}^2(x) \rangle$. We characterize each subhalo i with the location of its center x_i and mass m_i. Density due to all the subhalos at a coordinate x is written as a sum of the density profile around the *seed* of each subhalo, that is,

$$\rho_{\text{sh}}(x) = \int dm' \int d^3x' \sum_i \delta_D(m' - m_i)\delta_D^3(x' - x_i) m' u_{\text{sh}}(x - x'|m'), \tag{5}$$

where δ_D^N is the N-dimensional Dirac delta function, and $u_{\text{sh}}(r|m)$ defines the density profile of the subhalo with mass m and is normalized to one after the volume integral.[2] We define the ensemble average of the product of these delta functions as

$$\frac{dn_{\text{sh}}(x,m)}{dm} = \left\langle \sum_i \delta_D(m - m_i)\delta_D^3(x - x_i) \right\rangle, \tag{6}$$

---

[1] We note, however, that the concentration c_{vir} has a scatter, which is often characterized by a log-normal distribution, whose mean $\bar{c}_{\text{vir}}$ is the function of M and z. We will include this in the latter sections.
[2] For the sake of simplicity for analytic expressions, we assume that the suhbalo mass is the only parameter characterizing its density profile. One can introduce many more parameters to make the model more realistic.

its volume integral over the host halo as

$$\frac{dN_{\rm sh}}{dm} = \int d^3x \frac{dn_{\rm sh}(x,m)}{dm}, \quad (7)$$

and call both $dn_{\rm sh}/dm$ and $dN_{\rm sh}/dm$ the subhalo mass function. We also obtain the mass fraction in the subhalos as

$$f_{\rm sh}(M,z) = \frac{1}{M}\int dm\, m \frac{dN_{\rm sh}}{dm}. \quad (8)$$

By multiplying Equation (5) by itself and taking both the ensemble and the volume averages, we have

$$\begin{aligned}
\langle \rho_{\rm sh}^2(x)\rangle_{M,z} &\equiv \frac{1}{V}\int d^3x \langle \rho_{\rm sh}^2(x)\rangle \\
&= \frac{1}{V}\int d^3x \int dm' \int d^3x' \int dm'' \int d^3x'' m' u_{\rm sh}(x-x'|m') m'' u_{\rm sh}(x-x''|m'') \\
&\quad \times \left\langle \sum_i \delta_D(m'-m_i)\delta_D^3(x'-x_i) \sum_j \delta_D(m''-m_j)\delta_D^3(x''-x_j) \right\rangle \\
&= \frac{1}{V}\int d^3x \int dm' \int d^3x' \frac{dn_{\rm sh}(x',m')}{dm'} m'^2 u_{\rm sh}^2(x-x'|m') \\
&= \frac{4\pi}{3V}\int dm \frac{dN_{\rm sh}}{dm} \rho_{\rm s,sh}^2 r_{\rm s,sh}^3 \left[1 - \frac{1}{(1+c_{t,\rm sh})^3}\right],
\end{aligned} \quad (9)$$

where at the last equality we adopted the NFW function for the subhalo density profile $mu_{\rm sh}(r|m)$ with the scale radius $r_{\rm s,sh}$, characteristic density $\rho_{\rm s,sh}$, and tidal truncation radius $r_{t,\rm sh} \equiv c_{t,\rm sh} r_{\rm s,sh}$ beyond which the subhalo density abruptly decreases to zero. At the third equality of Equation (9), we ignored the term arising from $j \neq i$ as we evaluate the quantity at one point x and assume that subhalos do not overlap. We note, however, that such a term becomes relevant for obtaining the two-point correlation function, or the power spectrum; see References [27,38] for more details.

We define the subhalo boost factor as the ratio of the total luminosity from dark matter annihilation in the subhalos and that from the smooth component *in the case that there is no substructure*. By comparing Equations (4) and (9), and remembering that the luminosity is proportional to the volume integral of the density squared, the boost factor is simply written as

$$B_{\rm sh}(M,z) = \frac{1}{L_{\rm host,0}(M,z)} \int dm \frac{dN_{\rm sh}(m|M,z)}{dm} L_{\rm sh}(m), \quad (10)$$

where the subscript 0 shows that this is a quantity in the case of no subhalo contributions. Equation (10) is also valid for any other spherically symmetric density profiles than the NFW.

Finally, we evaluate the last cross-correlation term in Equation (2). See also References [47,59–62]. Following a similar procedure as in Equation (9), we have

$$\begin{aligned}
2\langle \rho_{\rm sm}(x)\rho_{\rm sh}(x)\rangle_{M,z} &= \frac{2}{V}\int d^3x \langle \rho_{\rm sm}(x)\rho_{\rm sh}(x)\rangle \\
&= \frac{2}{V}\int d^3x \rho_{\rm sm}(x) \int dm' \int d^3x' m' u_{\rm sh}(x-x'|m') \frac{dn_{\rm sh}(x',m')}{dm'} \\
&\approx \frac{2}{V}\int d^3x \rho_{\rm sm}(x) \int dm\, m \frac{dn_{\rm sh}(x,m)}{dm},
\end{aligned} \quad (11)$$

where in the last equality, we first used the fact that the subhalo density profile is much more sharply peaked than their spatial distribution, and take $dn_{\rm sh}/dm$ out of x' integration adopting $x' \approx x$ as its spatial variable. Second, we performed volume integral for $u(x-x'|m)$ over x' variable, which simply returns one, to reach the last expression of Equation (11). Then we assume that the spatial distribution of the subhalos is independent of their masses:

$$\frac{dn_{\rm sh}(x,m)}{dm} = P_{\rm sh}(x)\frac{dN_{\rm sh}(m)}{dm}, \qquad (12)$$

where $P_{\rm msh}(x)d^3x$ represents the probability of finding a subhalo in a volume element d^3x around x. With this and Equation (8), we have

$$2\langle \rho_{\rm sm}(x)\rho_{\rm sh}(x)\rangle_{M,z} = \frac{2f_{\rm sh}M}{V}\int d^3x \rho_{\rm sm}(x) P_{\rm sh}(x). \qquad (13)$$

For simplicity, we assume that the subhalos are distributed following the smooth NFW component. In this case, we have $\rho_{\rm sm}(x) = (1-f_{\rm sh})MP_{\rm sh}(x)$, and

$$2\langle \rho_{\rm sm}(x)\rho_{\rm sh}(x)\rangle_{M,z} = \frac{2f_{\rm sh}}{1-f_{\rm sh}}\langle \rho_{\rm sm}^2(x)\rangle = \frac{8\pi f_{\rm sh}(1-f_{\rm sh})}{3V}\rho_s^2 r_s^3\left[1 - \frac{1}{(1+c_{\rm vir})^3}\right]. \qquad (14)$$

The luminosity from the smooth component in the presence of the subhalos ($L_{\rm sm}$) is related to the host luminosity in the subhalos' absence via $L_{\rm sm} = (1-f_{\rm sh})^2 L_{\rm host,0}$, because the density in the smooth component gets depleted by a factor of $1-f_{\rm sh}$, if there are subhalos. Thus, the total luminosity from both the smooth component and the subhalos are given by

$$\begin{aligned} L_{\rm total} &= L_{\rm sm} + L_{\rm sh} + L_{\rm cross} \\ &= \left[(1-f_{\rm sh})^2 + B_{\rm sh} + 2f_{\rm sh}(1-f_{\rm sh})\right]L_{\rm host,0} \\ &= \left(1 - f_{\rm sh}^2 + B_{\rm sh}\right)L_{\rm host,0} \end{aligned} \qquad (15)$$

Often, $L_{\rm total}/L_{\rm host,0}$ is also referred to as the subhalo boost factor in the literature. Note, however, that we have not included the effect of sub-subhalos (and beyond) yet in this formalism. In order to accommodate it, in the right-hand side of Equation (10), we need to include the sub-subhalo boost to the subhalo luminosity $L_{\rm sh}$. Thus, we replace $L_{\rm sh}$ with $(1-f_{\rm ssh}^2 + B_{\rm ssh})L_{\rm sh}$, where the subscript "ssh" represents the contribution from the sub-subhalos. If the subhalo mass fraction $f_{\rm sh}$ and the boost factor $B_{\rm sh}$ depend only on the host mass, then one can assume $f_{\rm ssh}(m) = f_{\rm sh}(m)$ and $B_{\rm ssh}(m) = B_{\rm sh}(m)$, and repeat the calculations in an iterative manner. See, however, Section 4.1 for a more realistic treatment.

2.2. Characterization of Dark Matter Halos

We shall discuss the density profile of dark matter halos that are characterized by the virial radius $r_{\rm vir}$, the scale radius r_s, and the characteristic density ρ_s. The halo is virialized when a mean density within a region reaches some critical value times the critical density of the Universe at that time: $\Delta_{\rm vir}(z)\rho_c(z)$, where $\rho_c(z) = \rho_{c,0}[\Omega_m(1+z)^3 + \Omega_\Lambda]$, $\rho_{c,0} = 3H_0^2/(8\pi G)$ is the present critical density, $H_0 = 100h$ km s^{-1} Mpc^{-1} is the Hubble constant, Ω_m and Ω_Λ are the density parameters for matter and the cosmological constant, respectively. In CDM cosmology with the cosmological constant, this critical value is given as [63]

$$\Delta_{\rm vir}(z) = 18\pi^2 + 82d(z) - 39d^2(z), \qquad (16)$$

where $d(z) = \Omega_m(1+z)^3/[\Omega_m(1+z)^3 + \Omega_\Lambda] - 1$. Given the virial mass M and the redshift z of the halo of interest, $r_{\rm vir}$ is therefore obtained by solving

$$M = \frac{4\pi}{3}\Delta_{\rm vir}(z)\rho_c(z)r_{\rm vir}^3. \qquad (17)$$

Alternatively, one can define M_{200} and r_{200} via

$$M_{200} = \frac{4\pi}{3} 200 \rho_c(z) r_{200}^3. \tag{18}$$

M_{200} is often adopted to define halo masses in N-body simulations.

The concentration parameter $c_{\rm vir} \equiv r_{\rm vir}/r_s$ (or $c_{200} \equiv r_{200}/r_s$) has been studied with numerical simulations and found to be a function of M and z. It follows a log-normal distribution with the mean of $\bar{c}_{\rm vir}(M,z)$ (e.g., Reference [64]) and the standard deviation of $\sigma_{\log c} \approx 0.13$ [65]. The mean $\bar{c}_{\rm vir}$ has been calibrated at both large (galaxies, clusters) and very small (of Earth-mass size) halos, and found to decreases as a function of M and z. Once $c_{\rm vir}$ is drawn from the distribution, it is used to obtain $r_s = r_{\rm vir}/c_{\rm vir}$. Finally, ρ_s is obtained through the condition of having mass M within $r_{\rm vir}$:

$$M = \int_0^{r_{\rm vir}} dr 4\pi r^2 \rho(r) = 4\pi \rho_s r_s^3 f(c_{\rm vir}), \tag{19}$$

$$f(x) = \ln(1+x) - \frac{x}{1+x}, \tag{20}$$

where the second equality of Equation (19) holds in the case of the NFW profile.

In the case of the subhalos, the procedures above cannot be adopted. This is because they are subject to tidal effects from the host, which strip masses away from the subhalos. However, the regions well inside the scale radius r_s—because of strong self-gravity—is resilient against the tidal force and hence the annihilation rate hardly changes. These tidal processes, therefore, make the subhalos more concentrated and hence effectively brighter compared with the field halos of the same mass. In many analytical studies in the literature [29,32,44,66], however, the effect of tidal stripping was ignored and the concentration-mass relation of the field halos was adopted, which resulted in underestimate of the annihilation boost factor. This has been pointed out by Reference Bartels and Ando [46] and will be discussed in Section 4 (see also References [48,49]).

3. Estimates of Annihilation Boost with Numerical Simulations

In order to assess the annihilation boosts, one has to have reasonably good ideas on the density profiles $\rho(r)$, the concentration-mass relation,[3] and the subhalo mass function. Cosmological N-body simulations have been a powerful tool for probing all of them because once a halo collapses from initial density fluctuations, it evolves under a strongly nonlinear environment. They have indeed demonstrated that there are a large amount of surviving subhalos (see Figure 1) in halos and halos have cuspy density profiles.

---

[3] The concentration-mass relation is defined as the average concentration parameter as a function of halo mass.

Figure 1. Dark matter distribution of a Milky-Way-size halo taken from a high-resolution cosmological N-body simulation [67].

3.1. Subhalo Abundance

Cosmological N-body simulations predict that there are many surviving subhalos in host halos as a consequence of hierarchical structure formation. Klypin et al. [68] and Moore et al. [69] performed high-resolution cosmological N-body simulations for the formation and evolution of galaxy-scale halos. They demonstrated that too many subhalos existed in simulated halos in comparison with the number of observed dwarf galaxies in the Local Group. This discrepancy is known as the "missing satellite problem" that has been investigated by a number of follow-up simulation studies (e.g., References [70–72]). Even though it triggered many studies attempting to reduce small-scale structures by imposing other non-CDM candidates such as warm dark matter [73] and self-interacting dark matter [74], it is also possible to solve it with standard baryonic physics including early reionization [75] and self-regulation of star formation in low-mass halos [76–79]. Hence, it is no longer regarded as a serious problem of the CDM model.

These studies suggest a large number of "dark satellites" exist in halos, which do not contain optically visible components such as gases and stars. The population of dark satellites is more abundant in host halos than visible satellite galaxies and could enhance annihilation boosts significantly. To estimate subhalo boosts to annihilation signals accurately, understanding abundance of subhalos as well as their density structure is crucial.

A number of studies have calculated the subhalo mass function in halos using cosmological N-body simulations (e.g., References [29,37,51,80]), indicating that it obeys a power law

$$\frac{dN_{\rm sh}}{dm} \propto m^{-\alpha}, \qquad (21)$$

where the slope $-\alpha$ ranges from -2 to -1.8, although no consensus has yet emerged. There is also a large halo-to-halo scatter for the subhalo abundances [80,81]. The subhalo abundance at a fixed mass halo depends on their accretion history. Namely, it increases with the mass of halo and decreases with the halo concentration (e.g., References [65,80–87]).

Due to the limitation of currently available computational resources, simulations cannot resolve the full hierarchy of subhalos from the smallest to the most massive scales, which ranges more than twenty orders of magnitude in the mass. Even in the highest resolution simulations for galaxy-scale halos, the smallest resolved subhalo mass is around $\sim 10^5 M_\odot$ [37,51], which is still more than ten orders of magnitude more massive than that of the cutoff scale. To study subhalo boosts to annihilation signals, a single-power-law subhalo mass function, Equation (21), is traditionally extrapolated beyond the resolution.

Another approach is to use some analytical models (e.g., References [46,50,85,88–90]), which can shed light on the resolution issue. Hiroshima et al. [50] developed a model of the subhalo evolution calibrated with cosmological N-body simulations and found that the power-law index of the subhalo mass function is in a rather narrow range between -2 and -1.8 with a vast range of subhalo mass from $z = 0$ to 5. This picture is more or less consistent with the assumption of the subhalo mass function of the single power law. More details on the analytic approach are discussed in the following section.

Note that the annihilation boost factors strongly depend on the underlying subhalo mass function [44,45,48]. Assuming that ∼10% of the halo mass is within subhalos, the difference of the boost factors in the Milky-Way-size halo could be as large as a factor of ten between the slope of $-\alpha = -2$ and -1.9 [44]. More extensive simulations are needed to obtain the subhalo mass function in wide mass ranges and also to compare with analytic models [50].

3.2. Density Profile of Dark Matter Halos

By the end of the 1980's, it was already known in both analytic [91] and numerical [92] studies that the density profiles were described by power-law functions. Reference Dubinski and Carlberg [93] studied the density profiles of dark matter halos using cosmological N-body simulations and argued that the profiles were well described by a Hernquist model [94].

Navarro et al. [57,58] simulated the structures of CDM halos systematically with masses in the range of galaxy to rich cluster size. They claimed that the radial density profile $\rho(r)$ could be described by a simple universal profile, Equation (3), the so-called NFW profile. They also claimed that the shape of the profile was universal, independent of cosmological parameters, the primordial power spectrum, and the halo mass. Today, the NFW profile has been extensively used to model halos analytically for various purposes.

After the work of NFW, A number of subsequent studies (e.g., References [95,96]) performed simulations with better mass resolutions. Whereas previous studies [57,58] used only ∼10,000 particles, they used ∼1,000,000 particles for a halo, and found that the slope was steeper than -1. In the original results of the NFW, the numerical two-body relaxation effects due to the small number of particles affected the structures of central regions and led to form a shallower cusp. Higher resolution simulations could resolve more inner structures of halos [97–104]. In most cases, the slope of density became shallower as the radius went inward. A different approach was adopted by Jing and Suto [105], who used the triaxial model for describing the central structures. Moore et al. [96] and Diemand et al. [106] considered a more general profile,

$$\rho(r) = \frac{\rho_s}{(r/r_s)^\gamma \left[1 + (r/r_s)^\eta\right]^{(\beta-\gamma)/\eta}}. \tag{22}$$

If $\beta = 3$, $\gamma = 1$, and $\eta = 1$, the profile is the same as NFW.

More recent studies [37,51,52] archived one of the highest resolution dark matter only simulations for galaxy-size halos with mass resolution better than $10^4 M_\odot$. Their results are in agreement in that the density slope cannot be described by a single power law and the slope is around -1 at the radius $\sim 0.001 r_{200}$. Besides, Springel et al. [37] and Stadel et al. [52] fitted the density using the Einasto profile [107]

$$\rho(r) = \rho_s \exp\left\{-\frac{2}{\alpha_E}\left[\left(\frac{r}{r_s}\right)^{\alpha_E} - 1\right]\right\}, \tag{23}$$

where α_E is a free parameter. Note that r_s and ρ_s are not the same parameters as those in Equation (3). Although we can obtain the density profile down to the radius $\sim 0.001 r_{200}$, the result does not converge to a single power law. In addition, the physical origin of this flattening towards the center is not understood at all. However, the importance of understanding the central structures is increasing. In particular, if we would like to detect signals from dark matter annihilation, the central structure of the dark matter halo is essential.

The most important parameter to describe the halo profile is the concentration parameter. Assuming the universal NFW profile regardless of the halo mass, the concentration-mass relation gives the annihilation rate as a function of the halo mass. Combined with assumed subhalo mass functions, they enable to estimate the annihilation boost factor. The concentration-mass relation of halos has been widely investigated and a number of fitting functions has been suggested [44,45,64,65,108–117]. The concentration shows a weak dependence on the halo mass. The average concentration at fixed halo mass becomes smaller with increasing halo mass because the central density is tightly correlated with the cosmic density at the halo formation epoch, reflecting the hierarchical structure formation [108,118].

Traditionally, the concentration-mass relation has been calibrated with cosmological N-body simulations for relatively massive halos ($10^{10} M_\odot \lesssim M_{200} \lesssim 10^{15} M_\odot$). Because the mass dependence of the concentration is weak for these mass halos, it is found that a single power-law function, $c \propto M_{200}^{-\alpha_c}$, with slope α_c in the range of 0.08 to 0.13, gives reasonable fits [108,110–112,114]. However, the dependence gradually becomes weaker toward less massive halos, and a clear flattening emerges [44,45,64,65,117,119,120], ruling out single power-law concentration-mass relation for the full hierarchy of subhalos.

These fitting functions are valid for the NFW density profile. More generally, the concentration can be defined independently of the density profile and the subhalo mass as (e.g., [37,48,121])

$$c_V = \frac{\bar{\rho}(< r_{\max})}{\rho_c} = 2 \left(\frac{V_{\max}}{H_0 r_{\max}} \right)^2, \qquad (24)$$

where $r_{\max}$ is the radius at which the circular velocity reaches its maximum value $V_{\max}$. This definition is also used to estimate the annihilation boost factor (e.g., [48]).

Even with the highest resolution simulations for galaxy-scale halos, the smallest resolved subhalo mass is around $\sim 10^5 M_\odot$ [37,51]. To estimate the annihilation boost from the full hierarchy of subhalos, we have to make some assumption of the concentration at unresolved scales, which has a significant impact on the result. One approach is extrapolating single power-law fittings to the smallest scale beyond the mass range calibrated with simulations, although the literature including the above cautions the risk of such extrapolations. With such extrapolations, the concentration of the smallest halo can reach more than 100, substantially enhancing the annihilation boost. A number of studies have computed the concentration in such a manner and the resulting boost factor is a few hundreds for Milky-Way halos [36], and ~ 1000 for cluster-scale halos [40,41,56].

Another approach is adopting analytic models or fitting functions that can reproduce flattening of the concentration-mass relation (e.g., [44,64,115,116]). In contrast to using the power-law extrapolation, the resulting boost factor is rather modest, three to a few tens [44,45,47] for Milky-Way halo, and less than ~ 100 for cluster-scale halos [44,45,56].

The density profile at fixed halo mass shows a significant halo-to-halo scatter [122], possibly making a big impact on the annihilation signal. Inferring from the cosmological Millennium simulation [123], the effect of this non-universality on the annihilation flux is a factor of ~ 3 [122], which indicates that the uncertainty of the concentration-mass relation for low-mass halos has a more significant effect.

These discussions are based on the universal density profile and the concentration-mass relation for field halos. There is a concern that whether or not we can apply the universal NFW profile for the full hierarchy of halos and subhalos beyond the range that cosmological simulations have been able to tackle. We discuss this issue in Section 3.4. More importantly, we have to use the concentration-mass relation for subhalos, not field halos. We also discuss this issue in the following section.

3.3. Density Profile of Dark Matter Subhalos

Density structures of subhalos are more challenging to be investigated than field halos because it requires much higher resolutions. Therefore, to evaluate the subhalo contribution to the annihilation signals, the universal NFW profile and the concentration-mass relation for field halos have been historically assumed to be the same for subhalos as a first approximation, although the underlying assumption is not well studied. Complex physical mechanisms relevant to subhalos could change their original density profiles, such as the tidal effect from host halos, the encounter with other subhalos, and denser environment than the field.

Cosmological simulations have been suggesting that the density profile of subhalos is cuspy in analogy with field halos. On the other hand, the average concentration of subhalos tend to be higher than those of field halos (e.g., [37,48,97,108,121]). For example, Bullock et al. [108] showed that subhalos and halos tend to be more concentrated in dense environments than in the field, and the scatter of concentrations is larger. This result was taken into account to estimate the gamma-ray flux from dark matter annihilation (e.g., [124]). Diemand et al. [121] showed that outer regions of subhalos tend to be tidally stripped by host halos, which gives higher concentrations. These results suggest that both earlier formation of halos/subhalos in dense environments and tidal effect are responsible for the increased concentration. Pieri et al. [125] derived the concentration-mass relation of subhalos in Milky-Way-size halos by analyzing high resolution cosmological simulations [37,51] and showed that it depends on the location of subhalos relative to host halos. Subhalos have considerably large concentrations near the center than at the edge of host halos. Moliné et al. [48] quantified the concentration of subhalos in Milky-Way-size halos as a function of not only subhalo mass but distance from host halo center, and found a factor 2–3 enhancement of the boost factor compared to the estimation that relied on the concentration-mass relation of field halos (see also Figure 2).

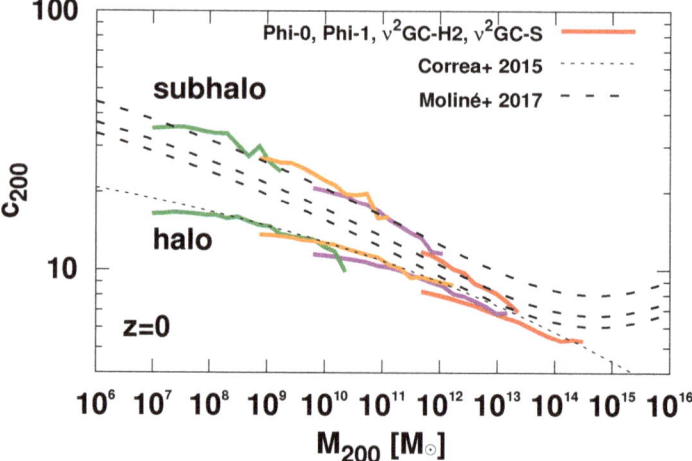

Figure 2. The mass-concentraion relation of halos and subhalos at $z = 0$, derived from high-resolution cosmological N-body simulations, Phi-0 [67], Phi-1, ν^2GC-H2, and ν^2GC-S [50,126,127] (green, orange, purple, and red, respectively). Dashed and dotted curves are fitting formulae proposed by [48,64], respectively. The dependence of the distance to host halo center gives three different dashed curves.

As shown in the literature in the above, higher concentrations of subhalos than field halos could have a big impact on the annihilation boost. However, van den Bosch and Ogiya [128] argued that subhalos even in state-of-the-art cosmological simulations suffer from excessive mass loss and artificial tidal disruption due to inadequately large force softening (see also [129,130]). If that is the case, it might be possible that subhalos have larger concentrations than those ever considered. These issues should be addressed by extremely high resolution cosmological N-body simulations and analytic models (e.g., [50]).

3.4. Density Profile of Dark Matter Halos Near Cutoff Scales

In the CDM framework, smaller halos collapse first, and then they merge into more massive halos. Since the smallest halos contain no subhalos, their central structures might entirely differ from that observed in more massive halos. If the dark matter particle is the lightest supersymmetric particle such as the neutralino, the smallest halo mass is predicted to be around the Earth mass [11,12,14]. Such halos are sometimes referred to as "microhalos."

The density profiles of the microhalos have been investigated using cosmological N-body simulations [45,66,131–133]. Diemand et al. [131] simulated the formation of Earth-mass microhalos by means of cosmological N-body simulation. They claimed that a single power law could describe the density profiles of microhalos, $\rho(r) \propto r^{-\gamma}$, with a slope γ in the range of 1.5 to 2. As a consequence of such steep slope, most microhalos could not be completely destructed by the Galactic tide and encounters with stars, even in the Galactic center.

Ishiyama et al. [132] have performed N-body simulations with much higher resolution and showed that the density profile of microhalos had steeper cusps than the NFW profile. The central density scales as $\rho(r) \propto r^{-1.5}$, which is supported by follow-up cosmological simulations [45,66,133] and cold-collapse simulations [134]. Ishiyama [45] has also shown that the cusp slope gradually becomes shallower with increasing halo mass. Major merger of halos is responsible for the flattening, indicating that the process of violent relaxation plays a key role (see also [133,135]). Similar density structures are observed in recent simulations of ultracompact minihalos [136–138] and warm dark matter [139]. The self-similar gravitational collapse models (e.g., [91,140–143]) can also give hints to understand the main physical origin of such steeper cusps, because the smallest halos do not contain smaller density fluctuations by definition and collapse from initially overdense patches.

Such microhalos with steep cusps can cause a significant effect on indirect dark matter searches. Ishiyama et al. [132] argued that the central parts of microhalos could survive against the encounters with stars except in the very Galactic center. The nearest microhalos could be observable via gamma rays from dark matter annihilation, with usually large proper motions of ∼0.2 deg yr^{-1}, which are, however, stringently constrained with the diffuse gamma-ray background [33]. Gravitational perturbations to the millisecond pulsars might be detectable with future observations by pulsar timing arrays [132,144–146]. Anderhalden and Diemand [66] have assumed a transition from the NFW to steeper cusps at scales corresponding to ∼100 times more massive than the cutoff and have found that such profiles can enhance moderately the annihilation boost of a Milky-Way-size halo by 5–12%. They also have found that concentrations of microhalos are consistent with a toy model proposed by Bullock et al. [108].

Ishiyama [45] showed that the steeper inner cusps of halos in the smallest scale and near the cutoff scale could increase the annihilation rate of a Milky-Way-size halo by 12–67%, compared with estimates adopting the universal NFW profile and an empirical concentration-mass relation [44] (see Figure 3). The value, however, depends strongly on the adopted subhalo mass function and concentration model. They have found that concentrations near the free-streaming scale show little dependence on the halo mass and corresponding conventional NFW concentrations are 60–70, consistent with the picture that the mass dependence is gradually becoming weaker toward less massive halos (e.g., [44,45,64,65,117,119]), ruling out a single power-law concentration-mass relation.

As shown in the literature above, steep density cusps of halos near the free-streaming scale have an impact on the annihilation boost. However, these studies rely on the density structure seen in field halos, not subhalos. It is also important to quantify the structures of subhalos near the free-streaming scale by larger simulations. Another concern is that the cutoff in the matter power spectrum should suppress the number of subhalos near the free-streaming scale, which should weaken the annihilation signal. However, the shape of the mass function near the free-streaming scale is not understood well for the neutralino dark matter. The structure of subhalos and the subhalo mass function near the free-streaming scale should be explored by larger volume cosmological *N*-body simulations.

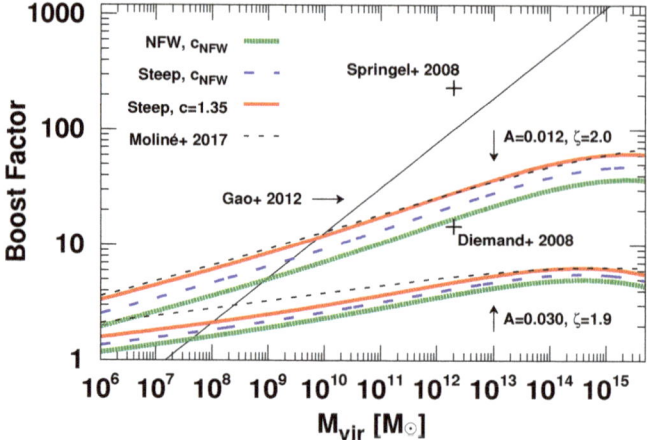

Figure 3. Boost factor as a function of halo virial mass. The data used in six thick curves are taken from Ishiyama [45]. Two subhalo mass functions, $dn/dm = A/M_{vir}(m/M_{vir})^{-\zeta}$, are used ($A = 0.012$, $\zeta = 2.0$, and $A = 0.030$, $\zeta = 1.9$ [44,48]). Thick dotted curves are for the NFW profile, where the empirical concentration-mass relation of field halos [44] are assumed for the full hierarchy of subhalos. Including the effect of steeper cusp of halos near the free streaming scale gives thick dashed curves. Besides, thick solid curves are results of incorporating the concentration of these halos derived from cosmological simulations [45]. For comparison, boost factors obtained in other studies are shown with thin dashed curves [48] (two subhalo mass functions are used), thin solid curves [41], and crosses [36,51].

4. Semi-Analytic Approaches

4.1. Models Based on Structure Formation and Tidal Evolution

In an analytical approach in Section 2, the subhalo luminosity L_{sh} is characterized with the mass of the subhalo and the redshift of interest; see, for example, Equation (10). The mass and redshift, however, are not the only quantities that fully characterize the subhalo properties. Indeed, they depend on the accretion history and mass loss after they fall onto their host halo, that is, two subhalos that have the identical mass could have formed with different masses and accreted at different redshifts, evolved down to $z = 0$ reaching the same mass. Bartels and Ando [46] and Hiroshima et al. [50] developed an analytical prescription to take these effects into account, which we follow in this section.

A subhalo is characterized with its mass and redshift when it accreted onto its host, (m_a, z_a). The concentration parameter c_a is drawn from the log-normal distribution with mean $\bar{c}_a(m_a, z_a)$ [64] and $\sigma_{\log c} = 0.13$ [65]. Since the subhalo was a *field halo* when it just accreted, one can use the relations in Section 2.2 to obtain $r_{s,a}$ and $\rho_{s,a}$ for the NFW profile.

After the accretion, the subhalos evolve by losing their mass through tidal forces. The mass-loss rate is typically characterized by a dynamical timascale at the redshift z,

$$\tau_{dyn}(z) = 1.628 h^{-1} \text{Gyr} \left[\frac{\Delta_{vir}(z)}{178} \right]^{1/2} \left[\frac{H(z)}{H_0} \right]^{-1}, \quad (25)$$

as follows [90]:

$$\dot{m}(z) = -A \frac{m(z)}{\tau_{\rm dyn}(z)} \left[\frac{m(z)}{M(z)} \right]^{\zeta}, \qquad (26)$$

where $H(z) = H_0[\Omega_m(1+z)^3 + \Omega_\Lambda]^{1/2}$, $m(z)$ and $M(z)$ are the subhalo and host-halo masses at z, respectively. Following Jiang and van den Bosch [90], Hiroshima et al. [50] adopted simple Monte Carlo simulations to estimate $\dot{m}$ based on the assumption that the subhalo loses all the masses beyond its tidal radius in one complete orbit at its peri-center passage. While Jiang and van den Bosch [90] found $A = 0.81$ and $\zeta = 0.04$, Hiroshima et al. [50] extended the mass and redshift ranges of applicability and found that these parameters are weakly dependent on both M and z:

$$\log A = \left\{ -0.0003 \log \left[\frac{M(z)}{M_\odot} \right] + 0.02 \right\} z + 0.011 \log \left[\frac{M(z)}{M_\odot} \right] - 0.354, \qquad (27)$$

$$\zeta = \left\{ 0.00012 \log \left[\frac{M(z)}{M_\odot} \right] + 0.0033 \right\} z + 0.0011 \log \left[\frac{M(z)}{M_\odot} \right] + 0.026. \qquad (28)$$

One can solve Equation (26) to obtain the subhalo mass at a redshift of interest z, $m(z)$, with a boundary condition of $m(z_a) = m_a$. For the evolution of the host, $M(z)$, Hiroshima et al. [50] adopted a fitting formula given by Correa et al. [147].

The subhalo density profile after accretion is also well described with the NFW profile with a sharp truncation at r_t:

$$\rho(r) = \begin{cases} \rho_s r_s^3 / [r(r+r_s)]^2, & \text{for } r < r_t, \\ 0, & \text{for } r \geq r_t. \end{cases} \qquad (29)$$

This is indeed a good approximation found in the simulations [37]. In addition to r_t, Peñarrubia et al. [129] found that the internal structure changes. If the inner profile is $\propto r^{-1}$ just like NFW, the maximum circular velocity $V_{\max}$ and its corresponding radius $r_{\max}$ evolve as

$$\frac{V_{\max}(z)}{V_{\max,a}} = \frac{2^{0.4}[m(z)/m_a]^{0.3}}{[1+m(z)/m_a]^{0.4}}, \qquad (30)$$

$$\frac{r_{\max}(z)}{r_{\max,0}} = \frac{2^{-0.3}[m(z)/m_a]^{0.4}}{[1+m(z)/m_a]^{-0.3}}, \qquad (31)$$

respectively. After computing $V_{\max}$ and $r_{\max}$ at z, one can convert them to ρ_s and r_s through

$$r_s = \frac{r_{\max}}{2.163}, \qquad (32)$$

$$\rho_s = \frac{4.625}{4\pi G} \left(\frac{V_{\max}}{r_s} \right)^2, \qquad (33)$$

which are valid for the NFW profile. Finally by solving the condition

$$m(z) = \int_0^{r_t} dr\, 4\pi r^2 \rho(r) = 4\pi \rho_s r_s^3 f(r_t/r_s), \qquad (34)$$

the truncation radius r_t is obtained. Hiroshima et al. [50] omitted subhalos with $r_t < 0.77 r_s$ from the subsequent calculations assuming that they were tidally disrupted [148]. This criterion, however, might be a numerical artifact [130]. Either case, Hiroshima et al. [50] checked that whether one implements this condition or not did not have impact on the results of, for example, subhalo mass functions.

Thus, given (m_a, z_a, c_a), one can obtain all the subhalo parameters after the evolution, (m, r_s, ρ_s, r_t), in a deterministic manner. The differential number of subhalos accreted onto a host with a mass m_a and at redshift z_a, $d^2N_{\rm sh}/(dm_a dz_a)$, is given by the excursion set or the extended Press-Schechter formalism [55]. Especially Yang et al. [89] obtained analytical formulation for the distribution that provides good fit to the numerical simulation data over a large range of m/M and z. Hiroshima et al. [50] adopted their model III.

The subhalo mass function is obtained as

$$\frac{dN_{\rm sh}(m|M,z)}{dm} = \int dm_a \int dz_a \frac{d^2 N_{\rm sh}}{dm_a dz_a} \int dc_a P(c_a|m_a,z_a)\delta(m - m(z|m_a,z_a,c_a)), \quad (35)$$

where $P(c_a|m_a,z_a)$ is the probability distrbution for c_a given m_a and z_a, for which Hiroshima et al. [50] adopted the log-normal distribution with the mean $\bar{c}_a(m_a,z_a)$ [64] and the standard deviation $\sigma_{\log c_a} = 0.13$ [65]. We show the subhalo mass functions obtained with Equation (35) for various values of M and z in Figure 4, where comparison is made with simulation results of similar host halos. Halos and subhalos formed in these simulations were identified with ROCKSTAR phase space halo finder [149]. The bound mass is used as the subhalo mass, which nearly corresponds to the tidal mass [48]. For all these halos, one can see remarkable agreement between the analytic model and the corresponding simulation results in resolved regimes. Successfully reproducing behaviors at resolved regimes, this analytic model is able to make reliable predictions of the subhalo mass functions below resolutions of the numerical simulations, without relying on extrapolating a single power-law functions, from which most of the previous studies in the literature had to suffer. The subhalo mass fraction is then obtained as

$$\begin{aligned} f_{\rm sh}(M,z) &= \frac{1}{M}\int dm\, m \frac{dN_{\rm sh}(m|M,z)}{dm} \\ &= \frac{1}{M}\int dm_a \int dz_a \frac{d^2 N_{\rm sh}}{dm_a dz_a} \int dc_a P(c_a|m_a,z_a) m(z|m_a,z_a,c_a), \end{aligned} \quad (36)$$

and is shown in Figure 4 (bottom right) for various values of redshifts. The subhalo mass fraction is found to increase as a function of M and z. At higher redshifts, since there is shorter time for the subhalos to experience tidal mass loss, $f_{\rm sh}$ is larger. Again, a good agreement in $f_{\rm sh}$ is found between the analytic model and the simulation results by Giocoli et al. [86].

The annihilation boost factor is then

$$B^{(0)}_{\rm sh}(M,z) = \frac{1}{L_{\rm host,0}(M,z)} \int dm_a \int dz_a \frac{d^2 N_{\rm sh}}{dm_a dz_a} \int dc_a P(c_a|m_a,z_a) L^{(0)}_{\rm sh}(m_a,z_a,c_a|M,z), \quad (37)$$

which is to be compared with Equation (10) that was derived with a simpler (and unrealistic) discussion. The superscript (0) represents the quantity *in the absense of sub-subhalos and beyond*. The subhalo luminosity, $L^{(0)}_{\rm sh}(m_a,z_a,c_a|M,z)$, is proportional to the volume integral of density squared $\rho^2_{\rm sh}(r)$ out to the truncation radius,

$$L^{(0)}_{\rm sh}(m_a,z_a,c_a|M,z) \propto \int d^3 x \rho^2_{\rm sh}(x) = \frac{4\pi}{3}\rho^2_{s,{\rm sh}} r^3_{s,{\rm sh}} \left[1 - \frac{1}{(1+r_{t,{\rm sh}}/r_{s,{\rm sh}})^3}\right], \quad (38)$$

where $\rho_{s,{\rm sh}}$, $r_{s,{\rm sh}}$ and $r_{t,{\rm sh}}$ are functions of (m_a,z_a,c_a) as well as M and z.

Then, the effect of sub^n-subhalos (for $n \geq 1$) can be estimated iteratively. At nth iteraction, when a subhalo accreted onto its host at z_a with m_a, it is assigned a sub-subhalo boost factor $B^{(n-1)}_{\rm sh}(m_a,z_a)$. After the accretion, the outer region of the subhalo is stripped away by the tidal force and thus all the sub-subhalos within this stripped region will disappear, reducing the sub-subhalo boost accordingly. Hiroshima et al. [50] assumed that the sub-subhalos were distributed within the subhalo following $n_{\rm ssh}(r) \propto [1 + (r/r_s)^2]^{-3/2}$. The luminosity due to sub-subhalos within a radius r is therefore proportional to their enclosed number

$$N_{\rm ssh}(<r|r_s) = \int_0^r dr' 4\pi r'^2 n_{\rm ssh}(r') \propto r_s^3 \left[\sinh^{-1}\left(\frac{r}{r_s}\right) - \frac{r}{\sqrt{r^2 + r_s^2}}\right], \quad (39)$$

and it gets suppressed by a factor of $N_{\text{ssh}}(< r_t|r_s)/N_{\text{ssh}}(< r_{\text{vir}}|r_{s,a})$ due to the tidal stripping.[4] The luminosity due to the smooth component also decreases as $L^{(0)}_{\text{sh}}(< r_t|\rho_s, r_s)/L^{(0)}_{\text{sh}}(< r_{\text{vir}}|\rho_{s,a}, r_{s,a})$, where

$$L^{(0)}_{\text{sh}}(< r|\rho_s, r_s) \propto \rho_s^2 r_s^3 \left[1 - \frac{1}{(1+r/r_s)^3}\right]. \tag{40}$$

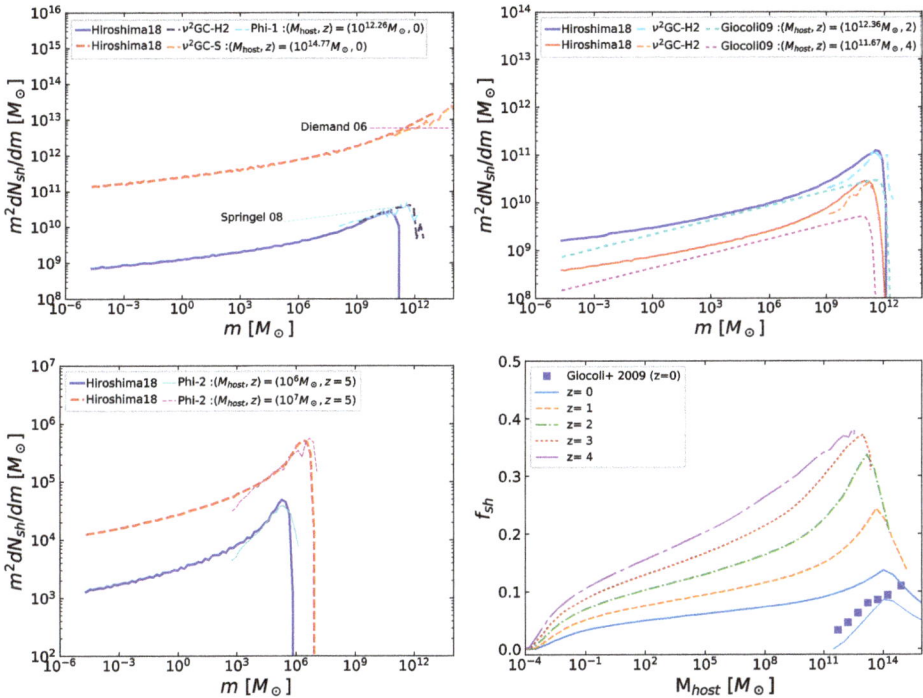

Figure 4. Subhalo mass function for galaxy ($M_{200} = 1.8 \times 10^{12} M_\odot$) and cluster ($M_{200} = 5.9 \times 10^{14} M_\odot$) halos at $z = 0$ (**top left**), halos with $2.3 \times 10^{12} M_\odot$ at $z = 2$ and $4.7 \times 10^{11} M_\odot$ at $z = 4$ both of which would evolve to $M_{200} = 10^{13} M_\odot$ at $z = 0$ (**top right**) and smaller halos of $M_{200} = 10^6 M_\odot$ and $10^7 M_\odot$ at $z = 5$ (**bottom left**). Results of the analytic models by Hiroshima et al. [50] are compared with those from the numerical simulations of similar halos and other fitting functions: Springel et al. [37], Diemand et al. [150] (**top left**), ν^2GC H2 [126,127], Giocoli et al. [86] (**top right**) and Phi-2 (Ishiyama et al., in preparation; **bottom left**). The bottom right panel shows the subhalo mass fraction f_{sh} as a function of the host mass M_{200} for various values of redshift z. The thin solid curve is for $z = 0$ but with lower mass threshold of $1.73 \times 10^{10} h^{-1} M_\odot$ to be compared with Giocoli et al. [86] results shown as the squares.

Thus the sub-subhalo boost after the nth iteration, $B^{(n)}_{\text{ssh}}$ is obtained by

$$B^{(n)}_{\text{ssh}}(m_a, z_a, c_a|M, z) = B^{(n-1)}_{\text{sh}}(m_a, z_a) \frac{N_{\text{ssh}}(< r_t|r_s)/N_{\text{ssh}}(< r_{\text{vir}}|r_{s,a})}{L^{(0)}_{\text{sh}}(< r_t|\rho_s, r_s)/L^{(0)}_{\text{sh}}(< r_{\text{vir}}|\rho_{s,a}, r_{s,a})}. \tag{41}$$

---

[4] We note that in estimating the effect of sub^n-subhalos in the boost factors, Reference Hiroshima et al. [50] ignored the changes of ρ_s and r_s and hence did not include the factor of r_s^3 in Equation (39) and $\rho_s^2 r_s^3$ in Equation (40). In addition, in Equation (43), they multiplied $L^{(0)}_{\text{sh}}$ by a factor of $1 + B^{(n)}_{\text{ssh}}$ instead of $1 - f_{\text{ssh}}^2 + B^{(n)}_{\text{ssh}}$. We correct for all these effects in this review.

Similarly, the sub-subhalo mass fraction $f_{\rm ssh}$ is obtained by

$$f_{\rm ssh}(m_a, z_a, c_a | M, z) = f_{\rm sh}(m_a, z_a) \frac{N_{\rm ssh}(<r_t | r_s)/N_{\rm ssh}(<r_{\rm vir}|r_{s,a})}{m_{\rm sm}(<r_t|\rho_s, r_s)/m_{\rm sm}(<r_{\rm vir}|\rho_{s,a}, r_{s,a})}, \quad (42)$$

where $f_{\rm sh}(m_a, z_a)$ is obtained with Equations (36) and $m_{\rm sm}(<r|\rho_s, r_s) \propto \rho_s r_s^3 f(r/r_s)$ is the enclosed mass within r of the smooth component of the subhalo. The subhalo boost factor after nth iteration is obtained with Equation (37) by replacing $L_{\rm sh}^{(0)}$ with $[1 - f_{\rm ssh}^2 + B_{\rm ssh}^{(n)}]L_{\rm sh}^{(0)}$ [see discussions below Equation (15)]:

$$\begin{aligned}
B_{\rm sh}^{(n)}(M, z) &= \frac{1}{L_{\rm host,0}(M, z)} \int dm_a \int dz_a \frac{d^2 N_{\rm sh}}{dm_a dz_a} \int dc_a P(c_a | m_a, z_a) \\
&\quad \times \left[1 - f_{\rm ssh}^2(m_a, z_a, c_a | M, z) + B_{\rm ssh}^{(n)}(m_a, z_a, c_a | M, z)\right] L_{\rm sh}^{(0)}(m_a, z_a, c_a | M, z). \quad (43)
\end{aligned}$$

The host luminosity in the absence of the subhalos $L_{\rm host,0}(M, z)$ is defined by marginalizing over the concentration parameter $c_{\rm vir}$:

$$L_{\rm host,0}(M, z) \propto \frac{4\pi}{3} \int dc_{\rm vir} P(c_{\rm vir}|M, z) \rho_s^2(M, z, c_{\rm vir}) r_s^3(M, z, c_{\rm vir}) \left[1 - \frac{1}{(1 + c_{\rm vir})^3}\right]. \quad (44)$$

with the log-normal distribution $P(c_{\rm vir}|M,z)$.

Figure 5 shows the subhalo boost factors $B_{\rm sh}$ as a function of the host mass M at various redshifts z (top left). The boost factors are on the order of unity, while it can be as larger as ~ 5 for cluster-size halos. It is also noted that they are larger at higher redshifts, because the subhalos have less time to be disrupted. The top right panel of Figure 5 shows the effect of sub^n-subhalos, which is saturated after the second iteration. The contribution to the boost factors due to sub-subhalos and beyond is $\lesssim 10\%$ for the hosts with $M_{\rm host} \geq 10^{13} M_\odot$. The bottom left panel of Figure 5 shows the luminosity ratio $L_{\rm total}/L_{\rm host,0} = 1 - f_{\rm sh}^2 + B_{\rm sh}$ (Equation (15)) as a function of the host masses for various values of the redshifts. The bottom right panel of Figure 5 shows comparison with the results of the other work [41,44,48]. We note that the analytic models do not rely on the subhalo mass function prepared separately, as the models can provide them in a self-consistent manner. The resulting boost factors are, however, found to be more modest than the previous results. This is mainly because the subhalo mass function adopted in the literature is larger than the predictions of the analytic models. However, they might be larger because of halo-to-halo variance. See discrepancy between predictions of the subhalo mass function for the $1.8 \times 10^{12} M_\odot$ halo by Hiroshima et al. [50] and the result of Springel et al. [37] shown in the top left panel of Figure 4.

Finally, for convenience of the reader who might be interested in using the results without going into details of the formalism, we provide fitting functions for both the subhalo mass functions and the annihilation boost factors. They are summarized in Appendix A.

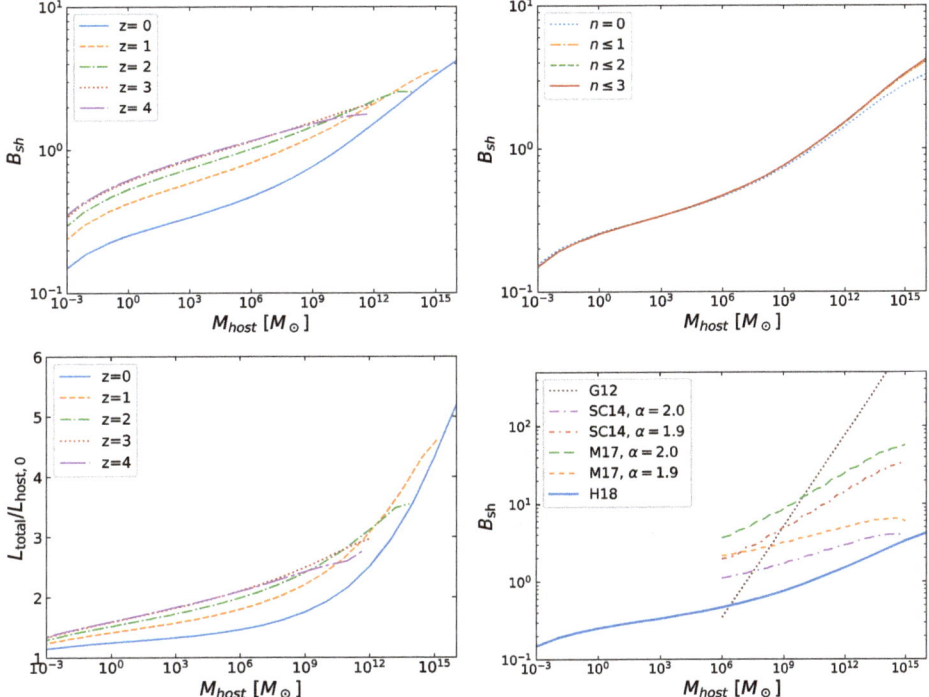

Figure 5. The subhalo boost factor B_{sh} as a function of the host mass M_{200} for various values of redshift z (**top left**) based on the analytic models by Hiroshima et al. [50]. The effect of sub^n-subhalos, up to $n = 3$, is shown in the **right panel** in the case of $z = 0$. Note that the three curves except for $n = 0$ overlap with each other. The **bottom left** panel shows the ratio between the total luminosity including the subhalo boost and the luminosity in absence of subhalos, $L_{total}/L_{host,0} = 1 - f_{sh}^2 + B_{sh}$. The **bottom right** panel shows comparison of B_{sh} between several models at $z = 0$: G12 [41], SC14 [44] and M17 [48] are based on N-body calculations while H18 [50] is on analytic calculations. The subhalo mass function for the N-body results is assumed to be $dN_{sh}/dm \propto m^{-\alpha}$.

4.2. Models for Self-Similar Subhalos

Assuming a self-similarity of the subhalos, Kamionkowski and Koushiappas [35] developed a fully analytic formulation for the probability distribution function of the dark matter density, $P(\rho)$. Then Kamionkowski et al. [39] applied the formulation to the result of cosmological N-body simulations, to obtain the fitting function for the Galactic local boost factor at Galactocentric radius r:

$$B_{sh}(r) = f_{sm}(r)e^{\delta_f^2} + [1 - f_{sm}(r)]\frac{1 + \alpha_K}{1 - \alpha_K}\left[\left(\frac{\rho_{max}}{\rho_\chi(r)}\right)^{1-\alpha_K} - 1\right], \quad (45)$$

where $f_{sm}(r)$ is the volume fraction occupied by the smooth component and ρ_{max} is the highest dark matter subhalo density. Through the calibration with the numerical simulations, Kamionkowski et al. [39] found $\delta_f = 0.2$, $\alpha_K = 0$ and that the subhalo fraction was given by

$$1 - f_{sm}(r) = \kappa \left[\frac{\rho_\chi(r)}{\rho_\chi(100 \text{ kpc})}\right]^{-0.26}, \quad (46)$$

where $\kappa = 0.007$. Fornasa et al. [42] then suggested a larger value of $\kappa = 0.15$–0.2 to obtain a larger boost consistent with earlier work [36,40,41]. The maximum subhalo density ρ_{max} is estimated as $\rho_{max} = [c^3/f(c)/12]200\rho_c(z_f)$, where c and z_f are the concentration parameter and collapse redshift of the smallest halos. Kamionkowski et al. [39] adopted $c = 3.5$ and $z_f = 40$ and $\rho_{max} = 80$ GeV cm^{-3}. On the other hand, Ng et al. [43] obtained a smaller $\rho_{max} \approx 20$ GeV cm^{-3} even with a very small cutoff masses of $M_{min} = 10^{-12} M_\odot$ Within the virial radius of the Milky-Way halo, the subhalo boost factor for dark matter annihilation is found to be no greater than ~ 10 [39].

4.3. Universal Clustering of Dark Matter in Phase Space

Zavala and Afshordi [49] investigated the behavior of dark matter particles that belong to the halo substructure in the phase space of distance and velocity. Reconstructing the phase-space distribution using the Aquarius numerical simulations [37], they found universality of the coarse-grained phase-space distribution ranging from dwarfs to clusters of galaxies. They developed physically motivated models based on the stable clustering hypothesis, spherical collapses and the tidal stripping of the subhalos and applied to the obtained phase-space distribution data from the simulations to find a good agreement. Then, they computed the nonlinear matter power spectrum based on the halo model [151] down to very small free-streaming cutoff scales. Based on the power spectrum, they obtained the subhalo boost factor greater than ~ 30–100 for the Milky-Way size halos, which is significantly larger than the values obtained with other analytic work [39,46,50]. This discrepancy might come from the treatment at very small scales, where it is very hard to calibrate the analytic models against the results of the numerical simulations.

5. Conclusions

It is established that dark matter halos are made up with lots of substructures. Especially in the cold dark matter scenario, small structures form first and merge and accrete to create larger halos. If the dark matter is made of weakly interacting massive particles such as the supersymmetric neutralino, the smallest halos can be as light as or even lighter than the Earth. The rate of dark matter annihilation and hence its signatures such as gamma-ray fluxes are proportional to dark matter density squared and therefore, having small-scale "clumpy" subhalos will *boost* the signals.

It is, however, of an extraordinary challenge to estimate this subhalo boost factor, that is, the ratio of luminosity from dark matter annihilation in the subhalos to that in the smoothly distributed main component. This is mainly because subhalos of all the mass scales ranging from Earth to galaxy masses can contribute to the boost factor nearly equally per decade in mass. In this review, we cover recent progress to overcome this issue to obtain realistic and unbiased estimates on the subhalo boost factor that will impact on interpretation of the measurements on particle physics parameters such as the annihilation cross section. While cosmological N-body simulations provide the most accurate avenue to study structures in highly nonlinear regime, it is inevitably limited by the numerical resolution. Even the state-of-the-art N-body simulations [37,150] can resolve subhalos ranging for only several decades, which is still more than ten orders of magnitude in short to resolve all the subhalos. Therefore, the boost estimates have to rely on extrapolation of the subhalo properties such as its mass function and concentration parameter, which are often well described with power-law functions. Danger of extrapolating trends found in resolved regime for other many orders of magnitude had been widely acknowledged but nevertheless, it was found that the estimates based on such extrapolations tended to give very large amount of boost factor of ~ 100 (~ 1000) for galaxy (cluster) size halos [36,41].

As a complementary approach, analytic models have been investigated. They are based on self similar propertiese of the subhalos [35,39], universal phase-space distribution [49] and extended Press-Schechter formalism combined with tidal stripping modeling [46,50]. (More recent numerical approach also adopts the concentration-mass relation calibrated for the subhalos in order to take the tidal effects into account [48].) Most importantly, these are all calibrated with the cosmological N-body simulations at resolved regimes and proven to reproduce the simulation results such as the subhalo mass functions. For example, the most recent analytic models by Reference Hiroshima et al. [50] predict the subhalo mass functions for various host masses and redshifts, which are found to be in good agreement with the simulation results [37,86,126,150]. The annihilation boost factors based on these analytic models tend to be more modest, $\mathcal{O}(1)$ for galaxy-size halos and $\lesssim \mathcal{O}(10)$ for cluster-size halos. However, none of these models have been tested against simulations at very small host halos that are less massive than $10^6 M_\odot$. Simulations of microhalos with $10^{-6} M_\odot$ suggest cuspier profiles towards halo centers such as $r^{-1.5}$ [45] and if this is the case for the subhalos too, it would boost the annihilation rate further.

It is known that including baryons in the simulations affects properties of subhalos such as spatial distribution and density profiles (e.g., References [152–155]) and hence there might be some effect on the annihilation boost factors. This, however, remains largely unexplored and has to wait for future progress. However, since the subhalos of all masses ranging down to about the Earth mass contribute to the boost factors and the baryons will likely affect only halos of dwarf galaxies or larger, we anticipate that it is not a very important effect for the annihilation boost factors.

The subhalo boosts directly impact the obtained upper limits on the dark matter annihilation cross section from the extragalactic halo observations. Therefore, to obtain the most accurate estimates of the boost factor by reducing uncertainties on structure formation at small scales as well as the physics of tidal stripping is of extreme importance for the indirect searches for particle dark matter through self-annihilation with the current and near future observations of high-energy gamma-rays, neutrinos and charged cosmic rays.

Author Contributions: S.A. wrote most of the review sections except for Section 3, which was contributed by T.I. including Figures 1–3. N.H. prepared Figures 4 and 5 and wrote Appendix A.

Funding: This work was supported by JSPS KAKENHI Grant Numbers JP17H04836 (S.A.), JP18H04340 (S.A. and N.H.), JP18H04578 (S.A.), JP15H01030 (T.I.), JP17H04828 (T.I.), JP17H01101 (T.I.) and JP18H04337 (T.I.). T.I. has been supported by MEXT as "Priority Issue on Post-K computer" (Elucidation of the Fundamental Laws and Evolution of the Universe) and JICFuS.

Acknowledgments: Numerical computations were partially carried out on the K computer at the RIKEN Advanced Institute for Computational Science (Proposal numbers hp150226, hp160212, hp170231, hp180180), and Aterui supercomputer at Center for Computational Astrophysics, CfCA, of National Astronomical Observatory of Japan.

Conflicts of Interest: The authors declare no conflict of interest.

Appendix A. Fitting Formulae

In this appendix, we provide fitting functions obtained with the analytical calculation in Hiroshima et al. [50] that covers more than twenty orders of magnitude in the mass range and the redshift up to ~ 10. The mass function of the subhalo $m^2 dN_{\rm sh}/dm$, the luminosity of the subhalo separated from the particle physics factors $L_{\rm sh}$, and the boost factor $B_{\rm sh} = L_{\rm sh}/L_{\rm host,0}$ [see Equation (10)] as functions of the host mass and the redshift are provided here. We also summarize fitting functions for the boost factor $B_{\rm sh}$ at $z = 0$ in the literature. Note that the host mass is always measured in units of the solar mass ($M_\odot$) in this appendix.

Appendix A.1. Subhalo Mass Function

The fitting formula is written in the follwing form:

$$m^2 \frac{dN}{dm} = (a + bm^\alpha) \exp\left[-\left(\frac{m}{m_c}\right)^\beta\right], \tag{A1}$$

introducing a cutoff mass m_c; a, b, and m_c in Equation (A1) are functions of the host mass and the redshift:

$$a(M_{\text{host}}, z) = 6.0 \times 10^{-4} M_{\text{host}}(z) \left(2 - 0.08 \log_{10} [M_{\text{host}}(z)]\right)^2 \left(\log_{10}\left[\frac{M_{\text{host}}(z)}{10^{-5}}\right]\right)^{-0.1} \quad \text{(A2)}$$
$$\times (1+z),$$

$$b(M_{\text{host}}, z) = 8.0 \times 10^{-5} M_{\text{host}}(z) \left(2 - 0.08 \log_{10} [M_{\text{host}}(z)]\right) \left(\log_{10}\left[\frac{M_{\text{host}}(z)}{10^{-5}}\right]\right)^{-0.08z} \quad \text{(A3)}$$
$$\times \left(\log_{10}\left[\frac{M_{\text{host}}(z)}{10^{-8}}\right]\right)^{-1} \left(\log_{10}\left[\frac{M_{\text{host}}(z)}{10^{18}}\right]\right)^2,$$

$$m_c = 0.05 (1+z) M_{\text{host}}(z), \quad \text{(A4)}$$
$$\alpha = 0.2 + 0.02z, \quad \text{(A5)}$$
$$\beta = 3. \quad \text{(A6)}$$

In Figure A1, we show the comparison between the mass function obtained in analytical calculations [50] and Equation (A1). By integrating Equation (A1), we obtain the mass fraction shown in Figure A2.

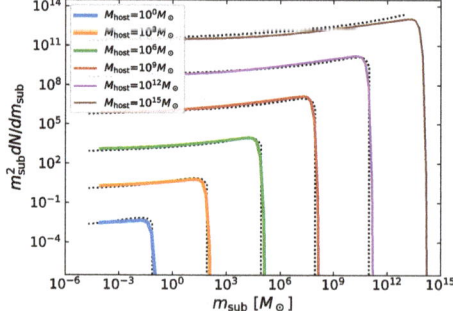

Figure A1. Subhalo mass function $m^2 dN/dm$ at $z=0$. Each line corresponds to a different host halo mass. The fitting formula is applicable for the host mass from $M_{\text{host}} \simeq 10^{-4}$ to $10^{14} M_\odot$ and redshifts up to ~ 6.

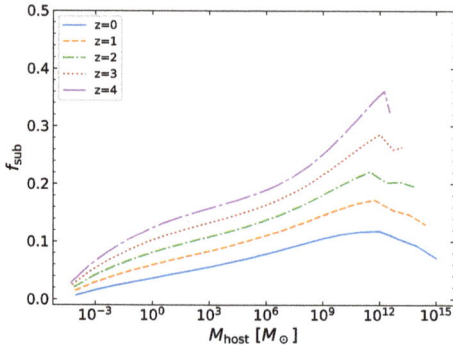

Figure A2. The subhalo mass fraction obtained by integrating Equation (A1).

Appendix A.2. Subhalo Luminosity

The luminosity of a subhalo is written as

$$L_{\text{host}}(M) \propto \rho_s^2 r_s^3 \left[1 - \frac{1}{(1+c_{\text{vir}}^3)}\right], \tag{A7}$$

assuming a subhalo of NFW profile. In this section, the characteristic density ρ_s and the scale radius r_s are measured in units of g/cm^3 and cm, respectively. For simplicity, we do not show the integration over the distribution of the virial concentration parameter $P(c_{\text{vir}}|M,z)$ in the above expression. The constant of proportionality is detemined by fixing the particle physics model. However, we do not include it in the following expression since it cancels in the calculation of the boost factor B_{sh} by taking the ratio of the host and subhalo luminosity $L_{\text{sh}}/L_{\text{host}}$.

The fitting formula for the luminosity takes the form of

$$\log_{10} L_{\text{sh}} = b + a \log_{10} m, \tag{A8}$$

where

$$a = (-0.025z + 0.18)\left(\log_{10}\left[\frac{M_{\text{host}}(z)}{10^{-5}}\right]\right)^{0.3} + (0.06z + 0.53), \tag{A9}$$

$$b = -0.95 \log_{10}[M_{\text{host}}(z)] + (0.1 - 0.015z)\log_{10}\left[\frac{M_{\text{host}}(z)}{10^4}\right] + 0.07. \tag{A10}$$

Appendix A.3. Annihilation Boost Factor

The boost factor is sensitive to the models of the concentration-mass relation. We provide fitting functions for two different concentration-mass relation models here. One coresponds to the canonical model in Reference [50] which assumes the concentration-mass relation derived in Reference [64]. The other corresponds to the concentration-mass relation in Reference [156]. For both cases, the fitting function of the boost factor is written in a combination of two sigmoid functions, $f(x) = (1 + e^{-ax})^{-1}$,

$$\log_{10} B_{\text{sh}} = \frac{X(z)}{1 + e^{-a(z)(\log_{10}[M_{\text{host}}] - m_1(z))}} + c(z)\left(1 + \frac{Y(z)}{1 + e^{-b(z)(\log_{10}[M_{\text{host}}] - m_2(z))}}\right). \tag{A11}$$

Funcitons X, Y, a, b, c, m_1, and m_2 depend on the redshift but they do not on the host mass.

- For Correa's concentration [64]

$$X(z) = 2.7e^{-0.2z} + 0.15, \tag{A12}$$
$$Y(z) = 0.4 + (-0.224z + 0.56)e^{-0.8z}, \tag{A13}$$
$$a(z) = 0.10 + 0.095e^{-0.5z}, \tag{A14}$$
$$b(z) = 0.03z^2 - 0.08z - 0.83, \tag{A15}$$
$$c(z) = 0.004z^2 - 0.04z - 0.6, \tag{A16}$$
$$m_1 = -3.17z + 17.4, \tag{A17}$$
$$m_2 = -(0.2z - 1)^5 - 4. \tag{A18}$$

- For Okoli's concentration [156]

$$X(z) = 2.2e^{-0.75z} + 0.67, \quad (A19)$$
$$Y(z) = 2.5e^{-0.005z} + 0.8, \quad (A20)$$
$$a(z) = 0.1e^{-0.5z} + 0.22, \quad (A21)$$
$$b(z) = 0.8e^{-0.5(z-12)^4} - 0.24, \quad (A22)$$
$$c(z) = -0.0005z^3 - 0.032z^2 + 0.28z - 1.12, \quad (A23)$$
$$m_1 = -2.6z + 8.2, \quad (A24)$$
$$m_2 = 0.1e^{-3z} - 12. \quad (A25)$$

All of these formulae are applicable for hosts at arbitrary redshifts up to $z \sim 7$.

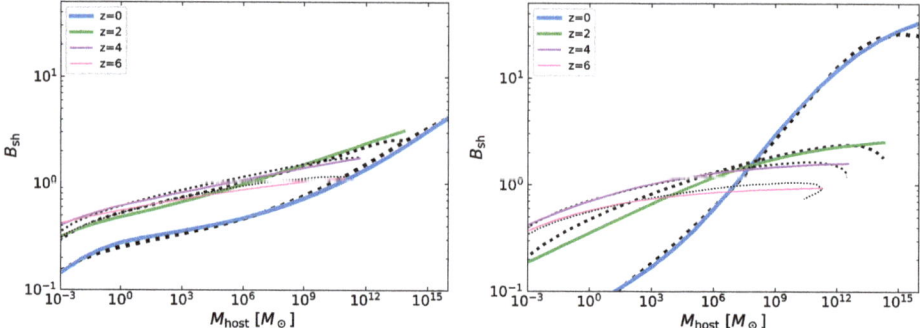

Figure A3. Comparisons between the boost factor from our calculations in Reference [50] and the fitting functions in this section. The left panel is the result assuming the concentration-mass relation in Reference [64] while the right panel assuming the relation in Reference [156].

Appendix A.4. Fitting Functions for the Boost Factor in the Literature

Several works provide the fitting function for the boost factor at $z = 0$. We summarize functions provided in Gao et al. [41], Sánchez-Conde and Prada [44], and Moliné et al. [48] here.

- Gao et al. [41] have analyzed cluster scale halo of $M_{\text{host}} = [5, 20] \times 10^{14} h^{-1} M_\odot$ in the Phoenix project [157]. Subhalos down to $m \sim 10^6$ can be resolved in their calculations.

$$B_{\text{sh}} = 1.6 \times 10^{-3} \left(\frac{M_{\text{host}}}{M_\odot} \right)^{0.39}. \quad (A26)$$

- Sánchez-Conde and Prada [44] derive the boost factor based on the concentration-mass relation in Reference [115]. The fitting function is provided for their fiducial model assuming the minimum halo mass to be $M_{\text{min}} = 10^{-6} M_\odot$ and the subhalo mass function $dN/dm \propto m^{-2}$. Each subhalo is assumed to be a field halo.

$$\log_{10} B_{\text{sh}}(z=0) = \sum_{i=0}^{5} b_i \left(\ln \frac{M_{\text{host}}}{M_\odot} \right)^i \quad (A27)$$

with

$$b_0 = -0.442 \tag{A28}$$
$$b_1 = 0.0796 \tag{A29}$$
$$b_2 = -0.0025 \tag{A30}$$
$$b_3 = 4.77 \times 10^{-6} \tag{A31}$$
$$b_4 = 4.77 \times 10^{-6} \tag{A32}$$
$$b_5 = -9.69 \times 10^{-8} \tag{A33}$$

- Moliné et al. [48] derive the boost factor taking the dependence of the survived halo properties on the distance from the host, that is, the host potential. The form of the function is similar to that of Sánchez-Conde and Prada [44] but adopitng different log bases:

$$\log_{10} B_{\rm sh}(z=0) = \sum_{i=0}^{5} b_i \left(\log_{10} \frac{M_{\rm host}}{M_\odot} \right)^i \tag{A34}$$

and parameters b_i are

$$b_0 = -0.186 \tag{A35}$$
$$b_1 = 0.144 \tag{A36}$$
$$b_2 = -8.8 \times 10^{-3} \tag{A37}$$
$$b_3 = 1.13 \times 10^{-3} \tag{A38}$$
$$b_4 = -3.7 \times 10^{-5} \tag{A39}$$
$$b_5 = -2 \times 10^{-7} \tag{A40}$$

for $\alpha = 2$ and

$$b_0 = -6.8 \times 10^{-2} \tag{A41}$$
$$b_1 = 9.4 \times 10^{-2} \tag{A42}$$
$$b_2 = -9.8 \times 10^{-3} \tag{A43}$$
$$b_3 = 1.05 \times 10^{-3} \tag{A44}$$
$$b_4 = -3.4 \times 10^{-5} \tag{A45}$$
$$b_5 = -2 \times 10^{-7} \tag{A46}$$

for $\alpha = 1.9$.

References

1. Bertone, G.; Hooper, D.; Silk, J. Particle dark matter: Evidence, candidates and constraints. *Phys. Rept.* **2005**, *405*, 279–390. [CrossRef]
2. Bringmann, T.; Weniger, C. Gamma Ray Signals from Dark Matter: Concepts, Status and Prospects. *Phys. Dark Univ.* **2012**, *1*, 194–217. [CrossRef]
3. Jungman, G.; Kamionkowski, M.; Griest, K. Supersymmetric dark matter. *Phys. Rept.* **1996**, *267*, 195–373. [CrossRef]
4. Hooper, D.; Profumo, S. Dark matter and collider phenomenology of universal extra dimensions. *Phys. Rept.* **2007**, *453*, 29–115. [CrossRef]
5. Boveia, A.; Doglioni, C. Dark Matter Searches at Colliders. *Ann. Rev. Nucl. Part. Sci.* **2018**, *68*, 429–459. [CrossRef]

6. Steigman, G.; Dasgupta, B.; Beacom, J.F. Precise Relic WIMP Abundance and its Impact on Searches for Dark Matter Annihilation. *Phys. Rev.* **2012**, *D86*, 023506. [CrossRef]
7. Hofmann, S.; Schwarz, D.J.; Stoecker, H. Damping scales of neutralino cold dark matter. *Phys. Rev.* **2001**, *D64*, 083507. [CrossRef]
8. Green, A.M.; Hofmann, S.; Schwarz, D.J. The power spectrum of SUSY—CDM on sub-galactic scales. *Mon. Not. R. Astron. Soc.* **2004**, *353*, L23. [CrossRef]
9. Loeb, A.; Zaldarriaga, M. The Small-scale power spectrum of cold dark matter. *Phys. Rev.* **2005**, *D71*, 103520. [CrossRef]
10. Bertschinger, E. The Effects of Cold Dark Matter Decoupling and Pair Annihilation on Cosmological Perturbations. *Phys. Rev.* **2006**, *D74*, 063509. [CrossRef]
11. Profumo, S.; Sigurdson, K.; Kamionkowski, M. What mass are the smallest protohalos? *Phys. Rev. Lett.* **2006**, *97*, 031301. [CrossRef] [PubMed]
12. Bringmann, T. Particle Models and the Small-Scale Structure of Dark Matter. *New J. Phys.* **2009**, *11*, 105027. [CrossRef]
13. Cornell, J.M.; Profumo, S. Earthly probes of the smallest dark matter halos. *J. Cosmol. Astropart. Phys.* **2012**, *1206*, 011. [CrossRef]
14. Diamanti, R.; Catalan, M.E.C.; Ando, S. Dark matter protohalos in a nine parameter MSSM and implications for direct and indirect detection. *Phys. Rev.* **2015**, *D92*, 065029. [CrossRef]
15. Silk, J.; Stebbins, A. Clumpy cold dark matter. *Astrophys. J.* **1993**, *411*, 439–449. [CrossRef]
16. Bergstrom, L.; Edsjo, J.; Gondolo, P.; Ullio, P. Clumpy neutralino dark matter. *Phys. Rev.* **1999**, *D59*, 043506. [CrossRef]
17. Bergstrom, L.; Edsjo, J.; Ullio, P. Possible indications of a clumpy dark matter halo. *Phys. Rev.* **1998**, *D58*, 083507. [CrossRef]
18. Calcaneo-Roldan, C.; Moore, B. The Surface brightness of dark matter: Unique signatures of neutralino annihilation in the galactic halo. *Phys. Rev.* **2000**, *D62*, 123005. [CrossRef]
19. Tasitsiomi, A.; Olinto, A.V. The Detectability of neutralino clumps via atmospheric Cherenkov telescopes. *Phys. Rev.* **2002**, *D66*, 083006. [CrossRef]
20. Stoehr, F.; White, S.D.M.; Springel, V.; Tormen, G.; Yoshida, N. Dark matter annihilation in the halo of the Milky Way. *Mon. Not. R. Astron. Soc.* **2003**, *345*, 1313. [CrossRef]
21. Koushiappas, S.M.; Zentner, A.R.; Walker, T.P. The observability of gamma-rays from neutralino annihilations in Milky Way substructure. *Phys. Rev.* **2004**, *D69*, 043501. [CrossRef]
22. Baltz, E.A.; Taylor, J.E.; Wai, L.L. Can Astrophysical Gamma Ray Sources Mimic Dark Matter Annihilation in Galactic Satellites? *Astrophys. J.* **2007**, *659*, L125–L128. [CrossRef]
23. Ando, S. Can dark matter annihilation dominate the extragalactic gamma-ray background? *Phys. Rev. Lett.* **2005**, *94*, 171303. [CrossRef] [PubMed]
24. Oda, T.; Totani, T.; Nagashima, M. Gamma-ray background from neutralino annihilation in the first cosmological objects. *Astrophys. J.* **2005**, *633*, L65–L68. [CrossRef]
25. Pieri, L.; Branchini, E.; Hofmann, S. Difficulty of detecting minihalos via gamm rays from dark matter annihilation. *Phys. Rev. Lett.* **2005**, *95*, 211301. [CrossRef] [PubMed]
26. Koushiappas, S.M. Proper motion of gamma-rays from microhalo sources. *Phys. Rev. Lett.* **2006**, *97*, 191301. [CrossRef] [PubMed]
27. Ando, S.; Komatsu, E.; Narumoto, T.; Totani, T. Dark matter annihilation or unresolved astrophysical sources? Anisotropy probe of the origin of cosmic gamma-ray background. *Phys. Rev.* **2007**, *D75*, 063519. [CrossRef]
28. Pieri, L.; Bertone, G.; Branchini, E. Dark Matter Annihilation in Substructures Revised. *Mon. Not. R. Astron. Soc.* **2008**, *384*, 1627. [CrossRef]
29. Diemand, J.; Kuhlen, M.; Madau, P. Dark matter substructure and gamma-ray annihilation in the Milky Way halo. *Astrophys. J.* **2007**, *657*, 262–270. [CrossRef]
30. Berezinsky, V.; Dokuchaev, V.; Eroshenko, Y. Anisotropy of dark matter annihilation with respect to the Galactic plane. *J. Cosmol. Astropart. Phys.* **2007**, 011. [CrossRef]
31. Lavalle, J.; Yuan, Q.; Maurin, D.; Bi, X.J. Full Calculation of Clumpiness Boost factors for Antimatter Cosmic Rays in the light of Lambda-CDM N-body simulation results. Abandoning hope in clumpiness enhancement? *Astron. Astrophys.* **2008**, *479*, 427–452. [CrossRef]

32. Siegal-Gaskins, J.M. Revealing dark matter substructure with anisotropies in the diffuse gamma-ray background. *J. Cosmol. Astropart. Phys.* **2008**, 040. [CrossRef]
33. Ando, S.; Kamionkowski, M.; Lee, S.K.; Koushiappas, S.M. Can proper motions of dark-matter subhalos be detected? *Phys. Rev.* **2008**, *D78*, 101301. [CrossRef]
34. Lee, S.K.; Ando, S.; Kamionkowski, M. The Gamma-Ray-Flux Probability Distribution Function from Galactic Halo Substructure. *J. Cosmol. Astropart. Phys.* **2009**, *0907*, 007. [CrossRef]
35. Kamionkowski, M.; Koushiappas, S.M. Galactic substructure and direct detection of dark matter. *Phys. Rev.* **2008**, *D77*, 103509. [CrossRef]
36. Springel, V.; White, S.D.M.; Frenk, C.S.; Navarro, J.F.; Jenkins, A.; Vogelsberger, M.; Wang, J.; Ludlow, A.; Helmi, A. Prospects for detecting supersymmetric dark matter in the Galactic halo. *Nature* **2008**, *456*, 73–76. [CrossRef] [PubMed]
37. Springel, V.; Wang, J.; Vogelsberger, M.; Ludlow, A.; Jenkins, A.; Helmi, A.; Navarro, J.F.; Frenk, C.S.; White, S.D.M. The Aquarius Project: The subhalos of galactic halos. *Mon. Not. R. Astron. Soc.* **2008**, *391*, 1685–1711. [CrossRef]
38. Ando, S. Gamma-ray background anisotropy from galactic dark matter substructure. *Phys. Rev.* **2009**, *D80*, 023520. [CrossRef]
39. Kamionkowski, M.; Koushiappas, S.M.; Kuhlen, M. Galactic Substructure and Dark Matter Annihilation in the Milky Way Halo. *Phys. Rev.* **2010**, *D81*, 043532. [CrossRef]
40. Pinzke, A.; Pfrommer, C.; Bergstrom, L. Prospects of detecting gamma-ray emission from galaxy clusters: Cosmic rays and dark matter annihilations. *Phys. Rev.* **2011**, *D84*, 123509. [CrossRef]
41. Gao, L.; Frenk, C.S.; Jenkins, A.; Springel, V.; White, S.D.M. Where will supersymmetric dark matter first be seen? *Mon. Not. R. Astron. Soc.* **2012**, *419*, 1721. [CrossRef]
42. Fornasa, M.; Zavala, J.; Sanchez-Conde, M.A.; Siegal-Gaskins, J.M.; Delahaye, T.; Prada, F.; Vogelsberger, M.; Zandanel, F.; Frenk, C.S. Characterization of Dark-Matter-induced anisotropies in the diffuse gamma-ray background. *Mon. Not. R. Astron. Soc.* **2013**, *429*, 1529–1553. [CrossRef]
43. Ng, K.C.Y.; Laha, R.; Campbell, S.; Horiuchi, S.; Dasgupta, B.; Murase, K.; Beacom, J.F. Resolving small-scale dark matter structures using multisource indirect detection. *Phys. Rev.* **2014**, *D89*, 083001. [CrossRef]
44. Sánchez-Conde, M.A.; Prada, F. The flattening of the concentration–mass relation towards low halo masses and its implications for the annihilation signal boost. *Mon. Not. R. Astron. Soc.* **2014**, *442*, 2271–2277. [CrossRef]
45. Ishiyama, T. Hierarchical Formation of Dark Matter Halos and the Free Streaming Scale. *Astrophys. J.* **2014**, *788*, 27. [CrossRef]
46. Bartels, R.; Ando, S. Boosting the annihilation boost: Tidal effects on dark matter subhalos and consistent luminosity modeling. *Phys. Rev.* **2015**, *D92*, 123508. [CrossRef]
47. Stref, M.; Lavalle, J. Modeling dark matter subhalos in a constrained galaxy: Global mass and boosted annihilation profiles. *Phys. Rev.* **2017**, *D95*, 063003. [CrossRef]
48. Moliné, Á.; Sánchez-Conde, M.A.; Palomares-Ruiz, S.; Prada, F. Characterization of subhalo structural properties and implications for dark matter annihilation signals. *Mon. Not. R. Astron. Soc.* **2017**, *466*, 4974–4990. [CrossRef]
49. Zavala, J.; Afshordi, N. Universal clustering of dark matter in phase space. *Mon. Not. R. Astron. Soc.* **2016**, *457*, 986–992. [CrossRef]
50. Hiroshima, N.; Ando, S.; Ishiyama, T. Modeling evolution of dark matter substructure and annihilation boost. *Phys. Rev.* **2018**, *D97*, 123002. [CrossRef]
51. Diemand, J.; Kuhlen, M.; Madau, P.; Zemp, M.; Moore, B.; Potter, D.; Stadel, J. Clumps and streams in the local dark matter distribution. *Nature* **2008**, *454*, 735–738. [CrossRef] [PubMed]
52. Stadel, J.; Potter, D.; Moore, B.; Diemand, J.; Madau, P.; Zemp, M.; Kuhlen, M.; Quilis, V. Quantifying the heart of darkness with GHALO—A multi-billion particle simulation of our galactic halo. *Mon. Not. R. Astron. Soc.* **2009**, *398*, L21–L25. [CrossRef]
53. Berezinsky, V.S.; Dokuchaev, V.I.; Eroshenko, Y.N. Small-scale clumps of dark matter. *Phys. Usp.* **2014**, *57*, 1–36; reprinted in *Usp. Fiz. Nauk* **2014**, *184*, 3. [CrossRef]
54. Press, W.H.; Schechter, P. Formation of galaxies and clusters of galaxies by selfsimilar gravitational condensation. *Astrophys. J.* **1974**, *187*, 425–438. [CrossRef]

55. Bond, J.R.; Cole, S.; Efstathiou, G.; Kaiser, N. Excursion set mass functions for hierarchical Gaussian fluctuations. *Astrophys. J.* **1991**, *379*, 440. [CrossRef]
56. Anderson, B.; Zimmer, S.; Conrad, J.; Gustafsson, M.; Sánchez-Conde, M.; Caputo, R. Search for Gamma-Ray Lines towards Galaxy Clusters with the Fermi-LAT. *J. Cosmol. Astropart. Phys.* **2016**, *1602*, 026. [CrossRef]
57. Navarro, J.F.; Frenk, C.S.; White, S.D.M. The Structure of cold dark matter halos. *Astrophys. J.* **1996**, *462*, 563–575. [CrossRef]
58. Navarro, J.F.; Frenk, C.S.; White, S.D.M. A Universal density profile from hierarchical clustering. *Astrophys. J.* **1997**, *490*, 493–508. [CrossRef]
59. Charbonnier, A.; Combet, C.; Maurin, D. CLUMPY: A code for γ-ray signals from dark matter structures. *Comput. Phys. Commun.* **2012**, *183*, 656–668. [CrossRef]
60. Nezri, E.; White, R.; Combet, C.; Maurin, D.; Pointecouteau, E.; Hinton, J.A. gamma-rays from annihilating dark matter in galaxy clusters: Stacking vs single source analysis. *Mon. Not. R. Astron. Soc.* **2012**, *425*, 477. [CrossRef]
61. Ando, S.; Komatsu, E. Constraints on the annihilation cross section of dark matter particles from anisotropies in the diffuse gamma-ray background measured with Fermi-LAT. *Phys. Rev.* **2013**, *D87*, 123539. [CrossRef]
62. Hütten, M.; Combet, C.; Maier, G.; Maurin, D. Dark matter substructure modelling and sensitivity of the Cherenkov Telescope Array to Galactic dark halos. *J. Cosmol. Astropart. Phys.* **2016**, *1609*, 047. [CrossRef]
63. Bryan, G.L.; Norman, M.L. Statistical properties of x-ray clusters: Analytic and numerical comparisons. *Astrophys. J.* **1998**, *495*, 80. [CrossRef]
64. Correa, C.A.; Wyithe, J.S.B.; Schaye, J.; Duffy, A.R. The accretion history of dark matter haloes—III. A physical model for the concentration–mass relation. *Mon. Not. R. Astron. Soc.* **2015**, *452*, 1217–1232. [CrossRef]
65. Ishiyama, T.; Makino, J.; Portegies Zwart, S.; Groen, D.; Nitadori, K.; Rieder, S.; de Laat, C.; McMillan, S.; Hiraki, K.; Harfst, S. The Cosmogrid Simulation: Statistical Properties of small Dark Matter Halos. *Astrophys. J.* **2013**, *767*, 146. [CrossRef]
66. Anderhalden, D.; Diemand, J. Density Profiles of CDM Microhalos and their Implications for Annihilation Boost Factors. *J. Cosmol. Astropart. Phys.* **2013**, *1304*, 009. Erratum in *J. Cosmol. Astropart. Phys.* **2013**, *1308*, E02. [CrossRef]
67. Ishiyama, T.; Sudo, K.; Yokoi, S.; Hasegawa, K.; Tominaga, N.; Susa, H. Where are the Low-mass Population III Stars? *Astrophys. J.* **2016**, *826*, 9. [CrossRef]
68. Klypin, A.A.; Kravtsov, A.V.; Valenzuela, O.; Prada, F. Where are the missing Galactic satellites? *Astrophys. J.* **1999**, *522*, 82–92. [CrossRef]
69. Moore, B.; Ghigna, S.; Governato, F.; Lake, G.; Quinn, T.R.; Stadel, J.; Tozzi, P. Dark matter substructure within galactic halos. *Astrophys. J.* **1999**, *524*, L19–L22. [CrossRef]
70. Diemand, J.; Moore, B.; Stadel, J. Velocity and spatial biases in CDM subhalo distributions. *Mon. Not. R. Astron. Soc.* **2004**, *352*, 535. [CrossRef]
71. Reed, D.; Governato, F.; Quinn, T.R.; Gardner, J.; Stadel, J.; Lake, G. Dark matter subhaloes in numerical simulations. *Mon. Not. R. Astron. Soc.* **2005**, *359*, 1537–1548. [CrossRef]
72. Kase, H.; Makino, J.; Funato, Y. Missing dwarf problem in galaxy clusters. *Publ. Astron. Soc. Jpn.* **2007**, *59*, 1071. [CrossRef]
73. Kamionkowski, M.; Liddle, A.R. The Dearth of halo dwarf galaxies: Is there power on short scales? *Phys. Rev. Lett.* **2000**, *84*, 4525–4528. [CrossRef] [PubMed]
74. Spergel, D.N.; Steinhardt, P.J. Observational evidence for selfinteracting cold dark matter. *Phys. Rev. Lett.* **2000**, *84*, 3760–3763. [CrossRef] [PubMed]
75. Susa, H.; Umemura, M. Effects of early cosmic reionization on the substructure problem in galactic halo. *Astrophys. J.* **2004**, *610*, L5–L8. [CrossRef]
76. Benson, A.J.; Frenk, C.S.; Lacey, C.G.; Baugh, C.M.; Cole, S. The effects of photoionization on galaxy formation. 2. Satellites in the local group. *Mon. Not. R. Astron. Soc.* **2002**, *333*, 177. [CrossRef]
77. Stoehr, F.; White, S.D.M.; Tormen, G.; Springel, V. The Milky Way's satellite population in a lambdaCDM universe. *Mon. Not. R. Astron. Soc.* **2002**, *335*, L84–L88. [CrossRef]
78. Kravtsov, A.V.; Gnedin, O.Y.; Klypin, A.A. The Tumultuous lives of Galactic dwarfs and the missing satellites problem. *Astrophys. J.* **2004**, *609*, 482–497. [CrossRef]
79. Okamoto, T.; Frenk, C.S.; Jenkins, A.; Theuns, T. The properties of satellite galaxies in simulations of galaxy formation. *Mon. Not. R. Astron. Soc.* **2010**, *406*, 208–222. [CrossRef]

80. Ishiyama, T.; Fukushige, T.; Makino, J. Variation of the subhalo abundance in dark matter halos. *Astrophys. J.* **2009**, *696*, 2115–2125. [CrossRef]
81. Mao, Y.Y.; Williamson, M.; Wechsler, R.H. The Dependence of Subhalo Abundance on Halo Concentration. *Astrophys. J.* **2015**, *810*, 21. [CrossRef]
82. Kravtsov, A.V.; Berlind, A.A.; Wechsler, R.H.; Klypin, A.A.; Gottloeber, S.; Allgood, B.; Primack, J.R. The Dark side of the halo occupation distribution. *Astrophys. J.* **2004**, *609*, 35–49. [CrossRef]
83. Gao, L.; White, S.D.M.; Jenkins, A.; Stoehr, F.; Springel, V. The Subhalo populations of lambda-CDM dark halos. *Mon. Not. R. Astron. Soc.* **2004**, *355*, 819. [CrossRef]
84. Zentner, A.R.; Berlind, A.A.; Bullock, J.S.; Kravtsov, A.V.; Wechsler, R.H. The Physics of galaxy clustering. 1. A Model for subhalo populations. *Astrophys. J.* **2005**, *624*, 505–525. [CrossRef]
85. Van den Bosch, F.C.; Tormen, G.; Giocoli, C. The Mass function and average mass loss rate of dark matter subhaloes. *Mon. Not. R. Astron. Soc.* **2005**, *359*, 1029–1040. [CrossRef]
86. Giocoli, C.; Tormen, G.; Sheth, R.K.; van den Bosch, F.C. The Substructure Hierarchy in Dark Matter Haloes. *Mon. Not. R. Astron. Soc.* **2010**, *404*, 502–517. [CrossRef]
87. Gao, L.; Frenk, C.S.; Boylan-Kolchin, M.; Jenkins, A.; Springel, V.; White, S.D.M. The statistics of the subhalo abundance of dark matter haloes. *Mon. Not. R. Astron. Soc.* **2011**, *410*, 2309. [CrossRef]
88. Giocoli, C.; Pieri, L.; Tormen, G. Analytical Approach to Subhaloes Population in Dark Matter Haloes. *Mon. Not. R. Astron. Soc.* **2008**, *387*, 689–697. [CrossRef]
89. Yang, X.; Mo, H.J.; Zhang, Y.; Bosch, F.C.V.D. An analytical model for the accretion of dark matter subhalos. *Astrophys. J.* **2011**, *741*, 13. [CrossRef]
90. Jiang, F.; van den Bosch, F.C. Statistics of dark matter substructure—I. Model and universal fitting functions. *Mon. Not. R. Astron. Soc.* **2016**, *458*, 2848–2869. [CrossRef]
91. Gunn, J.E.; Gott, J.R., III. On the Infall of Matter into Clusters of Galaxies and Some Effects on Their Evolution. *Astrophys. J.* **1972**, *176*, 1–19. [CrossRef]
92. Quinn, P.J.; Salmon, J.K.; Zurek, W.H. Primordial density fluctuations and the structure of galactic haloes. *Nature* **1986**, *322*, 329–335. [CrossRef]
93. Dubinski, J.; Carlberg, R.G. The Structure of cold dark matter halos. *Astrophys. J.* **1991**, *378*, 496. [CrossRef]
94. Hernquist, L. An Analytical Model for Spherical Galaxies and Bulges. *Astrophys. J.* **1990**, *356*, 359. [CrossRef]
95. Fukushige, T.; Makino, J. On the Origin of Cusps in Dark Matter Halos. *Astrophys. J.* **1997**, *477*, L9. [CrossRef]
96. Moore, B.; Quinn, T.R.; Governato, F.; Stadel, J.; Lake, G. Cold collapse and the core catastrophe. *Mon. Not. R. Astron. Soc.* **1999**, *310*, 1147–1152. [CrossRef]
97. Ghigna, S.; Moore, B.; Governato, F.; Lake, G.; Quinn, T.R.; Stadel, J. Density profiles and substructure of dark matter halos. Converging results at ultra-high numerical resolution. *Astrophys. J.* **2000**, *544*, 616. [CrossRef]
98. Jing, Y.P. The Density Profile of Equilibrium and Nonequilibrium Dark Matter Halos. *Astrophys. J.* **2000**, *535*, 30–36. [CrossRef]
99. Fukushige, T.; Makino, J. Structure of dark matter halos from hierarchical clustering. *Astrophys. J.* **2001**, *557*, 533. [CrossRef]
100. Klypin, A.; Kravtsov, A.V.; Bullock, J.; Primack, J. Resolving the structure of cold dark matter halos. *Astrophys. J.* **2001**, *554*, 903–915. [CrossRef]
101. Power, C.; Navarro, J.F.; Jenkins, A.; Frenk, C.S.; White, S.D.M.; Springel, V.; Stadel, J.; Quinn, T.R. The Inner structure of Lambda CDM halos. 1. A Numerical convergence study. *Mon. Not. R. Astron. Soc.* **2003**, *338*, 14–34. [CrossRef]
102. Fukushige, T.; Kawai, A.; Makino, J. Structure of dark matter halos from hierarchical clustering. 3. Shallowing of the Inner cusp. *Astrophys. J.* **2004**, *606*, 625–634. [CrossRef]
103. Hayashi, E.; Navarro, J.F.; Power, C.; Jenkins, A.R.; Frenk, C.S.; White, S.D.M.; Springel, V.; Stadel, J.; Quinn, T.R. The Inner structure of lambda-CDM halos. 2. Halo mass profiles and LSB rotation curves. *Mon. Not. R. Astron. Soc.* **2004**, *355*, 794–812. [CrossRef]
104. Kazantzidis, S.; Zentner, A.R.; Kravtsov, A.V. The robustness of dark matter density profiles in dissipationless mergers. *Astrophys. J.* **2006**, *641*, 647–664. [CrossRef]
105. Jing, Y.P.; Suto, Y. Density profiles of dark matter halo are not universal. *Astrophys. J.* **2000**, *529*, L69–L72. [CrossRef]

106. Diemand, J.; Moore, B.; Stadel, J. Convergence and scatter of cluster density profiles. *Mon. Not. R. Astron. Soc.* **2004**, *353*, 624. [CrossRef]
107. Einasto, J. On the Construction of a Composite Model for the Galaxy and on the Determination of the System of Galactic Parameters. *Tr. Astrofiz. Instituta Alma-Ata* **1965**, *5*, 87–100.
108. Bullock, J.S.; Kolatt, T.S.; Sigad, Y.; Somerville, R.S.; Kravtsov, A.V.; Klypin, A.A.; Primack, J.R.; Dekel, A. Profiles of dark haloes. Evolution, scatter, and environment. *Mon. Not. R. Astron. Soc.* **2001**, *321*, 559–575. [CrossRef]
109. Zhao, D.H.; Jing, Y.P.; Mo, H.J.; Borner, G. Mass and redshift dependence of dark halo structure. *Astrophys. J.* **2003**, *597*, L9–L12. [CrossRef]
110. Maccio, A.V.; Dutton, A.A.; van den Bosch, F.C.; Moore, B.; Potter, D.; Stadel, J. Concentration, Spin and Shape of Dark Matter Haloes: Scatter and the Dependence on Mass and Environment. *Mon. Not. R. Astron. Soc.* **2007**, *378*, 55–71. [CrossRef]
111. Neto, A.F.; Gao, L.; Bett, P.; Cole, S.; Navarro, J.F.; Frenk, C.S.; White, S.D.M.; Springel, V.; Jenkins, A. The statistics of lambda CDM Halo Concentrations. *Mon. Not. R. Astron. Soc.* **2007**, *381*, 1450–1462. [CrossRef]
112. Maccio, A.V.; Dutton, A.A.; Bosch, F.C.V.D. Concentration, Spin and Shape of Dark Matter Haloes as a Function of the Cosmological Model: WMAP1, WMAP3 and WMAP5 results. *Mon. Not. R. Astron. Soc.* **2008**, *391*, 1940–1954. [CrossRef]
113. Zhao, D.H.; Jing, Y.P.; Mo, H.J.; Boerner, G. Accurate universal models for the mass accretion histories and concentrations of dark matter halos. *Astrophys. J.* **2009**, *707*, 354–369. [CrossRef]
114. Klypin, A.; Trujillo-Gomez, S.; Primack, J. Halos and galaxies in the standard cosmological model: Results from the Bolshoi simulation. *Astrophys. J.* **2011**, *740*, 102. [CrossRef]
115. Prada, F.; Klypin, A.A.; Cuesta, A.J.; Betancort-Rijo, J.E.; Primack, J. Halo concentrations in the standard LCDM cosmology. *Mon. Not. R. Astron. Soc.* **2012**, *423*, 3018–3030. [CrossRef]
116. Klypin, A.; Yepes, G.; Gottlober, S.; Prada, F.; Hess, S. MultiDark simulations: The story of dark matter halo concentrations and density profiles. *Mon. Not. R. Astron. Soc.* **2016**, *457*, 4340–4359. [CrossRef]
117. Ludlow, A.D.; Bose, S.; Angulo, R.E.; Wang, L.; Hellwing, W.A.; Navarro, J.F.; Cole, S.; Frenk, C.S. The mass–concentration–redshift relation of cold and warm dark matter haloes. *Mon. Not. R. Astron. Soc.* **2016**, *460*, 1214–1232. [CrossRef]
118. Wechsler, R.H.; Bullock, J.S.; Primack, J.R.; Kravtsov, A.V.; Dekel, A. Concentrations of dark halos from their assembly histories. *Astrophys. J.* **2002**, *568*, 52–70. [CrossRef]
119. Hellwing, W.A.; Frenk, C.S.; Cautun, M.; Bose, S.; Helly, J.; Jenkins, A.; Sawala, T.; Cytowski, M. The Copernicus Complexio: A high-resolution view of the small-scale Universe. *Mon. Not. R. Astron. Soc.* **2016**, *457*, 3492–3509. [CrossRef]
120. Pilipenko, S.V.; Sánchez-Conde, M.A.; Prada, F.; Yepes, G. Pushing down the low-mass halo concentration frontier with the Lomonosov cosmological simulations. *Mon. Not. R. Astron. Soc.* **2017**, *472*, 4918–4927. [CrossRef]
121. Diemand, J.; Kuhlen, M.; Madau, P. Formation and evolution of galaxy dark matter halos and their substructure. *Astrophys. J.* **2007**, *667*, 859–877. [CrossRef]
122. Reed, D.S.; Koushiappas, S.M.; Gao, L. Non-universality of halo profiles and implications for dark matter experiments. *Mon. Not. R. Astron. Soc.* **2011**, *415*, 3177–3188. [CrossRef]
123. Springel, V.; White, S.D.M.; Jenkin, A.; Frenk, C.S.; Yoshida, N.; Navarro, J.; Thacker, R.; Croton, D.; Helly, J.; Peacock, J.A; et al. Simulating the joint evolution of quasars, galaxies and their large-scale distribution. *Nature* **2005**, *435*, 629–636. [CrossRef]
124. Ullio, P.; Bergstrom, L.; Edsjo, J.; Lacey, C.G. Cosmological dark matter annihilations into gamma-rays—A closer look. *Phys. Rev.* **2002**, *D66*, 123502. [CrossRef]
125. Pieri, L.; Lavalle, J.; Bertone, G.; Branchini, E. Implications of High-Resolution Simulations on Indirect Dark Matter Searches. *Phys. Rev.* **2011**, *D83*, 023518. [CrossRef]
126. Ishiyama, T.; Enoki, M.; Kobayashi, M.A.R.; Makiya, R.; Nagashima, M.; Oogi, T. The ν^2GC simulations: Quantifying the dark side of the universe in the Planck cosmology. *Publ. Astron. Soc. Jpn.* **2015**, *67*, 61. [CrossRef]

127. Makiya, R.; Enoki, M.; Ishiyama, T.; Kobayashi, M.A.R.; Nagashima, M.; Okamoto, T.; Okoshi, K.; Oogi, T.; Shirakata, H. The New Numerical Galaxy Catalog (ν^2GC): An updated semi-analytic model of galaxy and active galactic nucleus formation with large cosmological N-body simulations. *Publ. Astron. Soc. Jpn.* **2016**, *68*, 25. [CrossRef]
128. Van den Bosch, F.C.; Ogiya, G. Dark Matter Substructure in Numerical Simulations: A Tale of Discreteness Noise, Runaway Instabilities, and Artificial Disruption. *Mon. Not. R. Astron. Soc.* **2018**, *475*, 4066–4087. [CrossRef]
129. Peñarrubia, J.; Benson, A.J.; Walker, M.G.; Gilmore, G.; McConnachie, A.; Mayer, L. The impact of dark matter cusps and cores on the satellite galaxy population around spiral galaxies. *Mon. Not. R. Astron. Soc.* **2010**, *406*, 1290. [CrossRef]
130. Van den Bosch, F.C.; Ogiya, G.; Hahn, O.; Burkert, A. Disruption of Dark Matter Substructure: Fact or Fiction? *Mon. Not. R. Astron. Soc.* **2018**, *474*, 3043–3066. [CrossRef]
131. Diemand, J.; Moore, B.; Stadel, J. Earth-mass dark-matter haloes as the first structures in the early Universe. *Nature* **2005**, *433*, 389–391. [CrossRef]
132. Ishiyama, T.; Makino, J.; Ebisuzaki, T. Gamma-ray Signal from Earth-mass Dark Matter Microhalos. *Astrophys. J.* **2010**, *723*, L195. [CrossRef]
133. Angulo, R.E.; Hahn, O.; Ludlow, A.; Bonoli, S. Earth-mass haloes and the emergence of NFW density profiles. *Mon. Not. R. Astron. Soc.* **2017**, *471*, 4687–4701. [CrossRef]
134. Ogiya, G.; Hahn, O. What sets the central structure of dark matter haloes? *Mon. Not. R. Astron. Soc.* **2018**, *473*, 4339–4359. [CrossRef]
135. Ogiya, G.; Nagai, D.; Ishiyama, T. Dynamical evolution of primordial dark matter haloes through mergers. *Mon. Not. R. Astron. Soc.* **2016**, *461*, 3385–3396. [CrossRef]
136. Gosenca, M.; Adamek, J.; Byrnes, C.T.; Hotchkiss, S. 3D simulations with boosted primordial power spectra and ultracompact minihalos. *Phys. Rev.* **2017**, *D96*, 123519. [CrossRef]
137. Delos, M.S.; Erickcek, A.L.; Bailey, A.P.; Alvarez, M.A. Are ultracompact minihalos really ultracompact? *Phys. Rev.* **2018**, *D97*, 041303. [CrossRef]
138. Delos, M.S.; Erickcek, A.L.; Bailey, A.P.; Alvarez, M.A. Density profiles of ultracompact minihalos: Implications for constraining the primordial power spectrum. *Phys. Rev.* **2018**, *D98*, 063527. [CrossRef]
139. Polisensky, E.; Ricotti, M. Fingerprints of the initial conditions on the density profiles of cold and warm dark matter haloes. *Mon. Not. R. Astron. Soc.* **2015**, *450*, 2172–2184. [CrossRef]
140. Gott, J.R., III. On the Formation of Elliptical Galaxies. *Astrophys. J.* **1975**, *201*, 296–310. [CrossRef]
141. Fillmore, J.A.; Goldreich, P. Self-similar gravitational collapse in an expanding universe. *Astrophys. J.* **1984**, *281*, 1–8. [CrossRef]
142. Bertschinger, E. Self-similar secondary infall and accretion in an Einstein-de Sitter universe. *Astrophys. J. Suppl.* **1985**, *58*, 39. [CrossRef]
143. Gurevich, A.V.; Zybin, K.P. Large-scale structure of the Universe. Analytic theory. *Phys. Uspekhi* **1995**, *38*, 687–722. [CrossRef]
144. Baghram, S.; Afshordi, N.; Zurek, K.M. Prospects for Detecting Dark Matter Halo Substructure with Pulsar Timing. *Phys. Rev.* **2011**, *D84*, 043511. [CrossRef]
145. Kashiyama, K.; Oguri, M. Detectability of Small-Scale Dark Matter Clumps with Pulsar Timing Arrays. *arXiv* **2018**, arXiv:1801.07847.
146. Dror, J.A.; Ramani, H.; Trickle, T.; Zurek, K.M. Pulsar Timing Probes of Primordial Black Holes and Subhalos. *arXiv* **2019**, arXiv:1901.04490.
147. Correa, C.A.; Wyithe, J.S.B.; Schaye, J.; Duffy, A.R. The accretion history of dark matter haloes—I. The physical origin of the universal function. *Mon. Not. R. Astron. Soc.* **2015**, *450*, 1514–1520. [CrossRef]
148. Hayashi, E.; Navarro, J.F.; Taylor, J.E.; Stadel, J.; Quinn, T.R. The Structural evolution of substructure. *Astrophys. J.* **2003**, *584*, 541–558. [CrossRef]
149. Behroozi, P.S.; Wechsler, R.H.; Wu, H.Y. The Rockstar Phase-Space Temporal Halo Finder and the Velocity Offsets of Cluster Cores. *Astrophys. J.* **2013**, *762*, 109. [CrossRef]
150. Diemand, J.; Kuhlen, M.; Madau, P. Early supersymmetric cold dark matter substructure. *Astrophys. J.* **2006**, *649*, 1–13. [CrossRef]
151. Seljak, U. Analytic model for galaxy and dark matter clustering. *Mon. Not. R. Astron. Soc.* **2000**, *318*, 203. [CrossRef]

152. Zhu, Q.; Marinacci, F.; Maji, M.; Li, Y.; Springel, V.; Hernquist, L. Baryonic impact on the dark matter distribution in Milky Way-sized galaxies and their satellites. *Mon. Not. R. Astron. Soc.* **2016**, *458*, 1559–1580. [CrossRef]
153. Errani, R.; Peñarrubia, J.; Laporte, C.F.P.; Gómez, F.A. The effect of a disc on the population of cuspy and cored dark matter substructures in Milky Way-like galaxies. *Mon. Not. R. Astron. Soc.* **2017**, *465*, L59–L63. [CrossRef]
154. Chua, K.T.E.; Pillepich, A.; Rodriguez-Gomez, V.; Vogelsberger, M.; Bird, S.; Hernquist, L. Subhalo demographics in the Illustris simulation: Effects of baryons and halo-to-halo variation. *Mon. Not. R. Astron. Soc.* **2017**, *472*, 4343–4360. [CrossRef]
155. Kelley, T.; Bullock, J.S.; Garrison-Kimmel, S.; Boylan-Kolchin, M.; Pawlowski, M.S.; Graus, A.S. Phat ELVIS: The inevitable effect of the Milky Way's disk on its dark matter subhaloes. *arXiv* **2018**, arXiv:1811.12413.
156. Okoli, C.; Afshordi, N. Concentration, Ellipsoidal Collapse, and the Densest Dark Matter haloes. *Mon. Not. R. Astron. Soc.* **2016**, *456*, 3068–3078. [CrossRef]
157. Gao, L.; Navarro, J.F.; Frenk, C.S.; Jenkins, A.; Springel, V.; White, S.D.M. The Phoenix Project: The Dark Side of Rich Galaxy Clusters. *Mon. Not. R. Astron. Soc.* **2012**, *425*, 2169. [CrossRef]

 © 2019 by the authors. Licensee MDPI, Basel, Switzerland. This article is an open access article distributed under the terms and conditions of the Creative Commons Attribution (CC BY) license (http://creativecommons.org/licenses/by/4.0/).

Review

Gamma-Ray Dark Matter Searches in Milky Way Satellites—A Comparative Review of Data Analysis Methods and Current Results

Javier Rico

Institut de Física d'Altes Energies (IFAE), The Barcelona Institute of Science and Technology (BIST), 08193 Barcelona, Spain; jrico@ifae.es

Received: 29 July 2019; Accepted: 24 February 2020; Published: 17 March 2020

Abstract: If dark matter is composed of weakly interacting particles with mass in the GeV-TeV range, their annihilation or decay may produce gamma rays that could be detected by gamma-ray telescopes. Observations of dwarf spheroidal satellite galaxies of the Milky Way (dSphs) benefit from the relatively accurate predictions of dSph dark matter content to produce robust constraints to the dark matter properties. The sensitivity of these observations for the search for dark matter signals can be optimized thanks to the use of advanced statistical techniques able to exploit the spectral and morphological peculiarities of the expected signal. In this paper, I review the status of the dark matter searches from observations of dSphs with the current generation of gamma-ray telescopes: Fermi-LAT, H.E.S.S, MAGIC, VERITAS and HAWC. I will describe in detail the general statistical analysis framework used by these instruments, putting in context the most recent experimental results and pointing out the most relevant differences among the different particular implementations. This will facilitate the comparison of the current and future results, as well as their eventual integration in a multi-instrument and multi-target dark matter search.

Keywords: dark matter; indirect searches; gamma rays; dwarf spheroidal satellite galaxies; statistical data analysis

1. Introduction

The existence of a dominant non-baryonic, neutral, cold matter component in the Universe, called dark matter, has been postulated in order to explain the kinematics of galaxies in galaxy clusters [1] and stars in spiral galaxies [2], as well as the power spectrum of temperature anisotropies of the cosmic microwave background [3]. In one of the most plausible and thoroughly studied theoretical scenarios, dark matter is composed of weakly interacting particles with mass in the range between tens of GeV and hundreds of TeV, generically referred to as WIMPs [4]. The Standard Model (SM) particles that could result from WIMP annihilation or decay would hadronize, radiate and/or decay, producing detectable stable particles such as photons, neutrinos, proton–antiproton pairs or electron–positron pairs [5]. Looking for unambiguous spectral and/or morphological signatures of dark matter annihilation or decay in the extra-terrestrial fluxes of those particles is usually referred to as indirect dark matter searches.

Gamma rays are promising messengers to search for WIMPs. Since they are electrically neutral, they are not deflected by magnetic fields and point back to their production site, and therefore could be used to determine the underlying dark matter spatial distribution. At non-cosmological scales, gamma rays are also essentially unaffected by energy losses, which would preserve the features expected for dark matter annihilation and/or decay spectra, which depend on the values of the dark matter mass and the branching ratios to the different annihilation/decay channels, which could thus be studied.

Finally, the gamma-ray signal intensity would depend on the annihilation cross-section or the decay lifetime, which could therefore be determined if we measured a signal from an astronomical site for which we have a good estimate of its dark matter content based on independent measurements and/or simulations.

N-body simulations predict the formation of cold dark matter haloes in a hierarchical clustering fashion [6]. dSphs form in dark matter galactic subhalos that contain enough baryonic matter to have activated stellar formation (pure dark matter halos should also exist, but they remain as of yet unidentified). They are irregular satellite galaxies with mass $\sim 10^7 M_\odot$ and the largest known ratios of dark to luminous matter. The extension of the expected gamma-ray emission from the Milky Way dSphs is typically between ~ 0.1–$0.5°$ [7], which is of the order of the angular resolution of most of the current-generation gamma-ray telescopes.

Gamma-ray telescopes of the current generation have performed extensive observational campaigns of dSphs in search for dark matter signals. Along the years, gamma-ray telescopes have progressively adopted state-of-the-art statistical analysis techniques for their dark matter searches, optimized to exploit the particular spectral and morphological features expected for the signal. All the instruments have converged into a general statistical analysis framework, albeit with some significant differences among the different implementations. Some of these differences are unavoidable, since they are needed to adapt the analysis to the different experimental scenarios, whereas others rather consist in choices of conventions, approximations, or simplifications. These latter ones include the methods for computing the spectral and morphological models for the expected gamma-ray signal and associated background, their use in the statistical analysis, and the treatment of the related statistical and systematic uncertainties. Understanding both the similarities and the differences among the various analysis implementations is fundamental in view of meaningful comparison and combination of the obtained results.

In this paper, I review the present status of indirect dark matter searches with observations of dSphs with gamma-ray telescopes. In Section 2 I summarize the formalism for the computation of the gamma-ray fluxes expected to be produced by dark matter processes in dSphs. In Section 3, I briefly introduce the current generation of gamma-ray telescopes, their working principles and main features. Section 4 is devoted to the detailed description of the common statistical data analysis framework used by all these instruments in their search for dark matter in dSphs. Finally, in Section 5, I perform a critical comparison of the particular analysis implementations, review and contextualize the latest experimental results published by the different instruments, and show the prospects for their near-future combination.

2. Gamma-Ray Signals From Dark Matter Processes in dSphs

dSphs are among the cleanest astronomical targets for indirect dark matter searches. They are thought to be highly dominated by dark matter (mass-to-light ratios of the order of 10^3 [8]), and they harbor no known astrophysical gamma-ray sources that could produce a relevant background. Furthermore, dSphs contain in general no significant amount of dark gas, which allows their dark matter distribution to be inferred with relatively good precision from the stellar motions, enabling in turn robust predictions of the intensity of the associated gamma-ray signals, generally within an accuracy of one order of magnitude [7]. Finally, given how most of the known dSphs sit on relatively clean interstellar environments (i.e., out of the Galactic plane, where the particle densities, cosmic ray fluxes and radiation fields are small), the expected gamma-ray signal would come from well-understood prompt processes. Secondary processes such as inverse Compton scattering of primary or secondary electrons, or gamma-ray cascading processes initiated by their interaction with radiation fields (hence depending on local details of those radiation fields), can be in general ignored when computing the gamma-ray flux expected from dark matter at dSphs. Therefore, since flux predictions rely on relatively few assumptions compared to other typical observational targets like e.g., the Galactic center or clusters of galaxies, the bounds on the WIMP properties that can be inferred from the presence or absence of a gamma-ray signal are also relatively robust.

If WIMPs (hereafter denoted by χ) concentrate with number density n_χ in a dSph, annihilating and/or decaying with a rate Γ_χ and an average isotropic gamma-ray spectrum $\frac{dN_\gamma}{dE}$, then the differential flux of gamma rays of energy E observable from Earth coming from direction $\hat{p}$, per unit energy and solid angle Ω, is given by the following expression:

$$\frac{d^2\Phi}{dEd\Omega}(E,\hat{p}) = \frac{1}{4\pi}\frac{dN_\gamma}{dE}(E)\int_{\text{los}(\hat{p})} dl\, n_\chi(\hat{p},l)\,\Gamma_\chi \quad , \qquad (1)$$

with l the distance from Earth and the corresponding integral running over the line of sight in the direction $\hat{p}$.

As explicitly noted in Equation (1), $\frac{dN_\gamma}{dE}$ contains all the spectral dependence of the gamma-ray flux, and therefore determines the probability density function (PDF) for the energy of the emitted gamma rays. On the other hand, all the morphological dependence is contained in the line-of-sight integral, which hence determines the PDF for the gamma-ray arrival direction. Given that we can make relatively reliable predictions about these two PDFs, they will constitute key ingredients in the maximum-likelihood data analysis, as we will see below in detail.

The expected primary products of the WIMP annihilation and decay processes are pairs of leptons, quarks or gauge bosons, which would produce secondary gamma-rays (among other stable products) through final-state radiation or hadronization+decay chains. It is straightforward to compute the contribution to $\frac{dN_\gamma}{dE}$ from the different annihilation/decay channels, for a given WIMP mass, using standard Monte Carlo simulation packages such as PYTHIA [9]. The spectral energy distribution of the gamma-ray continuum resulting from these processes peaks between one and two orders of magnitude below the WIMP mass, depending on the channel, as shown in Figure 1. The plots show that Fermi-LAT is the most sensitive instrument for searching for WIMPs up to a dark matter mass (m_χ) of few TeV in the case of $b\bar{b}$ channel and of few 100 GeV for the $\tau^+\tau^-$ channel. Cherenkov telescopes dominate the search between those masses and ~ 100 TeV for $b\bar{b}$ and few 10 TeV for $\tau^+\tau^-$, and HAWC for even higher WIMP masses. Primary gamma rays like, e.g., those from the $\chi[\chi] \to \gamma\gamma$ or $\chi[\chi] \to \gamma Z$ processes would be [quasi-]monochromatic. These would constitute the cleanest possible dark matter signal, given how there is no known astrophysical process able to produce such gamma-ray spectral lines, and that backgrounds affecting the measurement could be drastically reduced using spectral criteria. If detected, a gamma-ray line would by itself be considered a clear evidence for the presence of dark matter. However, due to parity conservation, primary gamma rays can only be produced via loop processes, which significantly reduces their associated rate Γ_χ.

It is useful to particularize the line of sight integral in Equation (1) for the annihilation and decay cases:

- For annihilation, $\Gamma_\chi = \frac{1}{k}n_\chi \langle\sigma v\rangle$, with $\langle\sigma v\rangle$ the average of the product of the WIMP velocity and annihilation cross section. The value of k depends on whether WIMPs are Majorana ($k=2$, to take into account that an annihilation involves two identical particles) or Dirac particles ($k=4$, reflecting the fact that particles can only annihilate with their—equally abundant—antiparticles). Including this into Equation (1), and writing the WIMP number density n_χ in terms of its mass and density (ρ), we obtain:

$$\frac{d^2\Phi_{\text{ann}}}{d\Omega\, dE}(E,\hat{p}) = \frac{1}{4\pi}\frac{\langle\sigma v\rangle}{k\, m_\chi^2}\frac{dJ_{\text{ann}}}{d\Omega}(\hat{p})\frac{dN_\gamma}{dE}(E) \quad , \qquad (2)$$

where we have defined the annihilation differential J-factor as:

$$\frac{dJ_{\text{ann}}}{d\Omega}(\hat{p}) = \int_{\text{los}(\hat{p})} dl\, \rho^2(\hat{p},l) \quad . \qquad (3)$$

- For decay, the rate is given simply by the inverse of the dark matter decay lifetime, i.e., $\Gamma_\chi = \tau_\chi^{-1}$, since each WIMP particle decays independently of each other. Including this into Equation (1), we get:

$$\frac{d^2\Phi_{dec}}{d\Omega\, dE}(E, \hat{p}) = \frac{1}{4\pi} \frac{1}{\tau_\chi m_\chi} \frac{dJ_{dec}}{d\Omega}(\hat{p}) \frac{dN_\gamma}{dE}(E) \quad, \tag{4}$$

where we have defined the decay differential J-factor as:

$$\frac{dJ_{dec}}{d\Omega}(\hat{p}) = \int_{los(\hat{p})} dl\, \rho(\hat{p}, l) \quad. \tag{5}$$

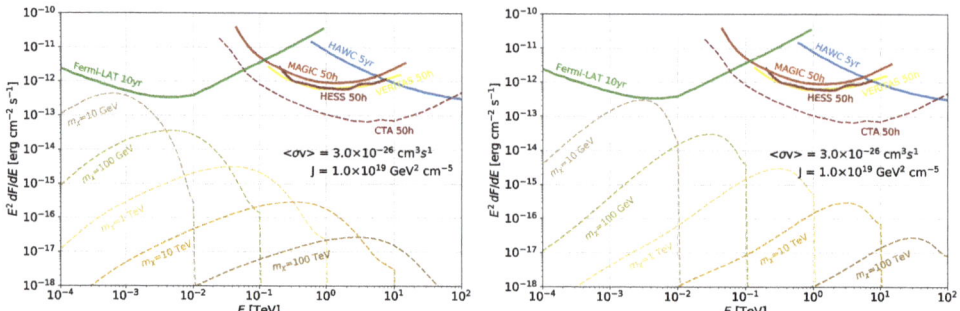

Figure 1. Expected gamma-ray spectral energy distribution for WIMPs of masses $m_\chi = 0.01, 0.1, 1, 10$ and 100 TeV annihilating with $\langle \sigma v \rangle = 3 \times 10^{-26}$ cm^3 s^{-1} into $b\bar{b}$ (left) and $\tau^+\tau^-$ (right) pairs in a dSph with associated J-factor $J_{ann} = 5 \times 10^{21}$ GeV^2 cm^{-5}; also shown are the sensitivity curves for the instruments considered in this paper. Fermi-LAT sensitivity curve [10] corresponds to observations of a point-like source at Galactic coordinates $(l, b) = (120°, 45°)$ for 10 years, analyzed using the latest (Pass8) data reconstruction tools; HESS [11], MAGIC [12] and VERITAS [13] curves correspond to 50 h of observations of a point-like source at low (Zd $\lesssim 30°$) zenith distance; HAWC curve [14] is for five years of observations of a point-like source at a declination of +22°N. The flux sensitivity for 50 h observations with the future Cherenkov Telescope Array [15] is shown for comparison.

The J-factor in a region of the sky $\Delta\Omega$ is given by:

$$J(\Delta\Omega) = \int_{\Delta\Omega} d\Omega\, \frac{dJ}{d\Omega} \quad, \tag{6}$$

both for J_{ann} and J_{dec}. It is convenient to define the total J-factor for a given dSph as:

$$\bar{J} \equiv J(\Delta\Omega_{tot}) \quad, \tag{7}$$

with $\Delta\Omega_{tot}$ a region of the sky containing the whole dSph dark matter halo. The differential J-factor can be written as:

$$\frac{dJ}{d\Omega}(\hat{p}) = \bar{J} \cdot \frac{d\mathcal{J}}{d\Omega}(\hat{p}) \quad, \tag{8}$$

where $\frac{d\mathcal{J}}{d\Omega}$ can be interpreted as the PDF for the arrival direction of gamma rays produced by dark matter processes in the dSph halo, since $\int_{\Delta\Omega_{tot}} d\Omega\, \frac{d\mathcal{J}}{d\Omega} = 1$. Using this notation, the differential gamma-ray flux per energy and solid angle can be written as:

$$\frac{d^2\Phi}{dEd\Omega}(E, \hat{p}) = a\bar{J}\, \frac{d\mathcal{J}}{d\Omega}(\hat{p}) \frac{dN_\gamma}{dE}(E) \quad, \tag{9}$$

(with a being either $a_{\text{ann}} \equiv \frac{1}{4\pi} \frac{\langle \sigma v \rangle}{k m_\chi^2}$ for annihilation or $a_{\text{dec}} \equiv \frac{1}{4\pi} \frac{1}{\tau_\chi m_\chi}$ for decay). The differential flux per unit energy is given by:

$$\frac{d\Phi}{dE}(E) \equiv \int_{\Delta\Omega_{\text{tot}}} d\Omega \frac{d^2\Phi}{dEd\Omega}(E,\hat{p}) = a \bar{J} \frac{dN_\gamma}{dE}(E) \quad . \tag{10}$$

The distribution of dark matter within the halo, $\rho(\hat{p}, l)$, is usually estimated by solving the spherical Jeans equation for the stellar kinematic data [16]. Using this technique, several authors have produced catalogues of J-factors for the different known dSphs. In general, the classical dSphs, with relatively large stellar populations ($O(100-1000)$), have relatively low associated J-factors (typically between 3×10^{17} and 7×10^{18} GeV^{2}cm^{-5} within an integrating angle of $0.5°$), with associated uncertainties also relatively low (typically below 50%), suitable for setting robust limits to dark matter properties. On the other hand, members of the ultra-faint population (those discovered by the Sloan Digital Sky Survey or later, with $O(10-100)$ members stellar populations) can have larger estimated J-factors (some above 10^{19} GeV^{2}cm^{-5}) but also larger uncertainties (some above a factor 10), therefore providing better prospects for discovery but less robust constraining power. A detailed review about the expected dark matter content and distribution of the known dSphs can be found elsewhere in this volume.

3. Gamma-Ray Telescopes

For WIMP indirect searches with gamma-rays, the relevant energy range spans from 100 MeV to 100 TeV (see Figure 1). Photons of these energies interact in the upper layers of the atmosphere, making impossible their direct detection from the ground. Several different experimental techniques have been developed to detect gamma rays, each optimized for a different energy range and hence for different dark matter masses.

At energies below ~100 GeV, we can efficiently measure gamma rays before their destructive interaction in the atmosphere by direct detection with balloon or satellite-borne detectors. Gamma rays interact within the detector, and convert into e^+e^- pairs, which are tracked to estimate the direction of the primary particle, and then stopped by a calorimeter to estimate its energy. This method is limited by the relatively small achievable collection area, corresponding essentially to the physical size of the detector. On the other hand, the technique presents the great advantages of ~100% duty cycle, large field of view, and that the much more abundant charged cosmic rays can be easily identified and therefore vetoed, resulting in virtually background-free gamma-ray measurements. Currently, the most advanced gamma-ray telescope using this detection technique is the Fermi-LAT. It consists of a large-field-of-view (2.4 sr), pair-conversion telescope, sensitive to gamma rays in the energy range between 20 MeV and about 300 GeV [17]. The latest Fermi-LAT source catalogue contains about 5000 sources [18], a third of which remain unassociated. Since its launch in June 2008, the LAT has primarily operated in survey mode, scanning the whole sky every 3 h. The exposure coverage of this observation mode is fairly uniform, with variations below 30% with respect to the average exposure. Thanks to this full-sky coverage, Fermi-LAT will be able to perform dark matter searches using its data archive should new dSphs be discovered in the future.

Above few tens of GeV, gamma-ray fluxes become too low for the relatively small collection area of Fermi-LAT, and it is advantageous to measure them indirectly through the detection of the secondary particles and/or the radiation present in the particle cascade resulting from their interaction in the atmosphere, which greatly increases the effective collection area.

Cherenkov telescopes measure the Cherenkov radiation emitted by the electrons and positrons of the cascade (which travel faster than light in the atmosphere), thus producing an image of such cascade. The intensity, orientation, and shape of Cherenkov images allow for the estimation of the energy and arrival direction of the primary particle, and provide some separation power between gamma rays and charged cosmic rays. Several nearby telescopes observing the same gamma-ray source may image the same cascade from different perspectives, increasing the precision of these measurements. The weak points of this technique are the small duty cycle (about 10–15%, since they operate only during night, with no or relatively dim moonlight and good atmospheric conditions), narrow fields of view of few degrees diameter at most, and the presence of the irreducible background produced by charged cosmic rays. Among its advantages, we find the large collection area, given by the size of the Cherenkov light pool projected on the plane of the telescope reflector (e.g., $\sim 10^5$ m^2 for 1 TeV gamma ray at low zenith distance). The resulting flux sensitivity achieved by this technique reaches currently around $\sim$1% of the Crab nebula in 25 h of observations. There are three main running Cherenkov observatories exploiting this detection technique: H.E.S.S, MAGIC and VERITAS. H.E.S.S is composed of four 12-m diameter telescopes operating since 2004, surrounding one 28-m diameter telescope since 2012, located in the Khomas Highland (Namibia). The energy threshold is 30 GeV and the field of view has a diameter of 5°. MAGIC is composed of two 17-m diameter telescopes, located at the Observatorio Roque de los Muchachos at La Palma, Canary Islands (Spain), in operation since 2004 in single-telescope mode and 2009 in two-telescope mode. MAGIC energy threshold is 30 GeV and the FoV is 3.5° diameter. Finally, VERITAS is composed of four 12-m diameter Cherenkov telescopes, located at the Fred Lawrence Whipple Observatory, Arizona (USA), operating since 2007. VERITAS has an energy threshold of 85 GeV and a FoV of 3.5° diameter.

Finally, water Cherenkov particle detectors measure the charged particles present in the cascades initiated by the primary gamma rays when interacting in the atmosphere. The amount of detected particles and their spatial distribution allow to measure the energy of the primary and to discriminate between gamma rays and cosmic rays, whereas the difference of detection time at different detectors allows to estimate the arrival direction. This technique is sensitive to gamma rays and cosmic rays between few hundred GeV and 100 TeV. It has the advantages of 100% duty cycle, plus large effective area and field of view, but a limited separation power between gamma rays and cosmic rays. The currently most advanced water Cherenkov gamma-ray detector is HAWC, composed of 300 water Cherenkov detectors located at an altitude of 4100 m at the Sierra Negra volcano, near Puebla (Mexico), covering 22,000 m^2. It is sensitive to gamma rays between 500 GeV and 100 TeV, with a field of view of 15% of the sky, and daily coverage of 8.4 sr, or 67% of the sky (a region where dark matter searches using the HAWC data archive will be possible should new dSphs will be discovered in the future). Partial HAWC operations started in 2013, and the full detector was completed in March 2015.

4. Statistical Data Analysis

Advanced searches for dark matter annihilation or decay in dSphs with gamma rays rely on the distinct spatial and spectral features of the expected signals. We expect dark matter signal to be distributed morphologically according to $\frac{dJ}{d\Omega}$, and spectrally according to $\frac{dN_\gamma}{dE}$, and those PDFs are in general clearly distinguishable from those expected for background processes.

Regarding the use of the morphological information, the spatial coincidence of the signal with the position of the dSph would provide strong discrimination power, because we do not expect that gamma rays can be produced at dSphs by any conventional astrophysical process. However, using the information of the morphology of the gamma-ray emission around the position of the dSph is more delicate, because such morphology is in general subject to relatively large uncertainties, and assuming an incorrect shape may bias the result of the search. In addition, the expected size of the dark matter halo is, for many of the known dSphs and for the considered gamma-ray instruments, consistent with point like sources, or at most slightly extended, which means that we can obtain no or little signal/background discrimination power from the use of the morphological information. All this is particularly true for dark matter annihilation, for which, due to the ρ^2 dependence of $\frac{dJ}{d\Omega}$, the expected signal is more compact and more affected by uncertainties on the details of the dark matter distribution within the halo. When looking for dark matter decay signal, on the other hand, such dependence is linear with ρ, which leads to less peaked and less uncertain morphologies.

The use of spectral information would be key for univocally attributing a dark matter origin to a detected gamma-ray signal, because in general, the features present in the spectra predicted for dark matter annihilation or decay cannot be produced by other conventional astrophysical processes. For instance, in the most extreme/luckiest case, the detection of gamma-ray spectral lines would be considered as unambiguous prove for the observation of dark matter annihilation or decay. Other processes, like creation of Standard Model particle pairs also produce distinct spectral features providing high discrimination power over backgrounds, such as the existence of sharp kinematic spectral cutoffs (see Figure 1). These considerations are general for all dark matter searches, independently of whether they are performed on dSphs or elsewhere. Searches in dSphs have the additional advantage that dark matter signals are, in principle, universal, any potential detection from a given dSph could be confirmed by looking for the same spectral features in the emission from other dSphs. Contrary to the case of $\frac{dJ}{d\Omega}$, uncertainties in $\frac{dN_\gamma}{dE}$ can be considered negligible for a given annihilation/decay channel. This is the main reason why gamma-ray instruments utilize the spectral information not only for reinforcing the credibility of an eventual future detection, but also to increase the sensitivity of the search and therefore provide more constraining bounds to the dark matter nature in case of no detection.

Current dark matter searches using gamma rays are based on different implementations of the likelihood-ratio test [19], which we use to quantify the compatibility of the measured data ($\mathcal{D}$) with different hypotheses, in particular with the null hypothesis (i.e., that no dark matter signal is present in $\mathcal{D}$), through the associated p-value. Finding a sufficiently low p-value (by convention in the field $p < 3 \times 10^{-7}$) for the observed data $\mathcal{D}$ under the null hypothesis assumption is usually referred to as detecting dark matter. In case of a positive detection, we can use the likelihood function to measure the dark matter physical parameters such as its mass, annihilation cross section, decay lifetime, and branching ratio to the different decay/annihilation channels (collectively represented here by the vector $\boldsymbol{\alpha}$). Conversely, if the null hypothesis cannot be excluded, we can use the likelihood function to set limits to the parameters $\boldsymbol{\alpha}$.

The likelihood function can be written in the following general form:

$$\mathcal{L}(\boldsymbol{\alpha}; \boldsymbol{\nu} | \mathcal{D}) \quad , \tag{11}$$

where, apart from its dependence on $\boldsymbol{\alpha}$ and $\mathcal{D}$, we have made explicit that $\mathcal{L}$ can also depend on other, so-called, nuisance parameters ($\boldsymbol{\nu}$), for which we only know their likelihood function (normally constrained using dedicated datasets). In general, nuisance parameters represent quantities used in the computation of $\boldsymbol{\alpha}$ and that are affected by some uncertainty, either of statistical or systematic nature, or both. Prototypical examples of nuisance parameters are the number of background events of certain estimated energy and arrival direction present in the signal region, or $\overline{J}$. One standard technique to eliminate the nuisance parameters when making statements about $\boldsymbol{\alpha}$ is using the profile likelihood ratio test:

$$\lambda_P(\boldsymbol{\alpha} | \mathcal{D}) = \frac{\mathcal{L}(\boldsymbol{\alpha}; \hat{\hat{\nu}} | \mathcal{D})}{\mathcal{L}(\hat{\boldsymbol{\alpha}}; \hat{\nu} | \mathcal{D})} \quad , \tag{12}$$

where $\hat{\boldsymbol{\alpha}}$ and $\hat{\nu}$ are the values maximizing $\mathcal{L}$, and $\hat{\hat{\nu}}$ the value that maximizes $\mathcal{L}$ for a given $\boldsymbol{\alpha}$. According to Wilks' theorem $-2 \ln \lambda_P(\boldsymbol{\alpha})$ is distributed, when $\boldsymbol{\alpha}$ are the true values, as a χ^2 distribution with number of degrees of freedom equal to the number of components of $\boldsymbol{\alpha}$, independent of the value of $\boldsymbol{\nu}$. It is an extended practice in indirect dark matter searches with gamma rays to decrease the n-dimensional vector $\boldsymbol{\alpha}$ of free parameters to a one-dimensional quantity α, by considering that gamma-ray production is dominated either by annihilation ($\alpha = \langle \sigma v \rangle$, i.e., the velocity-averaged annihilation cross section) or by decay ($\alpha = \tau_\chi^{-1}$, i.e., the decay rate), and scanning over values of the dark matter particle mass (m_χ) and pure annihilation/decay channels (i.e., considering at each iteration 100% branching ratio to one of the possible SM particle pairs). For each scanned combination, Equation (11) reduces to a likelihood function of just one purely free (i.e., non-nuisance) parameter.

In such a case, for instance, 1-sided 95% confidence level **upper limits** to α are taken as $\alpha^{\text{UL}_{95}} = \alpha_{2.71}$, with $\alpha_{2.71}$ found by solving the equation $-2\ln\lambda_P(\alpha_{2.71}) = 2.71$.

The data $\mathcal{D}$ can refer to N_{dSph} different dSphs, in which case it is convenient to write the joint likelihood function as:

$$\mathcal{L}(\alpha;\nu|\mathcal{D}) = \prod_{l=1}^{N_{\text{dSph}}} \mathcal{L}_\gamma(\alpha\bar{J}_l;\mu_l|\mathcal{D}_{\gamma_l}) \cdot \mathcal{L}_J(\bar{J}_l|\mathcal{D}_{J_l}) \quad , \tag{13}$$

where we have factorized the joint likelihood into the partial likelihood functions corresponding to each dwarf, and those subsequently into the parts corresponding to the gamma-ray observations ($\mathcal{L}_\gamma$) and J-factor measurement ($\mathcal{L}_J$), respectively; $\bar{J}_l$ is the total J-factor (see Equation (7)) of the l-th considered dSph, which, as we have made explicit, is a nuisance parameter degenerated with α in $\mathcal{L}_\gamma$; μ_l represents the additional nuisance parameters different from $\bar{J}_l$ affecting the analysis of the l-th dSph; $\mathcal{D}_{\gamma_l}$ represents the gamma-ray data of the l-th dSph, whereas $\mathcal{D}_{J_l}$ refers to the data constraining $\bar{J}_l$.

For each dSph, we may have N_{meas} independent measurements, each performed under different experimental conditions, by the same or different instruments. That is, we can factorize the $\mathcal{L}_\gamma$ term as:

$$\mathcal{L}_\gamma(\alpha\bar{J};\mu|\mathcal{D}_\gamma) = \prod_{k=1}^{N_{\text{meas}}} \mathcal{L}_{\gamma,k}(\alpha\bar{J};\mu_k|\mathcal{D}_{\gamma,k}) \quad , \tag{14}$$

where we have omitted the index l referring to the dSph for the sake of clarity, and with μ_k and $\mathcal{D}_{\gamma,k}$ representing the nuisance parameters and data, respectively, referred to the k-th measurement.

For each observation of a given dSph under certain experimental conditions, $\mathcal{L}_{\gamma,k}$ often consists of the product of $N_{E'} \times N_{\hat{p}'}$ Poissonian terms (P) for the observed number of gamma-ray candidate events (N_{ij}) in the i-th bin of reconstructed energy and j-th bin of reconstructed arrival direction, times the likelihood term for the μ nuisance parameters ($\mathcal{L}_\mu$), with $N_{E'}$ the number of bins of reconstructed energy and $N_{\hat{p}'}$ the number of bins of reconstructed arrival direction, i.e.:

$$\mathcal{L}_{\gamma,k}(\alpha\bar{J};\mu|\mathcal{D}_\gamma) = \prod_{i=1}^{N_{E'}}\prod_{j=1}^{N_{\hat{p}'}} P\left(s_{ij}(\alpha\bar{J};\mu) + b_{ij}(\mu)|N_{ij}\right) \cdot \mathcal{L}_\mu(\mu|\mathcal{D}_\mu) \quad , \tag{15}$$

where the indexes l and k referring to the dSph and the measurement have been removed for the sake of a clear notation. The parameter of the Poissonian term is $s_{ij} + b_{ij}$, where s_{ij} is the expected number of signal events in the i-th bin in energy and the j-th bin in arrival direction, computable using $\alpha\bar{J}$ as we will see below; and b_{ij} the corresponding contribution from background processes. $\mathcal{D}_\mu$ represents the data used to constrain the values of the nuisance parameters μ. We have made explicit that the uncertainties associated to μ can in principle affect both the computation of the signal and background contributions. For instance, uncertainties in the overall energy scale affect the computation of s_{ij}, whereas uncertainties in the background modeling affect the computation of b_{ij}. However, uncertainties affecting s_{ij} are usually considered to be largely dominated by the uncertainty in the J-factor and the dependence of s_{ij} on μ therefore ignored. Thus, s_{ij}, is given by:

$$s_{ij}(\alpha\bar{J}) = \int_{\Delta E'_i} dE' \int_{\Delta\hat{p}'_j} d\Omega' \int_0^\infty dE \int_{\Delta\Omega_{\text{tot}}} d\Omega \int_0^{T_{\text{obs}}} dt \, \frac{d^2\Phi(\alpha\bar{J})}{dE\,d\Omega} \, \text{IRF}(E',\hat{p}'|E,\hat{p},t) \quad , \tag{16}$$

where E', $\hat{p}'$, E and $\hat{p}$ are the estimated and true energies and arrival directions, respectively; $d\Omega'$ and $d\Omega$ infinitesimal solid angles containing $\hat{p}'$ and $\hat{p}$, respectively; T_{obs} the total observation time; t the time along the observations; and IRF the instrument response function, i.e. $\text{IRF}(E',\hat{p}'|E,\hat{p},t)\,dE'\,d\Omega'$ is the effective collection area of the detector times the probability for a gamma ray with true energy E and direction $\hat{p}$ to be assigned an estimated energy in the interval $[E', E'+dE']$ and $\hat{p}'$ in the solid angle $d\Omega'$ (see more details below), at the time t during the observations. The integrals over E and $\hat{p}$ perform the convolution of the gamma-ray spectrum with the instrumental response, whereas those over E' and $\hat{p}'$ compute the events observed within the i-th energy bin ($\Delta E'_i$) and the j-th arrival direction bin ($\Delta \hat{p}'_j$). It must be noted that, defining several spatial bins within the source produces relatively minor improvement in sensitivity to dark matter searches for not significantly extended sources (i.e., those well described by a point-like source, as it is the case for many dSphs) [20]. For significantly extended sources, on the other hand, using a too fine spatial binning makes the obtained result more sensitive to the systematic uncertainties in the dark matter spatial distribution within the dSph halo. Thus, a realistic optimization of $N_{\hat{p}'}$ based on sensitivity should balance the gain yielded by the use of more spatial information and the loss caused by the increase in the systematic uncertainty.

The IRF can be factorized as the product of the detector collection area A_{eff} ($T_{\text{obs}} \cdot A_{\text{eff}}$ is often referred to as exposure), times the PDFs for the energy (f_E) and incoming direction ($f_{\hat{p}}$) estimators, i.e.:

$$\text{IRF}(E', \hat{p}' | E, \hat{p}, t) = A_{\text{eff}}(E, \hat{p}, t) \cdot f_E(E' | E, t) \cdot f_{\hat{p}}(\hat{p}' | E, \hat{p}, t) \quad , \tag{17}$$

where, following the common practice, the (small) dependence of f_E with $\hat{p}$ has been neglected. $f_{\hat{p}}$ is often referred to as the point spread function (PSF).

Finally, the likelihood for the total J-factor is usually written as:

$$\mathcal{L}_J(\bar{J} | \bar{J}_{\text{obs}}, \sigma_J) = \frac{1}{\ln(10) \bar{J}_{\text{obs}} \sqrt{2\pi} \sigma_J} e^{-\left(\log_{10}(\bar{J}) - \log_{10}(\bar{J}_{\text{obs}})\right)^2 / 2\sigma_J^2} \quad ; \tag{18}$$

with $\log_{10} \bar{J}_{\text{obs}}$ and σ_J the mean and standard deviation of the fit of a log-normal function to the posterior distribution of the total J-factor [21]. Therefore, including $\mathcal{L}_J$ in the joint likelihood is a way to incorporate the statistical uncertainty of $\bar{J}$ in the estimation of α. It is worth noting that, because α and $\bar{J}$ are degenerate, in order to perform the profile of $\mathcal{L}$ with respect to $\bar{J}$ it is sufficient to compute $\mathcal{L}_\gamma$ vs α for a fixed value of $\bar{J}$, which facilitates significantly the computational needs of the profiling operation (see details in footnote 12 of reference [22]). Including $\bar{J}_{\text{obs}}$ systematic uncertainties is much more complex, since they depend mainly on our choice of the dark matter halo density profile function (e.g., NFW [23], Einasto [24], etc.), and there is no obvious way of assigning a PDF to that choice. Because of this, the impact of that uncertainty in the bounds in α are usually roughly quantified by performing the likelihood analysis several times, each assuming different fitting functions, and comparing the results obtained for each of them.

The PDF of the test statistic $-2 \ln \lambda_P$ for the no-dark matter null hypothesis, i.e., when the true value of α is given by $\alpha_{\text{true}} = 0$, is needed for evaluating the significance of a possible signal detection. Computing upper limits to α, on the other hand, consists in finding the value of α_{true} for which the integral of the PDF above $\hat{\alpha}$ corresponds to the required confidence level. Estimating the PDF for $-2 \ln \lambda_P$ with fast simulations is feasible (from a computational-demand standpoint) when the involved p-values are high enough so that they can be evaluated with a relatively low number of simulated datasets. In practice, however, results for dark matter searches using gamma-rays are generally computed assuming Wilks' theorem validity, and that $-2 \ln \lambda_P$ is distributed as a χ^2. The adoption of Wilks' theorem by all the experiments allows at least a direct comparison among their results. One should keep in mind, however, that the described statistical framework is also usually affected by the non-fulfillment of the conditions of validity of Wilks' theorem, at least because of two different reasons. First, because α is normally restricted to the physical region (i.e., to non-negative values), which produces over-coverage (i.e., the computed confidence interval contains the true value more often than the quoted confidence level) for negative background fluctuations, i.e., when the likelihood absolute maximum lies at the border of the physical region. This can be avoided by using the correct $-2 \ln \lambda_P$ PDF for this situation [25]. Another way commonly used to partially mitigate this problem is to show the obtained result (e.g., the upper limit to α) in comparison to its PDF for the no-dark matter ($\alpha_{\text{true}} = 0$) hypothesis. Such PDF is estimated using fast simulations and/or pure-background datasets (such as those obtained by considering randomly selected directions as potential DM targets), and it is normally characterized by its median (referred to as the sensitivity of the measurement) and the bounds for some predefined (e.g., 68%, 95%, etc.) symmetric containment quantiles. By such comparison one can evaluate whether the obtained result is significantly incompatible with the $\alpha_{\text{true}} = 0$ hypothesis. The second violation of Wilks' theorem validity conditions affects the computation of confidence intervals (i.e., the PDF of the test $-2 \ln \lambda_P$ for $\alpha_{\text{true}} > 0$). In this case, however, because α and $\bar{J}$ are degenerate in the likelihood function, the log-normal shape of the likelihood term $\mathcal{L}_J$ (see Equation (18)) results in the loss of Gaussianity of the likelihood for α required by the Wilks' theorem.

As we will see in the next Section, the most common simplifications adopted in gamma-ray data analyses consist in ignoring the statistical and/or systematic uncertainties in $\bar{J}$ or in the background contribution to the signal region. Omitting these relevant uncertainties in general improves artificially the reported sensitivity and bounds obtained by the analysis, which must be taken into account when comparing results obtained under different assumptions.

5. Results

None of the different gamma-ray telescopes has obtained a significant detection in their search for dark matter signals from dSphs. Therefore, they provide results in the form of upper limits to the annihilation cross section or lower limits to the decay lifetime. In this section, I summarize the results obtained by the different considered instruments. In addition, I highlight and motivate the main analysis choices adopted by the different experiments as well as the differences with respect to the general framework described in Section 4, also summarized in Table 1.

5.1. Fermi-LAT

The Fermi-LAT data are publicly available and several authors outside the Fermi-LAT Collaboration have searched for DM annihilation signals in dSphs (e.g., references [26–34]). The Fermi-LAT Collaboration has carried out several searches for dark matter signals from dSphs, corresponding respectively to 11 months observations of 14 dSphs [35], 24 months of observations of 10 dSphs [36], 4 years [37] and 6 years [21] of data of 25 dSphs. Here we concentrate on this latter work.

In their 6-year-data search, the Fermi-LAT Collaboration applied their most developed data (re-)analysis, known as Pass 8. They subsequently searched for gamma-ray signals individually in 25 dSphs (including the classical and the ultra-faint ones discovered by the Sloan Digital Sky Survey [38]), and combined the 15 targets with better determined dark matter content. The dark matter distribution in each dSph was parameterized using the Navarro-Frenk-White (NFW) profile [23], constrained using the prescription by Martinez (2015) [39]. The $\frac{dN_\gamma}{dE}$ average spectra for the different considered channels, on the other hand, were obtained from the PHYTIA-based [9] DMFIT package [40].

Table 1. Summary of dark matter searches with gamma-ray instruments. From left to right, columns show: Bibliographic reference; Instrument; Targets; Investigated decay and/or annihilation channels; dN/dE source; J-factor source; J-factor uncertainty, morphology of the source, restriction of α to physical region, statistical and systematic background uncertainties, determination of the true λ_P PDF; other relevant differences of the analyses with respect to the general framework. See main text for more details.

reference	Instrument	dSphs	Channels	dN/dE	J-Factor	ΔJ	Ext	$\alpha \geq 0$	Δ bkg sta	Δ bkg sys	PDF	Other
[21]	Fermi-LAT	Boötes 1, Canes Venatici II, Carina, Coma Berenices, Draco, Fornax, Hercules, Leo II, Leo IV, Sculptor, Segue 1, Sextans, Ursa Major II, Ursa Minor, Willman 1	$b\bar{b}$, $\tau^+\tau^-$, e^+e^-, $u\bar{u}$, $\mu^+\mu^-$, W^+W^- [annihilation]	PYTHIA 8.1 [9]	Following Martinez [39], assuming NFW [23]	✓	✓	✓	×	×	×	1
[41]	H.E.S.S	Carina, Coma Berenices, Fornax, Sagittarius, Sculptor	$b\bar{b}$, $\tau^+\tau^-$, $e^+e^-e^+e^-$, $\mu^+\mu^-$, $\mu^+\mu^-\mu^+\mu^-$, $W^+W^- + ZZ$ [annihilation]	Cembranos et al. [42]	Martinez [39] assuming NFW [23] and Burkert [43]	✓	×	✓	✓	×	×	2,3
[44]	H.E.S.S	Carina, Coma Berenices, Fornax, Sagittarius, Sculptor	$\gamma\gamma$ [annihilation]	trivial	Geringer-Sameth et al. [7]	✓	✓	✓	✓	×	×	2,3
[45]	MAGIC	Segue 1	$b\bar{b}$, $t\bar{t}$, $\mu^+\mu^-$, $\tau^+\tau^-$, W^+W^-, ZZ, [annihilation and decay], $\gamma\gamma$, $Z\gamma$, $\mu^+\mu^-\gamma$, $\tau^+\tau^-\gamma$ [annihilation], $\gamma\nu$, $\gamma\gamma\gamma$ [decay]	Cembranos et al. [42]	Essig et al. [24]	×	✓	×	×	×	×	—
[46]	MAGIC	Ursa Major II	$b\bar{b}$, $\mu^+\mu^-$, $\tau^+\tau^-$, W^+W^-, [annihilation]	PPC4DMID [47]	Geringer-Sameth et al. [7]	✓	✓	✓	✓	✓	×	—
[48]	VERITAS	Boötes 1, Draco, Segue 1, Ursa Minor, Willman 1	$u\bar{u}$, $d\bar{d}$, $s\bar{s}$, $b\bar{b}$, $t\bar{t}$, e^+e^-, $\mu^+\mu^-$, $\tau^+\tau^-$, W^+W^-, ZZ, hh [annihilation]	PPC4DMID [47]	Geringer-Sameth et al. [7]	×	✓	✓	×	×	✓	2,4

1 In Equation (19): assuming $d\phi/dE \propto E^{-2}$ and energy resolution and bias disregarded.
2 In Equation (16): A_{eff} dependence on $\vec{p}$ disregarded
3 In Equation (16): Effect of angular resolution disregarded (i.e., $f_{\vec{p}} \to \delta(\vec{p} - \vec{p}')$)
4 In Equation (28): f_s assumed radially symmetric with respect to the center of the dSph.

Table 1. Cont.

reference	Instrument	dSphs	Channels	dN/dE	J-Factor	ΔJ	Ext	α ≥ 0	Δ bkg sta	Δ bkg sys	PDF	Other
[49]	HAWC	Boötes 1, Canes Venatici I, Canes Venatici II, Coma Berenices, Draco, Hercules, Leo I, Leo II, Leo IV, Segue 1, Sextans, Triangulum II, Ursa Major I, Ursa Major II, Ursa Minor	$b\bar{b}$, $t\bar{t}$, $\mu^+\mu^-$, $\tau^+\tau^-$, W^+W^- [annihilation and decay]	PYTHIA 8.2 [50]	CLUMPY [51] assuming NFW [23]	×	×	✓	✓	×	×	–

Fermi-LAT data statistical analysis follows Equations (13) and (14), with $N_{\rm dSph} = 15$ observed dSphs and $N_{\rm meas} = 4$ referring to the four independent datasets, each containing events with one of the four possible event direction reconstruction quality level, and hence each described by different IRF. They consider bins of reconstructed energy in the range between 500 MeV and 500 GeV and bins of incoming direction in a region of interest of $10° \times 10°$ around the position of each dSph. The dominant background is produced by gamma rays from nearby sources (whose estimated energy and incoming direction are consistent, due to the finite angular resolution of the instrument, with being originated at the dSph dark matter halo), or by the diffuse gamma-ray component resulting from the interaction of cosmic rays with the interstellar medium or from unresolved sources of Galactic and extragalactic origin, depending on the particular dSphs being considered. The analysis does not explicitly treat the relevant background parameters μ in Equation (15) as nuisance parameters. Instead, the spectral parameters (e.g., normalization, photon index, etc.) of the different background sources are fixed using the following simplified method. The flux normalizations of the different background components are determined by means of a maximum-likelihood fit to the spacial and spectral distributions of the observed events, with the rest of spectral parameters fixed to the values listed in the updated third LAT source catalog [52]. Then, it is checked that the values of the background normalization factors obtained using this method do not change significantly by including an extra weak source at the locations of the dSph, which shows that the background are well-constrained by this procedure. Studies showed that the effect of the background uncertainty from this procedure contributed at a few percent of statistical uncertainty of the signal and are therefore safe to neglect.

In order to produce a result valid for arbitrary spectral shapes (i.e., arbitrary value of m_χ and of the branching ratios to the different annihilation/decay channels), the Fermi-LAT Collaboration computes, for each observed dSph and bin of estimated energy $\Delta E_i'$, the value of:

$$\mathcal{L}_{\gamma_i}(\bar{\Phi}_i) = \prod_{j=1}^{N_{\rho'}} P(s_{ij}(\bar{\Phi}_i) + b_{ij}|N_{ij}) \qquad (19)$$

as a function of $\bar{\Phi}_i$, that is, the sum over the spatial bins of the $\mathcal{L}_\gamma$ likelihood values within $\Delta E_i'$. In order to obtain a set of generic $\mathcal{L}_{\gamma_i}$ values, they compute $s_{ij}(\bar{\Phi}_i)$ using Equation (16) assuming a power-law gamma-ray spectrum ($\frac{d\Phi}{dE} \propto E^{-\Gamma}$) of spectral index $\Gamma = 2$. The spatial distribution of gammas (which does not depend on the energy) is considered known and fixed, and given by the $\frac{dJ}{d\Omega}$ curves obtained from the fit to the stellar kinematics to the different dSphs. Equation (15) can then be written in terms of the $\mathcal{L}_{\gamma_i}$ factors as:

$$\mathcal{L}_{\gamma,k}(\alpha\bar{J};\mu|\mathcal{D}_\gamma) = \prod_{i=1}^{N_{E'}} \mathcal{L}_{\gamma_i}(\bar{\Phi}_i(\alpha\bar{J})) \qquad . \qquad (20)$$

The values of $\mathcal{L}_{\gamma_i}$ vs $\bar{\Phi}_i$ for each of the analyzed dSphs were computed, tabulated and released by the Fermi-LAT Collaboration [53]. This allows any scientist to compute $\mathcal{L}_\gamma$ for the dark matter model of their choice by just selecting the corresponding values of α, m_χ, the total $\frac{dN_\gamma}{dE}$ and $\bar{J}$, and computing the corresponding $\bar{\Phi}_i$ values as:

$$\bar{\Phi}_i(\alpha\bar{J}) = \int_{\Delta E_i} dE \, \frac{d\Phi(\alpha\bar{J})}{dE} \qquad , \qquad (21)$$

with $\frac{d\Phi}{dE}$ given by Equation (10). We note that this approach allows to compute bounds on $\alpha\bar{J}$ with associated confidence level known only to a certain (unquantified) precision that depends on how similar are the investigated spectral shape and the one assumed when computing the values of $\mathcal{L}_{\gamma_i}$ (i.e., a power-law with $\Gamma = 2$ in the Fermi-LAT case). In addition, it should be stressed that such precision depends also on the PDF of the energy estimator and that, therefore, the range of investigated spectral shapes for which we can establish bounds within a certain precision using this technique is different for different instruments.

No significant gamma-ray signal from dSphs was found in the Fermi-LAT data, either individually in each dSph (the largest deviation from the null hypothesis is found for Sculptor, with $-2\ln\lambda_P = 4.3$), or in the combined analysis ($-2\ln\lambda_P = 1.3$). Some of the obtained exclusion limits are shown in Figure 2. This work represents the most constraining search for WIMP annihilation signals for the dark matter particle mass range below ~1 TeV. As shown in the figure, the limits exclude the thermal relic cross section for $m_\chi < 100$ GeV in the case of annihilation into $b\bar{b}$ or $\tau^+\tau^-$ pairs.

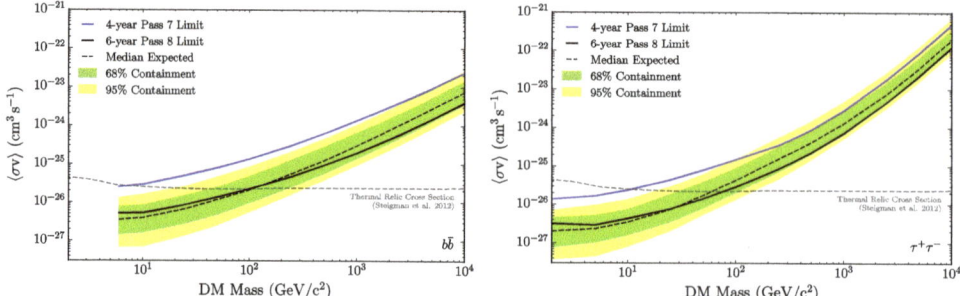

Figure 2. The 95% confidence-level upper limits to $\langle\sigma v\rangle$ for the $\chi\chi \to b\bar{b}$ (**left**) and $\chi\chi \to \tau^+\tau^-$ (**right**) annihilation channels derived from 6-year observations of 15 dSphs with Fermi-LAT. The dashed black line shows the median of the distribution of limits obtained from 300 simulated realizations of the null hypothesis using LAT observations of high-Galactic-latitude empty fields, whereas green and yellow bands represent the symmetric 68% and 95% quantiles, respectively. The dashed gray curve corresponds to the thermal relic cross-section [54]. Reprinted figure with permission from reference [37]; copyright (2014) by the American Physical Society.

These results were combined with MAGIC observations of Segue 1, into the first coherent search for dark matter using several gamma-ray instruments [22]. Details about this work are provided below.

In a later work, the Fermi-LAT and the Dark Energy Survey (DES) collaborations also used the data from 6 years of observations to look for dark matter signals over a sample of 45 stellar systems consistent with being dSphs [55]. The search was performed shortly after the discovery of 17 of the considered dSph candidates, for which no reliable estimate of the dark matter content was available at the time. Because of this, all considered candidates were assumed to be point-like sources, and the J-factors for the non-confirmed dSphs estimated from a purely empirical scaling relation based on their heliocentric distance. For four of the examined dSphs, a 2σ discrepancy with the null hypothesis was found, which does not contradict significantly such hypothesis, particularly once the number of investigated sources, channels and masses is considered. Overall, the strategy of observing a set of not fully confirmed dSphs candidates, for which no reliable estimate of the J-factor exists yet is justified since a solid positive gamma-ray signal from any of the observed targets would have been considered a strong experimental evidence of dark matter annihilation or decay. In absence of such signal, however, the obtained limits are less robust than those from the 15 confirmed dSphs described above, which remain the reference in the field for the sub-TeV mass range.

5.2. Cherenkov Telescopes

Dark matter searches with Cherenkov telescopes have evolved from simple event-counting analyses to more complex maximum-likelihood analyses of optimized sensitivity thanks to the inclusion of the expected spectral and morphological features of the dark matter signals [56].

In the most basic version of the likelihood function, the nuisance parameters μ (see Equation (15)) are the b_{ij} factors themselves. They are constrained by measurements in signal-free, background-control (or Off) regions with τ times the exposure of the signal (or On) region. A more complete analysis also includes the treatment of τ as a nuisance parameter, given that the latter is normally affected by significant statistic ($\sigma_{\tau,\text{stat}}$) and systematic ($\sigma_{\tau,\text{sys}}$) uncertainties. The statistical uncertainty comes from the fact that τ is often estimated from the data themselves (comparing the events observed in regions adjacent to the On and Off ones). The systematic uncertainty takes into account the residual differences of exposure between the Off and On regions, and it is normally assumed to be of the order of 1% for the current generation of Cherenkov telescopes [12]. It can be shown that this systematic uncertainty is the limiting factor to the sensitivity of the event-counting analyses for $N_{\text{On}} \gtrsim \frac{(\tau+1)\tau}{\Delta\tau^2}$, i.e., between $\sim 10^4$ and $\sim 2 \times 10^4$ events for τ in the typical range between 1 and 10, and 1% systematic uncertainty in τ (i.e., $\sigma_{\tau,\text{sys}} = 0.01\tau$). Once we reach this number of observed events in the signal region, increasing the statistics of the dataset does not longer contribute to improve the sensitivity of the search.

The gamma-ray likelihood function for Cherenkov telescopes can thus be written as the product of Poisson likelihoods for the On and Off region times a Gaussian likelihood for τ, i.e.:

$$\mathcal{L}_\gamma(\alpha\bar{J}; \{b_{ij}\}_{i=1,\ldots,N_{E'};j=1,\ldots,N_{\hat{p}'}}, \tau \mid \{N_{\text{On},ij}, N_{\text{Off},ij}\}_{i=1,\ldots,N_{E'};j=1,\ldots,N_{\hat{p}'}}) = \prod_{i=1}^{N_{E'}}\prod_{j=1}^{N_{\hat{p}'}} \left[P\left(s_{ij}(\alpha\bar{J}) + b_{ij} \mid N_{\text{On},ij}\right) \cdot P(\tau b_{ij} \mid N_{\text{Off},ij}) \right] \cdot G(\tau \mid \tau_{\text{obs}}, \sigma_\tau) \quad , \quad (22)$$

with $N_{\text{On},ij}$ and $N_{\text{Off},ij}$ the number of observed events in the On and Off regions, respectively, in the i-th bin of reconstructed energy and the j-th bin of reconstructed arrival direction; and G an (often neglected) Gaussian PDF with mean the measured value τ_{obs} and width $\sigma_\tau = \sqrt{\sigma_{\tau,\text{stat}}^2 + \sigma_{\tau,\text{sys}}^2}$.

The considered energy range depends on the instrument (e.g., larger reflectors provide lower thresholds) and the dSph observation conditions (e.g., higher zenith angle observations imply higher threshold). For the current instruments and observed dSphs, the lowest energy bin starts between 80 and 800 GeV, whereas the highest one can reach up to between 10 and 100 TeV.

In the analysis of Cherenkov telescope data, the convolution of $\frac{d^2\Phi}{dEd\Omega} \cdot A_{\text{eff}}$ with the PSF function $f_{\hat{p}}$ needed to compute s_{ij} according to Equation (16) is usually performed numerically through the analysis of Monte Carlo simulated events. We note that Equation (16) can be written as:

$$s_{ij} = \int_{\Delta E'_i} dE' \int_0^\infty dE \int_0^{T_{\text{obs}}} dt \, \frac{d\Phi}{dE}(E) \, \bar{A}_{\text{eff},j}\left(E, \frac{d\mathcal{J}}{d\Omega}, t\right) f_E(E'|E,t) \quad , \quad (23)$$

with $\bar{A}_{\text{eff},j}$ the signal morphology-averaged effective area within spatial bin j, defined as:

$$\bar{A}_{\text{eff},j}\left(E, \frac{d\mathcal{J}}{d\Omega}, t\right) = \frac{\int_{\Delta\hat{p}'_j} d\Omega' \int_{\Delta\Omega_{\text{tot}}} d\Omega \, \frac{d^2\Phi}{dEd\Omega}(E,\hat{p}) \, A_{\text{eff}}(E,\hat{p},t) \, f_{\hat{p}}(\hat{p}'|E,\hat{p},t)}{\frac{d\Phi}{dE}(E)}$$

$$= \int_{\Delta\hat{p}'_j} d\Omega' \int_{\Delta\Omega_{\text{tot}}} d\Omega \, \frac{d\mathcal{J}}{d\Omega}(\hat{p}) \, A_{\text{eff}}(E,\hat{p},t) \, f_{\hat{p}}(\hat{p}'|E,\hat{p},t) \quad . \quad (24)$$

$\bar{A}_{\text{eff},j}$ depends on the morphology of the gamma-ray emission ($\frac{d\mathcal{J}}{d\Omega}$), although not on its intensity, hence not on $\bar{J}$. Therefore, for point-like sources observed with constant IRF at a given fixed direction, $\bar{A}_{\text{eff},j}$ is only a function of the energy. As a matter of fact, what normally is referred to as the effective area of a given Cherenkov telescope is the value of $\bar{A}_{\text{eff},j}(E)$ for a circular spatial bin centered at the position of a point-like source (observed at low zenith distance under dark and good weather conditions), with radius optimized to maximize the signal-to-noise ratio. In practice, $\bar{A}_{\text{eff},j}(E)$ is computed with Monte Carlo simulations: Using a sample of simulated gamma rays with arrival directions distributed according to $\frac{d\mathcal{J}}{d\Omega}$ and trajectories impacting uniformly in a sufficiently large area (A_{tot}) around the telescope pointing axis, $\bar{A}_{\text{eff},j}$ is computed scaling A_{tot} by the ratio between events detected within spatial bin j and the total number of generated events. For reasons of economy of computational resources, $\bar{A}_{\text{eff},j}$ is computed in some of the analyses described here approximating $\frac{d\mathcal{J}}{d\Omega}$ by a point-like source (i.e., by a delta function), even for the analysis of moderately extended dSphs. This approximation is less accurate the more extended the source. The bias introduced in $\bar{A}_{\text{eff},j}$ becomes relevant when the source extension is comparable to or bigger than the region for which A_{eff} may be considered flat.

5.2.1. H.E.S.S

The first dark matter searches using observations of dSphs with the H.E.S.S telescopes were based on an event-counting analysis, with no attempt to use the expected spectral and morphological signatures in the search [57,58]. H.E.S.S also performed early searches on non-confirmed dSphs like Canis Major [59] or even globular clusters [60]. Their most recent searches use state-of-the-art analysis techniques like the one described in Section 3, and are based on observations of Sagittarius (~90 h), Coma Berenices (~9 h), Fornax (~6 h), Carina (~23 h) and Sculptor (~13 h) dSphs, where H.E.S.S has searched for both continuum [41] and line-like [44] dark matter spectra. We will concentrate in these two latter works.

There are significant differences in the high-level maximum-likelihood analyses used by H.E.S.S in their searches for continuum and spectral lines. In the search for continuum spectra the $\frac{dN_\gamma}{dE}$ was taken from analytical parameterizations [42], generally valid up to dark matter mass of 8 TeV; and the J-factors were estimated using the prescription by Martinez (2015) [39], assuming alternatively cuspy NFW [23], and cored Burkert [43] profiles. In the search for spectral lines, on the other hand, the $\frac{dN_\gamma}{dE}$ is trivial, and the J-factors were taken from the work by Geringer-Sameth et al. (2014) [7], which assumes a Zhao-Hernquist dark matter density profile [61]. The case of Sagittarius dSph was treated separately in both works, given that this galaxy is likely affected by tidal disruption [62], and therefore the J-factor calculation is subject to comparatively larger systematic uncertainties, not included in the likelihood analysis.

There are also slight differences in the likelihood function used by H.E.S.S in the continuum and spectral line dark matter searches. In both cases they use the likelihood function of Equation (22) without including the term accounting for the uncertainty in τ. In the case of continuum spectra, only one ($N_{\hat{p}'} = 1$) circular spatial bin centered at the position of each dSph was considered, whereas for spectral lines $N_{\hat{p}'} = 2$ or 3 (depending on the size of the considered dSph), concentric 0.1°-width ring-like spatial bins were used. The reason for this difference must be purely historical (given how the expected and measured spatial information is essentially common to both searches), probably in an attempt to increase the sensitivity by including more information in the likelihood analysis. The drawback of this approach, has already been discussed in Section 4: It can introduce a bias in s_{ij}, with an unquantified effect in the final sensitivity to α. In both analyses, for the computation of s_{ij} following Equation (16), the dependence of the effective area with $\hat{p}$ within the signal region and the effect of the PSF are ignored. We note that these two simplifications require opposite conditions: A_{eff} can be better approximated by a constant value for smaller signal regions, i.e., smaller dSphs, whereas the effect of the angular resolution in the distribution of measured events is smaller for larger dSphs. The effect in the final result of adopting these two simplifications is not quantified.

H.E.S.S found no significant gamma-ray signal in the observed dSphs, considered either individually or collectively, for any of the assumed emission spectra (continuum or spectral line). The maximum observed deviations from the null hypothesis are ~2.6σ for the continuum spectra search in Fornax, and ~1.2σ for the spectral-line search in Sagittarius. The exclusion limits for the annihilation cross-section for continuum spectra (see Figure 3-left) peak at dark matter masses of around 1–2 TeV, depending on the considered channel. Assuming a NFW density profile, the strongest constraint is provided by Sagittarius dwarf, with $\langle\sigma v\rangle^{\text{UL}_{95}} \sim 2 \times 10^{-23}$ cm^3 s^{-1} for a combination of W^+W^- and ZZ annihilation channels. The bounds resulting from the combination of all the observed dSphs are only marginally better because Sagittarius has, under the NFW-profile assumption, the largest by far J-factor among the considered dSph, and because it has been observed by H.E.S.S for significantly longer time than the rest of the dSphs. However, given that the value of the J-factor for Sagittarius is affected by large systematic uncertainties (on account of the possibility that the system is affected by tidal disruption), H.E.S.S has also provided constraints obtained from the combination of all the other dSphs, which results in the limit $\langle\sigma v\rangle^{UL_{95}} \sim 10^{-22}$ cm^3 s^{-1}, for the same annihilation channel. In the case of the search for spectral lines, limits do not depend significantly on the inclusion or not of Sagittarius (see Figure 3-right), since with the newer approach in the evaluation of the J-factor used in this work, the limits are dominated by Coma Berenices results. In the mass range between 400 GeV and 1 TeV, the obtained limit to the velocity-averaged cross section is $\langle\sigma v\rangle^{UL_{95}} \sim 3 \times 10^{-25}$ cm^3 s^{-1}.

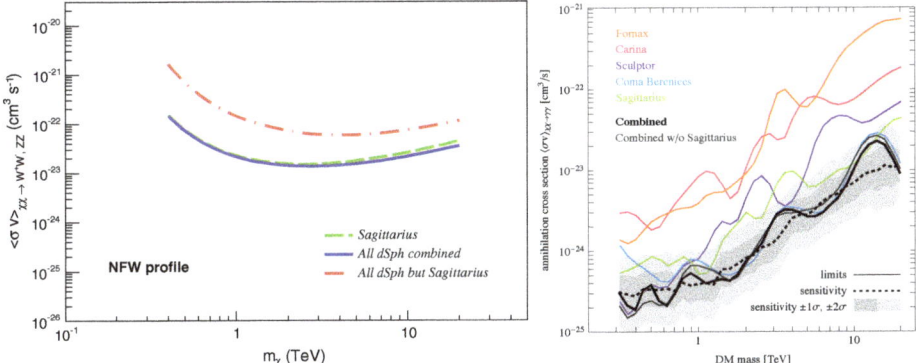

Figure 3. The 95% confidence level upper limits to the cross section of dark matter annihilating into a combination of W^+W^- and ZZ (**left**, reprinted figure with permission from reference [41]; copyright (2014) by the American Physical Society) and $\gamma\gamma$ pairs (**right**, reprinted figure with permission from reference [44], ©IOP Publishing Ltd. and Sissa Medialab; reproduced by permission of IOP Publishing; all rights reserved). Different lines show limits from individual dSphs and from their combination with and without Sagittarius. For the spectral line search, also the median of the distribution of limits obtained for simulated realizations of the null hypothesis is shown, together with the corresponding 1σ and 2σ symmetric quantiles.

5.2.2. MAGIC

MAGIC performed early dark matter searches using observations of the dSphs Draco [63], Willman 1 [64] and Segue 1 [65]. These searches had a relatively poor sensitivity, due to the fact that they were based on one-telescope observations and simple event-counting data analysis. With the addition of the second telescope, MAGIC dark matter search strategy was based on deep observations (~160 h) of the dSph with the highest J-factor known at that moment, namely Segue 1 [45], and the use for the first time by Cherenkov telescopes, of advanced maximum-likelihood analysis techniques. In addition, MAGIC Segue 1 observations were part of the aforementioned first multi-instrument combined search, together with data from Fermi-LAT [22], a work that I will discuss later in more detail. After that, MAGIC initiated a diversification of observed targets, starting by ~100 h of observations of the Ursa Major II dSph [46].

MAGIC dark matter searches in Segue 1 and Ursa Major II follow essentially the same data selection, calibration and processing procedures, but contain significant differences in several elements of their high-level analysis. The gamma-ray average spectra per annihilation reaction ($\frac{dN_\gamma}{dE}$) were obtained from the parameterization by Cembranos et al. (2011) [42] in the case of Segue 1, and the PPPC 4 DM ID computation [47] in the case of Ursa Major II. The spectra provided by these two works do not differ significantly for the considered energy range. The J-factor for Segue 1 was computed solving the Jeans equation assuming an Einasto density profile [24], and for Ursa Major II was taken from Geringer-Sameth et al. (2014) [7].

The likelihood function used by MAGIC for dark matter searches has also evolved over the years. For the observations of Segue 1 they used, instead of Equation (22), the following unbinned likelihood function:

$$\mathcal{L}_\gamma(\alpha \bar{J}; b \mid \{E'_i\}_{i=1,\ldots,N_{\mathrm{On}}}) = P\left(s(\alpha \bar{J}) + b \mid N_{\mathrm{On}}\right) \cdot \prod_{i=1}^{N_{\mathrm{On}}} f_{s+b}(E'_i) \quad, \tag{25}$$

where the uncertainty on τ is ignored, only one spatial bin is considered, and the energy-wise product of Poisson terms is substituted by a global Poisson term for the total number of observed events (N_{On}), times the joint likelihood for the observed values of estimated energies. The latter is computed as the product of the PDF for the reconstructed energy $f_{s+b}(E')$ evaluated at each observed E', where $f_{s+b} = \frac{1}{s+b}(s f_s + b f_b)$, with f_s and f_b the PDFs for the reconstructed energies for signal and background events, and s (the free parameter) and b (a nuisance parameter) the total expected number of signal and background events, respectively. f_s is the normalized convolution of the gamma-ray spectrum with the IRF, i.e.:

$$f_s(E') = \frac{T_{\mathrm{obs}}}{s} \int_0^\infty dE \, \frac{d\Phi}{dE} \, \bar{A}_{\mathrm{eff}}(E) \, f_E(E'|E) \quad, \tag{26}$$

with $\bar{A}_{\mathrm{eff}}(E)$ computed following Equation (24). f_b, on the other hand, is modeled using the data from one or several Off regions. This approach presents the drawback of neglecting the statistical and systematic uncertainties in the background spectral shape. In comparison, in the binned version of the likelihood function (Equation (22)) the statistical uncertainty is taken into account by the inclusion of the nuisance parameters b_{ij} and τ. This unbinned analysis hence typically produces results that are several tens of percent artificially more constraining than the binned one. Another important difference of the MAGIC Segue 1 analysis with respect to the general framework is that it does not include statistical uncertainties in the J-factor. This was justified by the fact that the bounds to α scale with $1/\bar{J}$, and therefore the provided results allow the computation of the limits for any other J-factor value (provided $\frac{dJ}{d\Omega}$ is kept fixed). This argument is valid only for single-target observations, but not for results obtained combining observations from different dSphs with different $\bar{J}$ values and uncertainties. Another main difference between this analysis and the general framework is the treatment of the cases when the value $\hat{\alpha}$ maximizing the likelihood lies outside the physical region, i.e., $\hat{\alpha} < 0$. For those cases, the 95% confidence limit on α was computed as $\alpha^{\mathrm{UL}_{95}} = \alpha_{2.71} - \hat{\alpha}$, with α unrestricted (i.e., allowed to take negative values) during the likelihood maximization process. With this prescription, the limit obtained for any negative fluctuation in the number of excess events is equal to the limit for zero excess events (i.e., the sensitivity), at the expense of some over-coverage (i.e., the bounds are conservative).

In the analysis of Ursa Major II data, MAGIC used the general analysis framework described in Section 4, with binned likelihood, statistical uncertainties in the J-factor considered, and α restricted to positive values. In addition, for the first time in the analysis of Cherenkov telescope data, the Off/On exposure ratio τ in Equation (22) was considered a nuisance parameter, taking into account both its statistical and systematic ($\sigma_{\tau,\mathrm{sys}} = 1.5\%$) uncertainties, thus providing more realistic results.

MAGIC found no significant gamma-ray signal in the observations of Segue 1 or Ursa Major II. This was translated into limits to the dark matter annihilation cross section (and decay lifetime), assuming different dark matter induced gamma-ray production mechanisms. Using Segue 1, MAGIC carried out a systematic search for annihilation and decay processes, looking for the continuum spectra from production of $b\bar{b}$, $t\bar{t}$, $\mu^+\mu^-$, $\tau^+\tau^-$, W^+W^- and ZZ pairs, spectral lines from $\gamma\gamma$ and γZ channels, and other spectral features such as those produced by virtual internal bremsstrahlung emission ($XX\gamma$) and gamma-ray "boxes" ($\Phi\Phi \to \gamma\gamma\gamma\gamma$). With Ursa Major II data, the searches were limited to annihilation into $b\bar{b}$, $\mu^+\mu^-$, $\tau^+\tau^-$ and W^+W^- pairs. Figure 4 shows the results for annihilation into $b\bar{b}$ pairs obtained from each of the observed dSphs (there is no MAGIC-only combined result). The obtained limits are in general within the 68% containment region expected for the null hypothesis, except for the low mass range $m_\chi \lesssim 300$ GeV in the case of Segue 1, where they stay nevertheless within the 95% containment region. 95% confidence level upper limits to the annihilation cross-section of dark matter particles into $b\bar{b}$ pairs reach $\langle\sigma v\rangle^{\text{UL}_{95}} \sim 5 \times 10^{-24}$ cm^3 s^{-1} for $m_\chi \sim 2$ TeV in the case of Segue 1, and $\langle\sigma v\rangle^{\text{UL}_{95}} \sim 2 \times 10^{-23}$ cm^3 s^{-1} in the case of Ursa Major II. Segue 1 observations were also used to constrain the lifetime of $m_\chi \sim 20$ TeV particles decaying into $b\bar{b}$ pairs to be larger than $\tau_\chi^{\text{LL}_{95}} \sim 3 \times 10^{25}$ s.

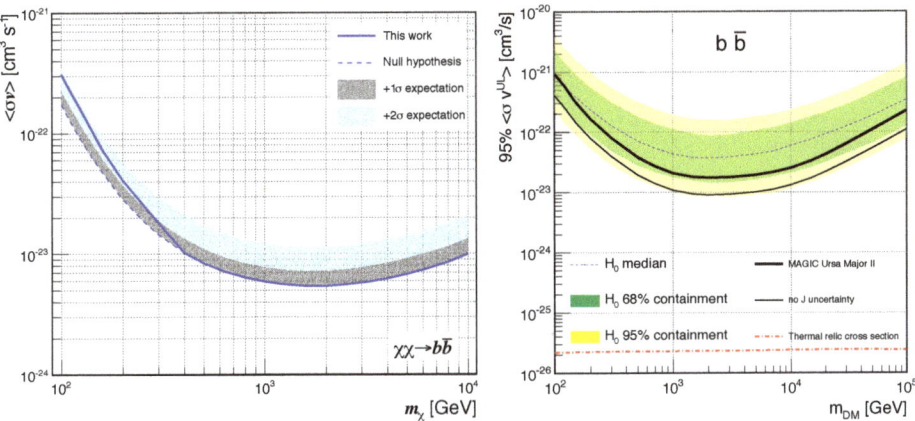

Figure 4. The 95% confidence level upper limits to $\langle\sigma v\rangle$ (solid line) for the process $\chi\chi \to b\bar{b}$ from observations of the dSphs Segue 1 (**left**, reprinted figure under CC BY license from reference [45]) and Ursa Major II (**right**, reprinted figure with permission from reference [46], ©IOP Publishing Ltd. and Sissa Medialab; reproduced by permission of IOP Publishing; all rights reserved) with MAGIC; also shown are the median of the distribution of limits for the null-hypothesis, and the limits of the symmetric 68% and 95% quantiles. For Ursa Major II, both the results with and without considering $\bar{J}$ statistical uncertainty are shown.

5.2.3. VERITAS

VERITAS has performed dark matter searches using observations of the dSphs Segue 1 (92 h), Draco (50 h), Ursa Minor (60 h), Boötes 1 (14 h) and Willman 1 (14 h). For early observations [66,67], they used a simple event counting analysis approach. More recently, they analyzed their full datasets and combined them using advanced analysis techniques [48].

In this latter work, the average gamma-ray spectra ($\frac{dN_\gamma}{dE}$) for the investigated dark matter annihilation channels were taken from the PPPC 4 DM ID computation [47], and the differential J-factors from Geringer-Sameth et al. (2014) [7]. For the high-level, statistical data analysis, VERITAS used a test statistic equivalent to the ratio of the following likelihood function [68], namely:

$$\mathcal{L}_\gamma(\alpha \mid \{E'_i, \theta'_i\}_{i=1,\ldots,N_{\text{On}}}) = \prod_{i=1}^{N_{\text{On}}} f_{s+b}(E'_i, \theta'_i) \quad . \tag{27}$$

This likelihood function is similar to the one used by MAGIC in the Segue 1 analysis (Equation (25)). They are both unbinned simplified versions of the general likelihood function for Cherenkov telescopes shown in Equation (22). With respect to the MAGIC Segue 1 likelihood function, in Equation (27) the external Poisson term for the total number of observed events is omitted, and the event-wise term consists in the evaluation of the 2-dimensional PDF for the measured energy E' and the angular separation θ' between the measured arrival direction and the dSph center. We remind the reader that $f_{s+b} = \frac{1}{s+b}(s f_s + b f_b)$. In the 2-dimensional case, assuming that the convolution of the gamma-ray distribution with the IRF is radially symmetric with respect to the center of the dSph (i.e., the dependence on $\hat{p}'$ reduces to a dependence on θ'), then $f_s(E', \theta')$ is given by:

$$f_s(E', \theta') = \frac{2\pi \theta' T_{\text{obs}}}{s} \int_0^\infty dE \int_{\Delta \Omega_{\text{tot}}} d\Omega \, \frac{d^2 \Phi}{dE \, d\Omega} \, \text{IRF}(E', \hat{p}' | E, \hat{p}) \quad . \tag{28}$$

Only events in an On region defined by a maximum distance of $\theta'_{\text{cut}} = 0.17°$ from the center of the dSphs are considered, and the dependence of the effective area A_{eff} on the arrival direction $\hat{p}$ for events passing such cut is ignored. The dependence of f_b on E' is modeled by smearing the distribution of E' measured for events of the background-control (Off) region, whereas the spatial distribution is assumed to be uniform within the On region. Both b and f_b are fixed during the likelihood maximization, i.e., no statistical or systematic uncertainties in the background estimation are considered. Moreover the J-factor uncertainty is not included in the likelihood. Instead, the effect of the uncertainty in J is quantified by repeating the limit calculation over an ensemble of dark matter halo realizations using, for each dSph, halo parameter values randomly chosen from their inferred PDFs, and reporting the 68% confidence level containment quantiles of the obtained distribution of results[5]. However, the main reported result in this case is still the median of such distribution, which is only sensitive to the central J-factor and not to its uncertainty, producing limits a factor ~ 2 more constraining than if $\bar{J}$ was considered a nuisance parameter.

A possible advantage of the use of the likelihood function of Equation (27) is that it allows a relatively simple estimation of the PDF for the associated $-2 \ln \lambda_P$ test statistic [68] directly from the data and without relying on the validity of the conditions of the Wilks' theorem. This is so, because $-2 \ln \lambda_P$ can be expressed as the sum of two random variables (those corresponding to the signal and background contributions to $-2 \ln \lambda_P$, respectively), which, for the likelihood function of Equation (27), are distributed according to a *compound Poisson* distribution. VERITAS results are hence robust in the sense that have a well determined confidence level under the assumption that the likelihood function was correct.

VERITAS has not found evidence of dark matter signals from neither of the four considered dSphs individually, or combined in a joint analysis. The null-hypothesis significance is well within the $\pm 2\sigma$ quantile, for all considered targets, annihilation channels ($u\bar{u}$, $d\bar{d}$, $s\bar{s}$, $b\bar{b}$, $t\bar{t}$, e^+e^-, $\mu^+\mu^-$, $\tau^+\tau^-$, W^+W^-, ZZ and hh) and m_χ values, except for $m_\chi \geq 5$ TeV dark matter particles annihilating into $\gamma\gamma$ in Draco dSph. In this latter case, a negative fluctuation slightly below -2σ is observed, which is not incompatible with purely statistical fluctuations, or could be alternatively explained by unaccounted systematic uncertainties in the background estimation. Figure 5 shows VERITAS limits to the annihilation cross-section into $b\bar{b}$ and $\tau^+\tau^-$ pairs, compared with other limits from dSph observations by other gamma-ray instruments. The constraints reach $\langle \sigma v \rangle^{\text{UL}_{95}} \sim 10^{-23}$ cm^3 s^{-1} at $m_\chi \sim 1$ TeV for $b\bar{b}$, and $\langle \sigma v \rangle^{\text{UL}_{95}} \sim 3 \times 10^{-24}$ cm^3 s^{-1} at ~ 300 GeV for $\tau^+\tau^-$ annihilation channels, respectively.

---

[5] That is: limits, which are one-sided confidence intervals, are provided with error bars, which are two-sided confidence intervals. Some authors [69] have described graphically the potentially pernicious consequences of extending this practice.

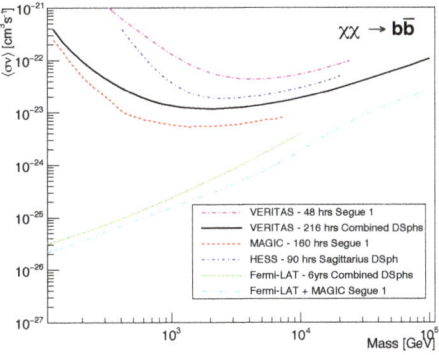 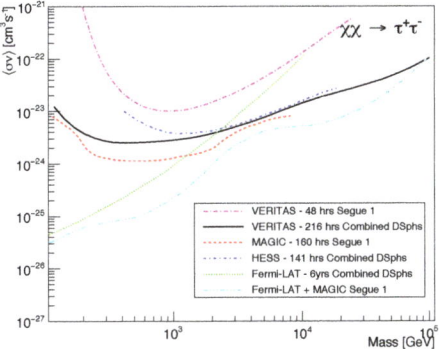

Figure 5. The 95% confidence-level upper limits to the dark matter annihilation cross-section into $b\bar{b}$ (**left**) and $\tau^+\tau^-$ (**right**) pairs, obtained from dSph observations by VERITAS (black solid line), compared with results from other gamma-ray instruments (see legend for the details). Reprinted figure with permission from reference [48]; copyright (2017) by the American Physical Society.

5.3. HAWC

HAWC has searched for dark matter annihilation and decay signals in 15 dSphs observed during 507 days between November 2014 and June 2016 [49]. They computed the average gamma-ray spectra per annihilation or decay event ($\frac{dN_\gamma}{dE}$) using the PYTHIA v8.2 simulation package [50], and the J-factors using the CLUMPY software package [51], assuming NFW [23] dark matter density profiles. The searches were carried out using the binned likelihood function described in Equation (15). Data were binned in reconstructed energy E' (referred to as f_{hit} in HAWC publications [14]) covering the range between 500 GeV and 100 TeV, and in reconstructed arrival direction $\hat{p}'$, covering an area of 5° radius around each of the analyzed dSphs. The computation of the signal events s_{ij} in each bin was performed using Monte Carlo simulations of the whole observations, assuming point-like sources and a reference value of α, and scaling the result for any other needed value, which is equivalent to using Equation (16).

No nuisance parameters accounting for uncertainties in the background estimation were considered, i.e., no $\mathcal{L}_\mu$ term was included in the Equation (15) likelihood function. The values b_{ij} were estimated from the measured number of events in the same bin of local (or detector) coordinates at times when such coordinates do not correspond to any of the analyzed dSphs or any known HAWC sources. Measured background rates at each local spatial bin were then normalized using the all-sky event rate measured in 2-h intervals. Using this method, the statistics used for background estimation correspond to an Off/On exposure ratio factor of $\tau = 30$–300 [70], and the related statistic uncertainties (included in the case of Cherenkov telescopes by the second Poisson term in Equation (22)), can therefore be safely neglected. However, the effect of the systematic uncertainty associated to this method is not quantified or taken into account in the analysis. In addition, similarly to the case of VERITAS, HAWC does also not include in the maximum likelihood analysis the statistical uncertainty in the J-factor, i.e., they ignore the $\mathcal{L}_J$ term in Equation (13). They do quantify the impact on the limits caused by the consideration of the dSphs as point-like sources and by several detector effects not perfectly under control in the Monte Carlo simulations used for calibrating the detector.

HAWC has not found gamma-rays associated to dark matter annihilation or decay from the examined dSphs, considered either individually or collectively. The significance of rejection of the null hypothesis for all considered targets, channels ($b\bar{b}$, $t\bar{t}$, $\tau^+\tau^-$, W^+W^- and $\mu^+\mu^-$), and m_χ values (between 1 and 100 TeV) is within 2σ, except for few marginally larger negative fluctuations. Figure 6 shows the limits to the annihilation cross section obtained by HAWC for the $b\bar{b}$ and $\tau^+\tau^-$ annihilation channels, compared to limits obtained by other gamma-ray instruments. Limits reach $\langle\sigma v\rangle^{UL_{95}} \sim 10^{-23}$ cm^3 s^{-1} at $m_\chi \sim 3$ TeV for $b\bar{b}$, and $\langle\sigma v\rangle^{UL_{95}} \sim 2 \times 10^{-24}$ cm^3 s^{-1} at ~ 1 TeV for $\tau^+\tau^-$ annihilation channels, respectively. For decay, lower limits to the decay lifetime were set to $\tau_\chi^{LL_{95}} \sim 3 \times 10^{26}$ s for the 100 TeV mass dark matter particle decaying into $b\bar{b}$ pairs or $\tau_\chi^{LL_{95}} \sim 10^{27}$ s for decaying into $\tau^+\tau^-$ pairs.

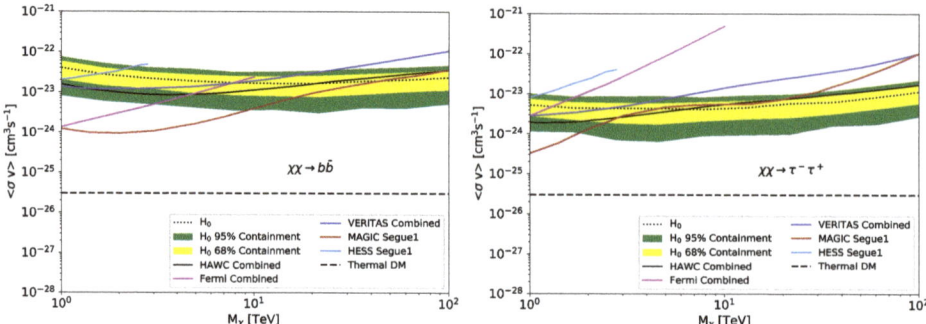

Figure 6. The 95% confidence level upper limits to the annihilation cross-section of dark matter particles annihilating into $b\bar{b}$ (**left**) and $\tau^+\tau^-$ (**right**) pairs, from HAWC observations of dSphs (black solid line). Results from other gamma-ray instruments are also shown (see legend for details), as well as the median and 65% and 95% symmetric quantiles of the distribution of limits obtained under the null hypothesis. Figure reproduced with permission from reference [49], ©AAS.

5.4. Multi-Instrument Searches

Following Equations (13) and (14), MAGIC and Fermi-LAT have computed a multi-target, multi-instrument, joint likelihood, producing the first coherent joint search for gamma-ray signals from annihilation of dark matter particles in the mass range between 10 GeV and 100 TeV [22]. The data used in this work correspond to the Fermi-LAT 6-years [21] and the MAGIC Segue 1 [45] observations discussed earlier in Sections 5.1 and 5.2.2, respectively. MAGIC analysis was slightly adapted to match LAT conventions, in the following aspects: (i) The determination of the J-factor; (ii) the treatment of the statistical uncertainty of $\bar{J}$ through the $\mathcal{L}_J$ term in Equation (13); and (iii) the treatment of the cases in which the limits lie outside the physical ($\alpha \geq 0$) region.

The MAGIC/Fermi-LAT combined search for dark matter did not produced a positive signal, but it allowed setting global limits to the dark matter annihilation cross section and, for the first time, a meaningful comparison of the individual results obtained with the two instruments. Figure 7 shows the 95% confidence level limits to the cross-section of dark matter particles of mass in the range between 10 GeV and 100 TeV annihilating into $b\bar{b}$ and $\tau^+\tau^-$ pairs. The obtained limits are the currently most constraining results from dSphs, and span the widest interval of masses, covering the whole WIMP range. In the regions of mass where Fermi-LAT and MAGIC achieve comparable sensitivities, the improvement of the combined result with respect to those from individual instruments reaches a factor ~ 2.

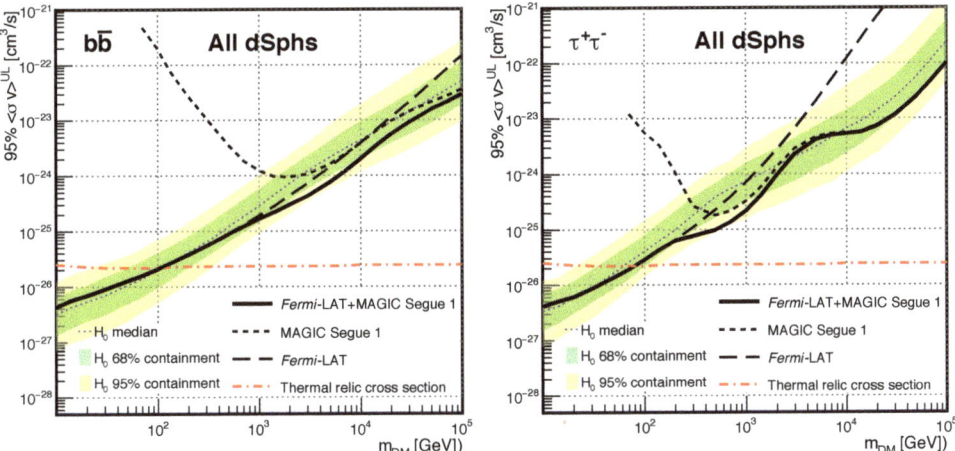

Figure 7. The 95% confidence level upper limits to the cross-section for dark matter particles annihilating into $b\bar{b}$ (**left**) and $\tau^+\tau^-$ (**right**) pairs. Thick solid lines show the limits obtained by combining Fermi-LAT observations of 15 dSphs with MAGIC observations of Segue 1. Dashed lines show the limit obtained individually by MAGIC (short dashes) and Fermi-LAT (long dashes), respectively. The thin-dotted line, green and yellow bands show, respectively, the median and the two-sided 68% and 95% symmetric quantiles for the distribution of limits under the null hypothesis. Reprinted figure with permission from reference [22], ©IOP Publishing Ltd. and Sissa Medialab; reproduced by permission of IOP Publishing; all rights reserved.

This approach is applicable to all the high-energy gamma-ray instruments (and also to high energy neutrino telescopes, with slight modifications in Equation (16) to account for the oscillations). The so-called Glory Duck working group has initiated an activity aimed at the combination of all dark matter searches performed with Fermi-LAT, H.E.S.S, MAGIC, VERITAS and HAWC using observations of dSphs [71]. Each collaboration will analyze their own datasets and will provide the likelihood values as a function of the free parameter α (i.e., the terms $\mathcal{L}_{\gamma,k}$ in Equation (14)) for the different considered annihilation channels and m_χ values, for their combination and J-factor profiling through Equation (13). Likelihood values from the different instruments will be computed using the same conventions for the computation of the gamma-ray spectra and the J-factors, as well as the same statistical treatment of the data, most notably a common consideration of all relevant uncertainties by the inclusion of the corresponding nuisance parameters in the likelihood functions. While in principle foreseen only for the combination of gamma-ray data in the search of annihilation signals, this work could pave the path for other combined searches, such as searches for decay signals, the inclusion of other kinds of targets or even extending the searches to include also results from neutrino telescopes. This approach will ensure that all the combined individual results will be directly comparable among them, and will produce the legacy result of the dark matter searches using the current generation of gamma-ray instruments.

Funding: This work is partially funded by grant FPA2017-87859-P from Ministerio de Economía, Industria y Competitividad (Spain).

Conflicts of Interest: The author declares no conflict of interest.

References

1. Zwicky, F. Die Rotverschiebung von extragalaktischen Nebeln. *Helv. Phys. Acta* **1933**, *6*, 110.
2. Babcock, H.W. The rotation of the Andromeda Nebula. *Lick Obs. Bull.* **1939**, *19*, 41. [CrossRef]

3. Aghanim, N.; Akrami, Y.; Arroja, F.; Ashdown, M.; Aumont, J.; Baccigalupi, C.; Ballardini, M.; Banday, A.J.; Barreiro, R.B.; Bartolo, N.; et al. Planck 2018 results. I. Overview and the cosmological legacy of Planck. *Astron. Astrophys.* **2020**, in press.
4. Hut, P. Limits on Masses and Number of Neutral Weakly Interacting Particles. *Phys. Lett.* **1977**, *B69*, 85. [CrossRef]
5. Bergstrom, L. Non-Baryonic Dark Matter—Observational Evidence and Detection Methods. *Rept. Prog. Phys.* **2000**, *63*, 793. [CrossRef]
6. Dubinski, J.; Carlberg, R.G. The structure of cold dark matter halos. *Astrophys. J.* **1991**, *378*, 496–503. [CrossRef]
7. Geringer-Sameth, A.; Koushiappas, S.M.; Walker, M. Dwarf galaxy annihilation and decay emission profiles for dark matter experiments. *Astrophys. J.* **2015**, *801*, 74. [CrossRef]
8. Strigari, L.E.; Bullock, J. S.; Kaplinghat, M.; Simon, J.D.; Geha, M.; Willman, B.; Walker, M.G. A common mass scale for satellite galaxies of the Milky Way. *Nature* **2008**, *454*, 1096. [CrossRef]
9. Sjöstrand, T.; Mrenna, S.; Skands, P.Z. A Brief Introduction to PYTHIA 8.1. *Comput. Phys. Commun.* **2007**, *178*, 852–867. [CrossRef]
10. Fermi LAT Performance. Available online: http://www.slac.stanford.edu/exp/glast/groups/canda/lat_Performance.htm (accessed on 3 March 2020).
11. Holler, M.M.; Balzer, A.; Chalmé-Calvet, R. de Naurois, M.; Zaborov, D. Photon Reconstruction for H.E.S.S. Using a Semi-Analytical Model. *PoS* **2015**, *ICRC2015*, 980.
12. Aleksić, J.; Ansoldi, S.; Antonelli, L.A.; Antoranz, P.; Babic, A.; Bangale, P.; Barcelo, M.; Barrio, J.A.; Becerra Gonzalez, J.; Bednarek, W.; et al. The major upgrade of the MAGIC telescopes, Part II: A performance study using observations of the Crab Nebula. *Astropart. Phys.* **2016**, *72*, 76-94 [CrossRef]
13. VERITAS Specifications. Available online: http://veritas.sao.arizona.edu/about-veritas-mainmenu-81/veritas-specifications-mainmenu-111 (accessed on 3 March 2020).
14. Abeysekara, A.U.; Albert, A.; Alfaro, R.; Alvarez, C.; Álvarez, J.D.; Arceo, R.; Arteaga-Velázquez, J.C.; Ayala Solares, H.A.; Barber, A.S.; Bautista-Elivar, N.; et al. Observation of the Crab Nebula with the HAWC Gamma-Ray Observatory. *Astrophys. J.* **2017**, *843*, 39 [CrossRef]
15. Acharyya, A.; Agudo, I.; Angüner, E.O.; Alfaro, R.; Alfaro, J.; Alispach, C.; Aloisio, R.; Alves Batista, R.; Amans, J.-P.; Amati, L.; et al. Monte Carlo studies for the optimisation of the Cherenkov Telescope Array layout. *Astropart. Phys.* **2019**, *111*, 35. [CrossRef]
16. Strigari, L.E.; Koushiappas, S.M.; Bullock, J.S.; Kaplinghat, M. Precise constraints on the dark matter content of Milky Way dwarf galaxies for gamma-ray experiments. *Phys. Rev.* **2007**, *D75*, 083526 [CrossRef]
17. Atwood, W.B.; Abdo, A.A.; Ackermann, M.; Althouse, W.; Anderson, B.; Axelsson, M.; Baldini, L.; Ballet, J.; Band, D.L.; Barbiellini, G.; et al. The Large Area Telescope on the Fermi Gamma-ray Space Telescope Mission. *Astrophys. J.* **2009**, *697*, 1071. [CrossRef]
18. Fermi-LAT Collaboration. Fermi Large Area Telescope Fourth Source Catalog. *arXiv* **2009**, arXiv:1902:10045.
19. Tanabashi, M.; Hagiwara, K.; Hikasa, K.; Nakamura, K.; Sumino, Y.; Takahashi, F.; Tanaka, J.; Agashe, K.; Aielli, G.; Amsler, C.; et al. (Particle Data Group), The Review of Particle Physics (2018). *Phys. Rev.* **2018**, *D98*, 030001.
20. Nievas-Rosillo, M.; Contreras, J.L. Extending the Li&Ma method to include PSF information. *Astropart. Phys.* **2016**, *74*, 51–57.
21. Ackermann, M.; Albert, A.; Anderson, B.; Atwood, W.B.; Baldini, L.; Barbiellini, G.; Bastieri, D.; Bechtol, K.; Bellazzini, R.; Bissaldi, E.; et al. Searching for Dark Matter Annihilation from Milky Way Dwarf Spheroidal Galaxies with Six Years of Fermi Large Area Telescope Data. *Phys. Rev. Lett.* **2015**, *115*, 231301. [CrossRef]
22. Ahnen, M.L.; Ansoldi, S.; Antonelli, L.A.; Antoranz, P.; Babic, A.; Banerjee, B.; Bangale, P.; Barres de Almeida, U.; Barrio, J.A.; Becerra González, J.; et al. Limits to dark matter annihilation cross-section from a combined analysis of MAGIC and Fermi-LAT observations of dwarf satellite galaxies. *J. Cosmol. Astropart. Phys.* **2016**, *1602*, 39. [CrossRef]
23. Navarro, J.F.; Frenk, C.S.; White, S.D.M. A Universal Density Profile from Hierarchical Clustering. *Astrophys. J.* **1997**, *490*, 493–508. [CrossRef]
24. Essig, R.; Sehgal, N.; Strigari, L.E.; Geha, M.; Simon, J.D. Indirect Dark Matter Detection Limits from the Ultra-Faint Milky Way Satellite Segue 1. *Phys. Rev.* **2010**, *D82*, 123503.
25. Chernoff, H. On the Distribution of the Likelihood Ratio. *Ann. Math. Stat.* **1954**, *25*, 573–578. [CrossRef]

26. Mazziotta, M.N.; Loparco, F.; de Palma, F.; Giglietto, N. A model-independent analysis of the Fermi Large Area Telescope gamma-ray data from the Milky Way dwarf galaxies and halo to constrain dark matter scenarios. *Astropart. Phys.* **2012**, *37*, 26–39. [CrossRef]
27. Drlica-Wagner, A.; Albert, A.; Bechtol, K.; Wood, M.; Strigari, L.; Sánchez-Conde, M.; Baldini, L.; Essig, R.; Cohen-Tanugi, J.; Anderson, B.; et al. Search for Gamma-Ray Emission from DES Dwarf Spheroidal Galaxy Candidates with Fermi-LAT Data. *Astrophys. J. Lett.* **2015**, *809*, 4. [CrossRef]
28. Baushev, A.N.; Federici, S.; Pohl, M. Spectral analysis of the gamma-ray background near the dwarf Milky Way satellite Segue 1: Improved limits on the cross section of neutralino dark matter annihilation. *Phys. Rev.* **2012**, *D86*, 063521. [CrossRef]
29. Cotta, R.C.; Drlica-Wagner, A.; Murgia, S.; Bloom, E.D.; Hewett, J.L.; Rizzo, T.G. Constraints on the pMSSM from LAT Observations of Dwarf Spheroidal Galaxies. *J. Cosmol. Astropart. P.* **2012**, *1204*, 016. [CrossRef]
30. Scott, P.; Conrad, J.; Edsjö, J.; Bergström, L.; Farnier, C.; Akrami, Y. Direct constraints on minimal supersymmetry from Fermi-LAT observations of the dwarf galaxy Segue 1. *J. Cosmol. Astropart. P.* **2010**, *1001*, 031. [CrossRef]
31. Hoof, S.; Geringer-Sameth, A.; Trotta, R. A Global Analysis of Dark Matter Signals from 27 Dwarf Spheroidal Galaxies using 11 Years of Fermi-LAT Observations. *J. Cosmol. Astropart. P.* **2020**, *2002*, 012. [CrossRef]
32. Li, S.; Liang, Y.-F.; Xia, Z.-Q.; Zu, L.; Duan, K.-K.; Shen, Z.-Q.; Feng, L.; Yuan, Q.; Fan, Y.-Z. Study of the boxlike dark matter signals from dwarf spheroidal galaxies with Fermi-LAT data. *Phys. Rev.* **2018**, *D97*, 083007. [CrossRef]
33. Zhao, Y.; Bi, X.-J.; Yin, P.-F.; Zhang, X. Constraint on the velocity dependent dark matter annihilation cross section from Fermi-LAT observations of dwarf galaxies. *Phys. Rev.* **2016**, *D93*, 083513. [CrossRef]
34. Baring, M.G.; Ghosh, T.; Queiroz, F.S.; Sinha, K. New Limits on the Dark Matter Lifetime from Dwarf Spheroidal Galaxies using Fermi-LAT. *Phys. Rev.* **2016**, *D93*, 103009. [CrossRef]
35. Abdo, A.A.; Ackermann, M.; Ajello, M.; Atwood, W.B.; Baldini, L.; Ballet, J.; Barbiellini, G.; Bastieri, D.; Bechtol, K.; Bellazzini, R.; et al. Observations of Milky Way Dwarf Spheroidal Galaxies with the Fermi-Large Area Telescope Detector and Constraints on Dark Matter Models. *Astrophys. J.* **2010**, *712*, 147–158. [CrossRef]
36. Ackermann, M.; Ajello, M.; Albert, A.; Atwood, W.B.; Baldini, L.; Ballet, J.; Barbiellini, G.; Bastieri, D.; Bechtol, K.; Bellazzini, R.; et al. Constraining Dark Matter Models from a Combined Analysis of Milky Way Satellites with the Fermi Large Area Telescope. *Phys. Rev. Lett.* **2011**, *107*, 241302. [CrossRef] [PubMed]
37. Ackermann, M.; Albert, A.; Anderson, B.; Baldini, L.; Ballet, J.; Barbiellini, G.; Bastieri, D.; Bechtol, K.; Bellazzini, R.; Bissaldi, E.; et al. Dark matter constraints from observations of 25 Milky Way satellite galaxies with the Fermi Large Area Telescope. *Phys. Rev. D* **2014**, *89*, 042001. [CrossRef]
38. Simon, J.D.; Geha, M. The Kinematics of the Ultra-Faint Milky Way Satellites: Solving the Missing Satellite Problem. *Astrophys. J.* **2007**, *670*, 313–331. [CrossRef]
39. Martinez, G.D. A Robust Determination of Milky Way Satellite Properties using Hierarchical Mass Modeling. *Mon. Not. R. Astron. Soc.* **2015**, *451*, 2524–2535. [CrossRef]
40. Jeltema, T.E.; Profumo, S. Fitting the Gamma-Ray Spectrum from Dark Matter with DMFIT: GLAST and the Galactic Center Region. *J. Cosmol. Astropart. Phys.* **2008**, *0811*, 003. [CrossRef]
41. Abramowski, A.; Aharonian, F.; Ait Benkhali, F.; Akhperjanian, A.G.; Angüner, E.; Backes, M.; Balenderan, S.; Balzer, A.; Barnacka, A.; Becherini, Y.; et al. Search for dark matter annihilation signatures in H.E.S.S. observations of dwarf spheroidal galaxies. *Phys. Rev. D.* **2014**, *90*, 112012. [CrossRef]
42. Cembranos, J.A.R.; de la Cruz-Dombriz, A.; Dobado, A. Lineros, R.A.; Maroto, A.L. Photon spectra from WIMP annihilation. *Phys. Rev.* **2011**, *D83*, 083507. [CrossRef]
43. Burkert, A. The Structure of Dark Matter Halos in Dwarf Galaxies. *Astrophys. J. Lett.* **1995**, *447*, L25–L28. [CrossRef]
44. Abdalla, H.; Aharonian, F.; Benkhali, F.A.; Angüner, E.O.; Arakawa, M.; Arcaro, C.; Armand, C.; Arrieta, M.; Backes, M.; Barnard, M.; et al. Searches for gamma-ray lines and 'pure WIMP' spectra from Dark Matter annihilations in dwarf galaxies with H.E.S.S. *J. Cosmol. Astropart. P* **2018**, *1811*, 037. [CrossRef]
45. Aleksić, J.; Ansoldi, S.; Antonelli, L.A.; Antoranz, P.; Babic, A.; Bangale, P.; Barres de Almeida, U.; Barrio, J.A.; Becerra González, J.; Bednarek, W.; et al. Optimized dark matter searches in deep observations of Segue 1 with MAGIC. *J. Cosmol. Astropart. Phys.* **2014**, *1406*, 008. [CrossRef]

46. Ahnen, M.; Ansoldi, S.; Antonelli, L.A.; Arcaro, C.; Baack, D.; Babić, A.; Banerjee, B.; Bangale, P.; Barres de Almeida, U.; Barrio, J.A.; et al. Indirect dark matter searches in the dwarf satellite galaxy Ursa Major II with the MAGIC Telescopes. *J. Cosmol. Astropart. Phys.* **2018**, *1803*, 009. [CrossRef]
47. Cirelli, M.; Corcella, G.; Hektor, A, Hütsi, G.; Kadastik, M.; Panci, P.; Raidal, M.; Sala, F.; and Strumia, A. PPPC 4 DM ID: A Poor Particle Physicist Cookbook for Dark Matter Indirect Detection. *J. Cosmol. Astropart. Phys.* **2011**, *1103*, 051. [CrossRef]
48. Archambault, S.; Archer, A.; Benbow, W.; Bird, R.; Bourbeau, E.; Brantseg, T.; Buchovecky, M.; Buckley, J.H.; Bugaev. V.; Byrum, K.; et al. Dark matter constraints from a joint analysis of dwarf Spheroidal galaxy observations with VERITAS. *Phys. Rev.* **2017**, *D95*, 082001. [CrossRef]
49. Albert, A.; Alfaro, R.; Alvarez, C.; Álvarez, J.D.; Arceo, R.; Arteaga-Velázquez, J.C.; Avila Rojas, D.; Ayala Solares, H.A.; Bautista-Elivar, N.; Becerril, A.; et al. Dark Matter Limits from Dwarf Spheroidal Galaxies with the HAWC Gamma-Ray Observatory. *Astrophys. J.* **2018**, *853*, 154. [CrossRef]
50. Sjöstrand, T.; Ask, S.; Christiansen, J.R.; Corke, R.; Desai, N.; Ilten, P.; Mrenna, S.; Prestel, S.; Rasmussen, C.O.; Skands, P.Z. An introduction to PYTHIA 8.2. *Comput. Phys. Commun.* **2015**, *191*, 159. [CrossRef]
51. Bonnivard, V.; Hütten, M.; Nezri, E.; Charbonnier, A.; Combet, E.; Maurin, D. CLUMPY: Jeans analysis, γ-ray and ν fluxes from dark matter (sub-)structures. *Comput. Phys. Commun.* **2016**, *200*, 336. [CrossRef]
52. Acero, F.; Ackermann, M.; Ajello, M.; Albert, A.; Atwood, W.B.; Axelsson, M.; Baldini, L.; Ballet, J.; Barbiellini, G.; Bastieri, D.; et al. Fermi Large Area Telescope Third Source Catalog. *Astrophys. J. Suppl. Ser.* **2015**, *218*, 23. [CrossRef]
53. Figures and Data Files Associated with the Fermi LAT Paper "Searching for Dark Matter Annihilation from Milky Way Dwarf Spheroidal Galaxies with Six Years of Fermi-LAT Data". Available online: https://www-glast.stanford.edu/pub_data/1048/ (accessed on 3 March 2020).
54. Steigman, G.; Dasgupta, B.; Beacom, J.F. Precise relic WIMP abundance and its impact on searches for dark matter annihilation. *Phys. Rev.* **2012**, *D86*, 023506. [CrossRef]
55. Albert, A.; Anderson, B.; Bechtol, K.; Drlica-Wagner, A.; Meyer, M.; Sánchez-Conde, M.; Strigari, L.; Wood, M.; Abbott, T.M.C.; Abdalla, F.B.; et al. Searching for Dark Matter Annihilation in Recently Discovered Milky Way Satellites with Fermi-LAT. *Astrophys. J.* **2017**, *834*, 110. [CrossRef]
56. Aleksić, J.; Rico, J.; Martinez, M. Optimized analysis method for indirect dark matter searches with imaging air Cherenkov telescopes. *J. Cosmol. Astropart. Phys.* **2012**, *1210*, 32. [CrossRef]
57. Aharonian, F.; Akhperjanian, A.G.; Bazer-Bachi, A.R.; Beilicke, M.; Benbow, W.; Berge, D.; Bernlöhr, K.; Boissont, C.; Bolza, O.; Borrel, V.; et al. Observations of the Sagittarius dwarf galaxy by the HESS experiment and search for a dark matter signal. *Astropart. Phys.* **2008**, *29*, 55–62. [CrossRef]
58. Abramowski, A.; Acero, F.; Aharonian, F.; Akhperjanian, A.G.; Anton, G.; Barnacka, A.; Barres de Almeida, U.; Bazer-Bachi, A.R.; Becherini, Y.; Becker, J.; et al. H.E.S.S. constraints on dark matter annihilations towards the sculptor and carina dwarf galaxies. *Astropart. Phys.* **2011**, *34*, 608–616. [CrossRef]
59. Aharonian, F.; Akhperjanian, A.G.; Barres de Almeida, U.; Bazer-Bachi, A.R.; Behera, B.; Benbow, W.; Bernlöhr, K.; Boisson, C.; Bochow, V.; Borrel, V.; et al. A Search for Dark Matter Annihilation Signal toward the Canis Major Overdensity with H.E.S.S. *Astrophys. J.* **2009**, *691*, 175–181. [CrossRef]
60. Abramowski, A.; Acero, F.; Aharonian, F.; Akhperjanian, A.G.; Anton, G.; Balzer, A.; Barnacka, A.; Barres de Almeida, U.; Bazer-Bachi, A.R.; Becherini, Y.; et al. H.E.S.S. Observations of the Globular Clusters NGC 6388 and M15 and Search for Dark Matter Signal. *Astrophys. J.* **2011**, *735*, 12. [CrossRef]
61. Zhao, H. Analytical models for galactic nuclei. *Mon. Not. R. Astron. Soc.* **1996**, *278*, 488. [CrossRef]
62. Ibata, R.; Irwin, M.; Lewis, G.F.; Stolte, A. Galactic Halo Substructure in the Sloan Digital Sky Survey: The Ancient Tidal Stream from the Sagittarius Dwarf Galaxy. *Astrophys. J. Lett.* **2001**, *547*, L133–L136. [CrossRef]
63. Albert, J.; Aliu, E.; Anderhub, H.; Antoranz, P.; Backes, M.; Baixeras, C.; Barrio, J.A.; Bartko, H.; Bastieri, D.; Becker, J.K.; et al. Upper limit for gamma-ray emission above 140 GeV from the dwarf spheroidal galaxy Draco. *Astrophys. J.* **2008**, *679*, 428. [CrossRef]
64. Aliu, E.; Anderhub, H.; A. Antonelli, L.; Antoranz, P.; Backes, M.; Baixeras, C.; Balestra, S.; Barrio, J.A.; Bartko, H.; Bastieri, D.; et al. MAGIC upper limits on the VHE gamma-ray emission from the satellite galaxy Willman 1. *Astrophys. J.* **2009**, *697*, 1299. [CrossRef]
65. Aleksić, J.; Alvarez, E.A.; Antonelli, L.A.; Antoranz, P.; Asensio, M.; Backes, M.; Barrio, J.A.; Bastieri, D.; Becerra González, J.; Bednarek, W.; et al. Searches for dark matter annihilation signatures in the Segue 1 satellite galaxy with the MAGIC-I telescope. *J. Cosmol. Astropart. Phys.* **2011**, *1106*, 035. [CrossRef]

66. Acciari, V.A.; Arlen, T.; Aune, T.; Beilicke, M.; Benbow, W.; Boltuch, D.; Bradbury, S.M.; Buckley, J.H.; Bugaev, V.; Byrum, K.; et al. VERITAS Search for the Gamma-ray Emission from Dwarf Spheroidal Galaxies. *Astrophys. J.* **2010**, *720*, 1174–1180. [CrossRef]
67. Aliu, E.; Archambault, S, Arlen, T.; Aune, T.; Beilicke, M.; Benbow, W.; Bouvier A.; Bradbury, S.M.; Buckley, J.H.; Bugaev, V.; et al. VERITAS deep observations of the dwarf spheroidal galaxy Segue 1. *Phys. Rev. D* **2012**, *85*, 062001. [CrossRef]
68. Geringer-Sameth, A.; Koushiappas, S.M.; Walker, M.G. A Comprehensive Search for Dark Matter Annihilation in Dwarf Galaxies. *Phys. Rev. D* **2015**, *91*, 083535. [CrossRef]
69. Error Bars. Available online: https://xkcd.com/2110/ (accessed on 3 March 2020).
70. Harding, P. *Private Communication*; Physics Division, Los Alamos National Laboratory: Los Alamos, NM, USA, 2019.
71. Oackes, L.; Armand, C.; Charles, E.; di Mauro, M.; Giuri, C.; Harding, J.P.; Kerszberg, D.; Miener, T.; Moulin, E.; Poireau, V.; et al. Combined Dark Matter searches towards dwarf spheroidal galaxies with Fermi-LAT, HAWC, HESS, MAGIC and VERITAS. In Proceedings of the 36th International Cosmic Ray Conference, Madison, WI, USA, 24 July–1 August 2019.

© 2020 by the authors. Licensee MDPI, Basel, Switzerland. This article is an open access article distributed under the terms and conditions of the Creative Commons Attribution (CC BY) license (http://creativecommons.org/licenses/by/4.0/).

Review

Radio-Frequency Searches for Dark Matter in Dwarf Galaxies

Geoff Beck

School of Physics, University of the Witwatersrand, Private Bag 3, Johannesburg WITS-2050, South Africa; geoffrey.beck@wits.ac.za or geoff.m.beck@gmail.com

Received: 29 November 2018; Accepted: 3 January 2019; Published: 13 January 2019

Abstract: Dwarf spheroidal galaxies have long been discussed as optimal targets for indirect dark matter searches. However, the majority of such studies have been conducted with gamma-ray instruments. In this review, we discuss the very recent progress that has been made in radio-based indirect dark matter searches. We look at existing work on this topic and discuss the future prospects that motivate continued work in this newly developing field that promises to become, in the light of the up-coming Square Kilometre Array, a prominent component of the hunt for dark matter.

Keywords: dark matter; indirect detection; dwarf spheroidal galaxies

1. Introduction

Dwarf spheroidal galaxies (dSphs) have long been known as highly Dark Matter (DM) dominated objects with little baryonic emission that would obscure indirect detection efforts [1]. This has lead to extensive searches for DM annihilation resulting in gamma-ray emissions with numerous telescopes. Early efforts focussed on the Draco dwarf [2,3] but later campaigns using the Fermi Large Area Telescope [4] (Fermi-LAT) [5–9], the High Energy Stereoscopic System (HESS) [10–13], and the High Altitude Water Cherenkov (HAWC) experiment [14] have greatly expanded the search to many other dwarf galaxy objects. However, radio continuum is another region of the spectra that few dSphs are detected in [15]. This suggests another possible avenue for hunting DM indirect emissions. For WIMP models, with masses above a few GeV [16], this emission would have to be in the form of long-lived leptons emitting synchrotron radiation. Therefore, the need to disentangle the magnetic field and DM contributions to this putative emission is a complicating factor that is not present in gamma-ray detection experiments. In addition, diffusion of the emitting electrons may substantially impact expected synchrotron emissions. To characterise the diffusive environment, we need detailed information about the diffuse baryon content and turbulent magnetic field structure within the dwarf galaxy [17]. This is, of course, considerably complicated by the low levels of expected diffuse baryonic content and weak emissions from target objects. These complications explain the historical preference for hunting indirect DM emission in gamma-rays. However, radio instruments have several points in their favour. Firstly, their angular resolution is vastly superior to that of gamma-ray experiments, especially when interferometry is employed (Atwood and others for the Fermi/LAT collaboration [4] vs. Perley et al. [18], for instance). This is important as it can be used to avoid the confusion of diffuse dark matter emission with that of unresolved point sources. In compliment to this, radio interferometers are entering a golden age of increasing sensitivity as embodied, in the GHz frequency range, by the Jansky Very Large Array (JVLA) [18], and the up-coming Square Kilometre Array (SKA) [19] and its precursor experiments MeerKAT [20] and the Australian Square Kilometre Array Pathfinder (ASKAP) [21]. In addition, lower frequency experiments such as the LOw Frequency ARray (LOFAR) [22] and the low-frequency SKA component are pushing the boundaries of minimum detectable fluxes to levels below 1 µJy. This is very promising for the indirect detection of DM as the advancement of radio astronomy techniques and technology will begin to overcome the traditional obstacles in the way of radio-based searches, allowing the strengths of radio instruments to make their impact on the hunt for DM.

This review therefore covers the progress that has been made towards the observation of diffuse radio emissions from dwarf spheroidal galaxies and the use of this to probe the parameter space of particle dark matter that produces electrons/positrons through annihilation or decay processes. The use of radio observation for indirect DM detection was prominently advocated in [23], using the magnetic field estimates on dSphs by Klein et al. [15] as motivation, with the Draco dwarf galaxy particularly in mind. This work spurred further searches with the Green Bank Telescope (GBT) [24,25] covering the Wilman I, Ursa Major II, and Coma Berenices. For this experiment, being a single dish, an external source catalogue was necessary (the NVSS was used [26]) to remove the contribution of point sources to the radio continuum map of the target objects. The authors of [24,25] noted the strong dependence of their results on the magnetic field scenario within the dSphs observed, a particular issue as the instruments used could not discern the magnetic field structure. This problem of source extraction can be obviated by using interferometers to make the radio maps of the dwarf galaxy in the first place. This was the approach that was followed in the subsequent works addressing three classical and three ultra-faint dSphs with the Australian Compact Telescope Array (ATCA) [27–29]. Higher achievable sensitivities could also have allowed for magnetic field estimation; however, it was found that ATCA was not sensitive enough to detect μG level fields via rotation measure, and polarimetry is complicated by low levels of dust and gas [28]. A follow-up observation was performed more recently targeting the Reticulum II dSph [30], following interest in this target by the gamma-ray DM community [7]. The direct use of interferometers allowed for greatly improved constraints on the DM annihilation cross-section over a wide mass range. Which are, in some cases, competitive with those obtained by a Fermi-LAT study of 15 dSph objects [6].

The results of these existing searches are presented here and compared to a literature benchmark of the Fermi-LAT dSph searches [6]. In addition, we follow Regis et al. [29] and present the future prospects of these radio searches by using estimation of sensitivity gains over ATCA by instruments such as JVLA, ASKAP, MeerKAT, and the SKA.

This review is structured as follows: Section 2 covers all the theoretical details needed to model the synchrotron emission from electrons resulting from DM annihilation/decay, including the handling of diffusion within the radio searches presented here. In Section 3, we go into detail on the approach taken to the deep radio searches, as well as instrumental details used, in [24,25,29,30]. In Section 4, we discuss the results of these aforementioned searches and compare them to our literature benchmark. Finally, in Section 5, we discuss the future prospects for deep radio searches with up-coming experiments and summarise the outlook in Section 6.

2. Radio Emissions from Dark Matter

In this section, we cover all necessary theoretical considerations needed to model potential radio emission that results from DM annihilation or decay.

2.1. Electron Source Functions from DM Annihilation/Decay

In general, we describe the production of some particle species i, via DM annihilation or decay, with a source function Q. This Q function gives the number of particles of type i produced per unit volume per unit time per unit energy. This function will depend upon both the position within the DM halo r and the energy of product i particles, E.

For annihilation, this is given by

$$Q_{i,A}(r,E) = \langle \sigma V \rangle \sum_f \frac{\mathrm{d}N_i^f}{\mathrm{d}E} B_f \mathcal{N}_\chi(r) , \qquad (1)$$

where $\langle \sigma V \rangle$ is the velocity-averaged DM annihilation cross-section at 0 K, the index f labels the states produced by annihilation with branching ratios B_f and i particle production spectra $\frac{\mathrm{d}N_i^f}{\mathrm{d}E}$, M_χ is the WIMP mass, and finally $\mathcal{N}_\chi(r) = \frac{\rho_\chi^2}{M_\chi^2}$ is the DM particle pair density at a given halo radius r.

The source function in the case of DM decay is given by

$$Q_{i,D}(r,E) = \Gamma \sum_f \frac{\mathrm{d}N_i^f}{\mathrm{d}E} B_f n_\chi(r) , \qquad (2)$$

where Γ is the decay rate of the DM particle, the spectra $\frac{\mathrm{d}N_i^f}{\mathrm{d}E}$ will match those used above but for annihilation cases where the DM particle mass is half of that used for studying decay processes, and $n_\chi(r) = \frac{\rho_\chi}{M_\chi}$ is the DM particle number density at a given halo radius r.

2.2. DM Halos of Dwarf Spheroidal Galaxies

There are three considered density profiles that are used in deep radio searches examined in this review. They are detailed as a function of the radial coordinate r below:

$$\rho_N(r) = \frac{\rho_s}{\frac{r}{r_s}\left(1+\frac{r}{r_s}\right)^2} ,$$

$$\rho_B(r) = \frac{\rho_s}{\left(1+\frac{1.52 r}{r_s}\right)\left(1+\left(\frac{1.52 r}{r_s}\right)^2\right)} , \qquad (3)$$

$$\rho_E(r) = \frac{1}{4}\rho_s e^{-\frac{2}{\alpha}\left(\left(\frac{r}{r_s}\right)^\alpha - 1\right)} ,$$

where ρ_s is characteristic density, which normalises the density profile to virial mass of the halo M_{vir}; r_s is the scale radius, related to the virial radius via $r_{vir} = r_s c_{vir}$ where c_{vir} is the virial concentration parameter; and α is the free Einasto parameter. In Equation (3), these density profiles are, in order, Navarro–Frenk–White (NFW) [31], the Burkert profile [32], and the Einasto profile [33]. Several profiles need to be considered when studying dwarf galaxies as there is some uncertainty as to their halo structure in the literature [34,35].

2.3. Diffusion of Secondary Electrons

It is vital in the discussion of DM-induced radio emission to consider the diffusion and energy-loss experienced by resultant electrons. This is because both the position and energy distributions of DM-produced electrons will influence the subsequent synchrotron emission. Particularly, it has been shown in [23,36,37] that the effect of diffusion on the emitted flux is highly significant in small structures such as dwarf galaxies. The diffusion equation for electrons within the halo is given by

$$\frac{\partial}{\partial t}\frac{\mathrm{d}n_e}{\mathrm{d}E} = \nabla\left(D(E,\mathbf{x})\nabla\frac{\mathrm{d}n_e}{\mathrm{d}E}\right) + \frac{\partial}{\partial E}\left(b(E,\mathbf{x})\frac{\mathrm{d}n_e}{\mathrm{d}E}\right) + Q_e(E,\mathbf{x}) , \qquad (4)$$

where $\frac{\mathrm{d}n_e}{\mathrm{d}E}$ is the electron spectrum, the spatial diffusion is characterised by $D(E,\mathbf{x})$, while $b(E,\mathbf{x})$ specifies the rate of energy-loss and $Q_e(E,\mathbf{x})$ is the electron source function from DM annihilation or decay. The solution that is sought from such an equation is the stable equilibrium electron distribution. Two main approaches exist in the literature, one being Crank–Nicolson method for discretising derivatives. This approach is used in publicly available cosmic-ray transport codes such as DRAGON and GALPROP [17,38] and is employed in [29,30]. To implement this method, the time derivative is discretised

$$\frac{\partial}{\partial t}\frac{dn_e}{dE} = \frac{\frac{dn_i}{dE}(t+\Delta t) - \frac{dn_i}{dE}(t)}{\Delta t}, \quad (5)$$

where i indicates a position r (assuming spherical symmetry) and we drop the e subscript for clarity. The Crank–Nicolson scheme is then

$$\frac{\frac{dn_i}{dE}(t+\Delta t) - \frac{dn_i}{dE}(t)}{\Delta t} = \frac{\alpha_1 \frac{dn_{i-1}}{dE}(t+\Delta t) - \alpha_2 \frac{dn_i}{dE}(t+\Delta t) + \alpha_3 \frac{dn_{i+1}}{dE}(t+\Delta t)}{2\Delta t} - \frac{\alpha_1 \frac{dn_{i-1}}{dE}(t) - \alpha_2 \frac{dn_i}{dE}(t) + \alpha_3 \frac{dn_{i+1}}{dE}(t)}{2\Delta t} + Q_i. \quad (6)$$

The α coefficients for the r discretisation are defined to match the form of Equation (4)

$$\frac{\alpha_1}{\Delta t} = \left[-\frac{D + \frac{\partial D}{\partial r}}{\Delta r} + \frac{D}{\Delta r^2} \right]\bigg|_{r=r_i}, \quad (7)$$

$$\frac{\alpha_2}{\Delta t} = \frac{2D(r = r_i)}{\Delta r^2}, \quad (8)$$

$$\frac{\alpha_3}{\Delta t} = \left[\frac{D + \frac{\partial D}{\partial r}}{\Delta r} + \frac{D}{\Delta r^2} \right]\bigg|_{r=r_i}. \quad (9)$$

The energy derivatives are discretised with the coefficients

$$\frac{\alpha_1}{\Delta t} = \frac{b_i(E)}{\Delta E}, \quad (10)$$

$$\frac{\alpha_2}{\Delta t} = \frac{b_i(E+\Delta E) - b_i(E-\Delta E)}{\Delta E} + 2, \quad (11)$$

$$\frac{\alpha_3}{\Delta t} = -\frac{b_i(E)}{\Delta E}. \quad (12)$$

The energy-loss function for inverse-Compton scattering of CMB photons and synchrotron emission is [28]

$$b(E, r) = 2.7 \times 10^{-17} \text{GeV s}^{-1} \left(1 + 0.095 \left(\frac{B(r)}{\mu G}\right)^2\right) \left(\frac{E}{\text{GeV}}\right)^2. \quad (13)$$

The diffusion function is [28]

$$D(E, r) = D_0 \left(\frac{B(r)}{1\mu G}\right)^{-\alpha} \left(\frac{E}{1\text{GeV}}\right)^\alpha, \quad (14)$$

where α is the slope of the magnetic field power spectrum and D_0 ranges between 10^{28} and 10^{30} cm^2 s^{-1}.

The other approach to solving Equation (4) employs a semi-analytical formalism via the use of Green's functions, as used in [23,36,37]. In this approach, it is assumed that the DM halo, and accompanying baryon distributions, have spherical symmetry. Additionally, it is assumed that the energy-loss and diffusion processes have no spatial dependence. Under these assumptions, the solution to diffusion equation takes the form

$$\frac{dn_e}{dE}(r, E) = \frac{1}{b(E)} \int_E^{M_\chi} dE' \, G(r, E, E') Q(r, E'), \quad (15)$$

with $G(r, E, E')$ being a Green's function. This is expressed as

$$G(r,E,E') = \frac{1}{\sqrt{4\pi\Delta v}} \sum_{n=-\infty}^{\infty} (-1)^n \int_0^{r_h} dr' \frac{r'}{r_n}$$
$$\times \left(\exp\left(-\frac{(r'-r_n)^2}{4\Delta v}\right) - \exp\left(-\frac{(r'+r_n)^2}{4\Delta v}\right) \right) \frac{Q(r')}{Q(r)}, \quad (16)$$

where r_h is the maximum radius considered for spatial diffusion, $r_n = (-1)^n r + 2nr_h$ are the image charge positions, and

$$\Delta v = v(u(E)) - v(u(E')), \quad (17)$$

with

$$v(u(E)) = \int_{u_{min}}^{u(E)} dx\, D(x),$$
$$u(E) = \int_E^{E_{max}} \frac{dx}{b(x)}. \quad (18)$$

These last equations constitute a similar change of variables to those used in [39,40] to solve Equation (4). Since we have assumed that diffusion and energy-loss do not depend on halo position r, we include their effects via average values for the field strength and thermal plasma density. These average values are defined as follows: $\bar{B} \equiv \sqrt{\langle B(r)^2 \rangle}$ and $\bar{n} \equiv \langle n(r) \rangle$, where the angular brackets indicate a radial average. We can then express the spatial diffusion coefficient in terms of these averages as [41]

$$D(E) = D_0 \left(\frac{\bar{B}}{1\mu G}\right)^{-\frac{1}{3}} \left(\frac{E}{1\text{GeV}}\right)^{\frac{1}{3}}, \quad (19)$$

where the turbulence has been assumed Kolmogorov distributed, and D_0 is the diffusion constant. Note that the radial dependence of the diffusion coefficient is very weak. This justifies the assumption that we can make use of only the averaged value of the magnetic field in the diffusion coefficient. The general electron energy-loss function is then

$$b(E) = b_{IC} E^2 (1+z)^4 + b_{sync} E^2 \bar{B}^2$$
$$+ b_{coul} \bar{n}(1+z)^3 \left(1 + \frac{1}{75} \log\left(\frac{\gamma}{\bar{n}(1+z)^3}\right)\right) \quad (20)$$
$$+ b_{brem} \bar{n}(1+z)^3 \left(\log\left(\frac{\gamma}{\bar{n}(1+z)^3}\right) + 0.36\right)$$

where $\gamma = \frac{E}{m_e c^2}$, $\bar{n}$ is given in cm^{-3} and b_{IC}, b_{synch}, b_{coul}, and b_{brem} are the inverse-Compton, synchrotron, Coulomb, and Bremsstrahlung energy loss factors, taken to be 0.25, 0.0254, 6.13, and 1.51, respectively, in units of 10^{-16} GeV s^{-1}. The energy E is expressed in GeV and the B-field is in terms of μG.

2.4. Synchrotron Emission

An electron of energy E, gyrating within a magnetic field of strength B produces synchrotron emission with frequency dependent power given by [42]:

$$P_{synch}(\nu, E, r, z) = \int_0^\pi d\theta\, \frac{\sin^2\theta}{2} 2\pi\sqrt{3} r_e m_e c \nu_g F_{synch}\left(\frac{\kappa}{\sin\theta}\right), \quad (21)$$

where ν is the observed frequency, z is the source redshift, the mass of the electron is given by m_e, the non-relativistic gyro-frequency of the electron is $\nu_g = \frac{eB}{2\pi m_e c}$, and $r_e = \frac{e^2}{m_e c^2}$ is the classical radius of the electron. Finally, κ and F_{synch} are defined as

$$\kappa = \frac{2\nu(1+z)}{3\nu_0 \gamma^2} \left(1 + \left(\frac{\gamma \nu_p}{\nu(1+z)}\right)^2\right)^{\frac{3}{2}},\tag{22}$$

and

$$F_{synch}(x) = x \int_x^\infty dy\, K_{5/3}(y) \simeq 1.25 x^{\frac{1}{3}} e^{-x} \left(648 + x^2\right)^{\frac{1}{12}}.\tag{23}$$

The synchrotron radiation emissivity at a position r within the halo can then be found to be

$$j_{synch}(\nu, r, z) = \int_{m_e}^{M_\chi} dE \left(\frac{dn_{e^-}}{dE} + \frac{dn_{e^+}}{dE}\right) P_{synch}(\nu, E, r, z).\tag{24}$$

This quantity is the basic ingredient in determining the flux seen by a distant observer. The flux density spectrum emitted within a radius r of the halo centre is found via

$$S_{synch}(\nu, z) = \int_0^r d^3 r'\, \frac{j_{synch}(\nu, r', z)}{4\pi D_L^2},\tag{25}$$

where D_L is the luminosity distance from observer to halo. Then, the azimuthally averaged surface brightness is given by

$$I_{synch}(\nu, \Theta, \Delta\Omega, z) = \int_{\Delta\Omega} d\Omega \int_{l.o.s} dl\, \frac{j_{synch}(\nu, l, z)}{4\pi},\tag{26}$$

where the integration regions $\Delta\Omega$ and $l.o.s$ define a cone of solid angle $\Delta\Omega$ around the line of sight ($l.o.s$). Note that this $l.o.s$ makes an angle Θ with the central axis of the halo.

3. Deep Radio Searches for Dark Matter Emissions

Dark matter radio emissions would be the result of synchrotron radiation emitted by electrons produced in annihilation/decay processes. This emission depends upon the DM density and so will be a truly diffuse component of the radio continuum, being on the scale of the DM halo or a few arcminutes in the case of nearby dSph targets. This means that there are two important sources of uncertainty in these searches. The first is the removal of contamination by point sources and the second is the dependence on the unknown magnetic field environment within the dwarf galaxy. The removal of point source contributions to the radio continuum data requires a fine enough resolution to resolve such objects. This means that radio interferometers form a necessary component of the hunt for diffuse radio emission. Two approaches have been considered in the literature: The first is the use of a single dish radio telescope (GBT) and extracting the sources via the use of a source catalogue produced by interferometer surveys (NVSS for instance). The second is to perform the observations directly with an interferometric array (ATCA is used in [29,30]).

In the single telescope approach of Spekkens et al. [24], a 40.5 deg^2 area of the sky is observed at 1.4 GHz with the GBT. The field in question contained the dSphs Wilman I, Ursa Major II, and Coma Berenices. The resolution attained is $10'$ and NVSS catalogue [26] is used to subtract the unresolved contributions of point sources to the continuum emissions. The final sensitivity attained in the source subtracted maps is around 7 mJy per beam. Limits on DM annihilation are then obtained via the surface brightness profile of the diffuse emission being compared to expected results for DM models with varying WIMP mass, halo density profile, and annihilation cross-section. The largest uncertainties in this work are the magnetic field profile and the diffusion of the synchrotron electrons (in this case the model from [23,36] was used). The DM limits are derived under the assumption of a fiducial scenario where the diffusion constant D_0 is taken to be 0.1 of that for the Milky Way, following scaling from [43], and the magnetic field is taken to be $B \sim 1\,\mu G$. The authors of these works [24,25] also showed how sensitive their limits are to the assumptions made in regards to B and D_0.

Regis et al. [29] and Regis et al. [30] followed a common methodology different to that used in [24,25]. In Refs. [27,29], the authors used the ATCA array to target the Carina, Fornax, Sculptor, Hercules, Segue 2, and Bootes II dSphs. Two mosaic regions of 1° and 0.5° were chosen containing three dSphs each. These regions were observed in a 2 GHz band around a central value of 2.1 GHz in the radio continuum. The H168 and H214 ATCA configurations were used, these having compact cores (baselines less than 100 m) and one long baseline around 4 km. A region of 20–30 arcminutes was targeted around each dSph for analysis with 10–17 h (varying by dSph) on these regions in question. This allowed a nominal sensitivity to 20–40 μJy to be reached. Data cleaning was performed via the MFCLEAN routine from Miriad [44]. Sources were extracted with two approaches: SExtractor [45] and SFIND within Miriad. The first of these approaches detects sources via their deviation in flux relative to the local background and the second uses a false detection rate method. These two methods were then tuned to match their source catalogues to a random position variation of around 1 arcsecond. The resulting high resolution maps have a synthesised 8 arcsecond beam size (with 10 beams per source), a confusion limit of 3 μJy, and an rms noise of 30–40 μJy. This is significant as it implies that source confusion will not be a factor in the analysis, as it lies below the rms sensitivity attained. A second set of maps was also produced with a Gaussian taper on the scale of 15 arcseconds. This Gaussian tapered case results in a larger 1 arcminute synthesised beam, which will be more suited to detecting fluxes on the scale of extended DM emission. This tapering also has the consequence of making the point sources easier to extract from the visibility plane prior to Fourier inversion, leaving an rms noise of 100 μJy due to confusion limitations (which are far more significant with the taper in use). The authors of [29,30] always presented the DM limits from the most constraining map, choosing between either the tapered case or the high resolution maps. The authors considered two additional sources of uncertainty: bandwidth smearing and clean bias. The small size of the observed frequency band was shown to result in no significant bandwidth smearing. Clean bias, resulting from incomplete UV coverage, involves flux from sources being redistributed to the noise during data cleaning. This was mitigated by following the approach suggested in [46] and stopping the cleaning process at a residual flux three times above the rms noise. The observations in question could not discern the magnetic field, being too insensitive to observe the rotation measure and the low dust and gas content of dSphs making polarimetry extremely challenging.

In Refs. [28,29], the Crank–Nicolson diffusion model (from Section 2.3) was implemented and the authors studied three diffusion schemes. The first case is an optimistic case (OPT) where there is no spatial diffusion of the DM-produced electrons and only energy-loss at injection is considered. This OPT scenario takes the DM halo of the target dwarf galaxies to have an Einasto density profile, and magnetic field strength is calculated via an assumption of local equipartition (yielding averaged values between 4 and 8 μG for the studied dSphs). The second diffusion scheme is situated between an optimistic or pessimistic scenario and is called AVG or average. This case assumes a diffusion constant given by $D_0 = 2 \times 10^{28}$ cm^2 s^{-1} with the diffusion function D experiencing an exponential increase over the scale of the stellar half-light radius r_*. The magnetic field in the AVG case is inferred from the rate of star formation (with the correlation normalised against data for the Large Magellanic cloud), which yields field strengths between 0.4 and 2.0 μG. The DM halo profile for the dSphs is assumed to be NFW in this AVG scenario. In the third pessimistic scenario (PES), the magnetic field is inferred from star formation but only by considering data for the last Gigayear of the history of each dSph target. The halo density profile is assumed to take a cored Burkert shape. In terms of diffusion functions, PES takes D to be of the same form as in the AVG case, but, with D_0 being smaller by two orders of magnitude. In both the AVG and PES cases, the magnetic field decays exponentially over the scale r_*.

In Ref. [30], the Reticulum II dSph was targeted with the ATCA telescope in a similar configuration to [27–29], complemented by large angular scale data from the KAT-7 array [47]. These ATCA observations involved a 23.7' region containing Reticulum II and attained a 10 μJy rms sensitivity, as the position of Reticulum II on the sky means that galactic foregrounds are less significant. The target was observed for 30 h in a 2 GHz band centred on 2.1 GHz. The synthesised beam is around $7.5'' \times 2.0''$ in size with well-imaged structures being above $3'$ in extent. The KAT-7 data came from 9 h with six antennae and 44 h with five antennae in a 400 MHz band centred on 1822 MHz. These data are used as a consistency check, as the lack of long-baselines means the source subtraction is not so well defined as with ATCA. Following kinematic estimates [48], the authors employed an Einasto density profile for the DM halo, a magnetic field model assuming $B_0 = 1$ µG with exponential decay on the scale r_* and the same diffusion function as the AVG scenario above.

4. Search Results

Preliminary work was done in this field in [24,25] motivated by arguments from [23,36]. Spekkens et al. [24] targeted the dSphs Wilman I, Coma Berenices, and Ursa Major II using data from the Greenbank telescope and the NVSS catalogue. The results of this study indicate that, for WIMPs with masses around 100 GeV, the DM annihilation cross-section was constrained below 10^{-25} cm^3 s^{-1}. These results are extended by the second study [25], where the authors targeted only Ursa Major II using data from Greenbank telescope and excluded (at 2σ confidence level) WIMP models with $m_\chi = 10$ GeV annihilating directly to electrons for cross-sections $> 10^{-26}$ cm^3 s^{-1} and those annihilating to b quarks with $m_\chi = 100$ GeV and $\langle \sigma \rangle > 10^{-24}$ cm^3 s^{-1}. The results in both studies [24,25] assume a constant magnetic field of 1 µG and diffusion consistent with [23]) and thus of similar magnitude to the AVG scenario in [28].

Now, we consider the more recent work of [29] (part of a trio of works [27–29] that contain all the observational and theoretical details of the study), where deep radio observations were performed with ATCA on the Carina, Fornax, Sculptor, Hercules Segue 2, and Bootes II dSphs. In this case, the authors studied three diffusion schemes detailed above in Section 3. Limits on the annihilation cross-section span around six or seven orders of magnitude between the three models with the largest gap being between OPT and AVG (AVG and PES differ by around 2 orders of magnitude). For individual galaxies, in the AVG scenario, the constraints are competitive with those found in [24] over a wide mass range (10–5000 GeV). However, a combined constraint produced in the AVG is considerably stronger (as can be seen for several annihilation channels in Figure 1).

Regis et al. [30] found no evidence of diffuse radio emission in Reticulum II and thus derived constraints on the WIMP annihilation cross-section or decay rate from this (following a model similar to AVG above). These constraints are displayed in Figure 1.

What is evident in Figure 1 is that the limits from non-observation of diffuse emission in Reticulum II from [30] are up to an order of magnitude stronger than those from [29] for all displayed annihilation channels (note that the diffusion scenarios are very similar in this comparison). Particularly, the $b\bar{b}$ channel is an order of magnitude better in [30] but other channels are far more similar. When these constraints are placed into literature context against a benchmark like the Fermi-LAT dwarf galaxy gamma-ray limits [6] we find that [29] is around an order of magnitude less stringent than Fermi. In the case of [30], we see that these limits are more competitive with those from gamma-rays than [29], being more stringent for low masses with the muon annihilation channel and within a order of magnitude of Fermi-LAT otherwise. It is worth noting that we are comparing a single dwarf galaxy with [30], and six galaxies in [29], against a combined 15 galaxy analysis in [49], which indicates the competitive potential of the radio approach.

In Figure 2, we display analogous results for the scenario of decaying dark matter particles. These are compared against the Fermi-LAT dwarf galaxy limits from [49]. In this case, we plot the limited channels presented in [29] for this particular study. What is evident is that the limits from [30] make some improvements over those from gamma-rays. This increase is more than an order of magnitude for masses below 1 TeV in the case of the muon channel (where [29] is also superior to Fermi-LAT by around an order of magnitude), and factor 2 improvements at mass between 100 GeV and 10 TeV for $b\bar{b}$, and similar increases between 20 and 1000 GeV for the tau-lepton channel. The W-boson channel is very similar for both gamma-ray and radio studies.

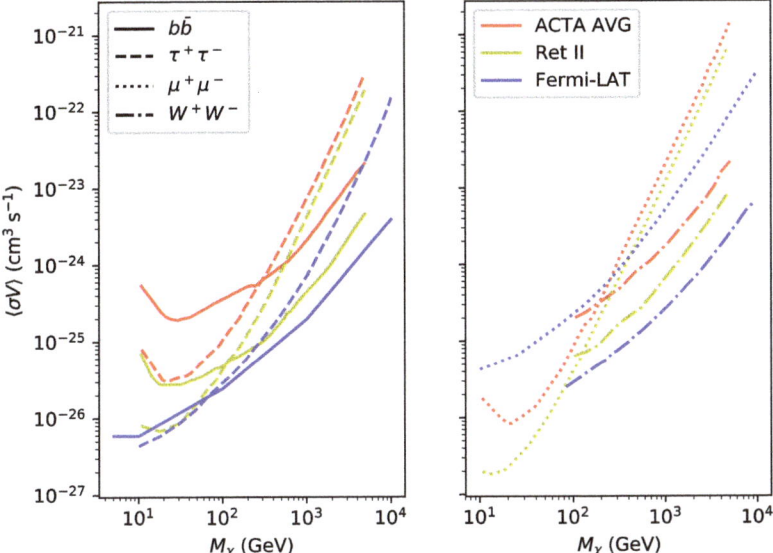

Figure 1. Cross-section upper limits from existing searches, presented at a confidence level of 95%. Four annihilation channels are covered and we display results from Fermi-LAT [6], the AVG diffusion scenario from [29], and Reticulum II results from [30]. The left panel shows the b-quark and τ-lepton channels. The right panel shows W-boson and muon channels.

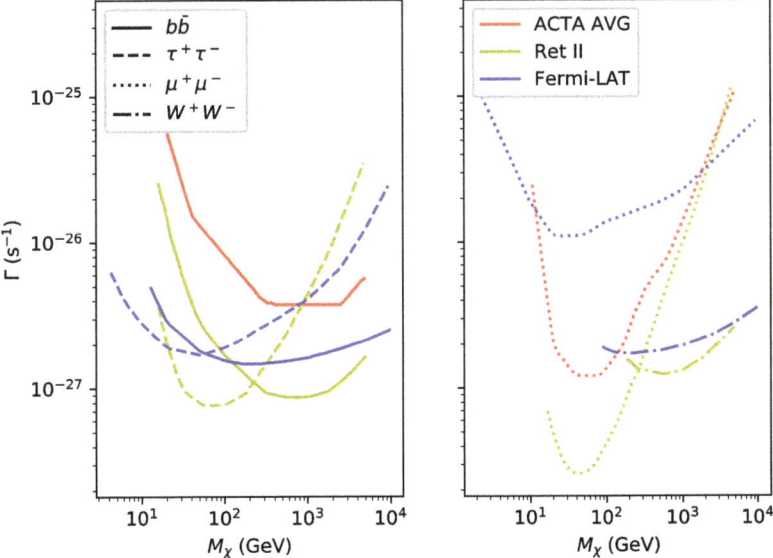

Figure 2. Decay rate limits from existing searches, presented at a confidence level of 95%. Four annihilation channels are covered and we display results from Fermi-LAT [49], the AVG diffusion scenario from [29] (limited to channels quoted in the reference), and Reticulum II results from [30]. The left panel shows the b-quark ($b\bar{b}$) and τ-lepton channels. The right panel shows W-boson and muon channels.

5. Future Prospects

Many new generation radio observatories are either coming online presently or are expected within the near future. We cover those that operate in a similar bandwidth to [24,25,29,30] in detail but do not discuss those experiments that operate outside this frequency band as their projections are not easily comparable to the results presented by the aforementioned studies. We note, however, that it has been argued that LOFAR [22] may have some potential in indirect DM detection [50,51], however neither of these studies directly addressed the dSph scenario considered here.

A particular example of improvements over the results from [29,30] could be drawn from deep radio searches with the existing JVLA [18] telescope. In particular, making use of the D configuration with baselines between 1 km and 35 m to observe both the large scale diffuse emission and perform source extraction. This instrument is capable of an rms sensitivity in the GHz range of around 10 µJy for 1 h per pointing which can provide a substantial advantage over the ATCA observations used in [29,30]. Despite this choice of optimal instrumental configuration, JVLA data would still require even longer baseline observations to remove point sources. This is because the D configuration confusion limit approaches 90 µJy and will thus impact on the potential to probe faint diffuse radio fluxes. Thus, overall, the JVLA may produce as much as factor of 2 improvement on the results of [29,30], as shown for the [29] targets in Figure 3. Such an improvement would make the limits from Reticulum II in [30] very competitive with the gamma-ray case presented for a study of 15 dSphs by Fermi-LAT and shown in the same plot.

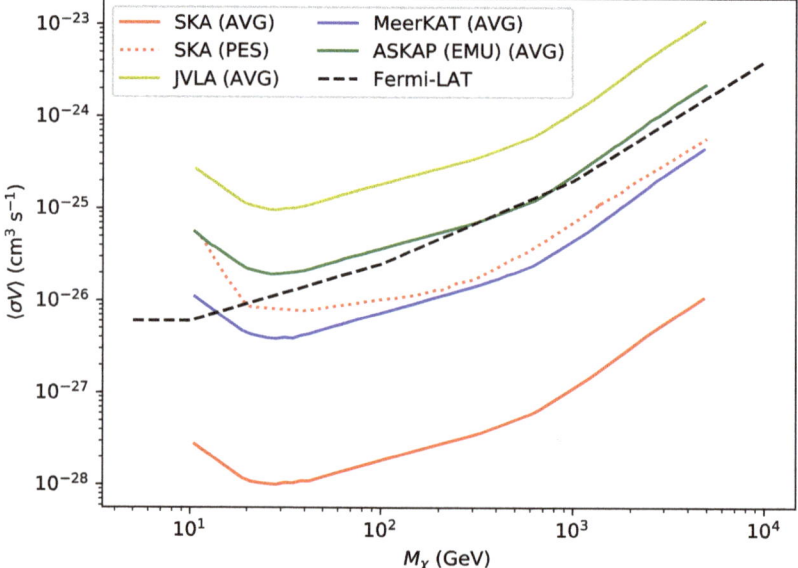

Figure 3. Cross-section upper limit prospects for the JVLA, ASKAP EMU, MeerKAT, and SKA from [29] compared to Fermi-LAT limits [6]. These limits are displayed at 95% confidence interval and only for the b-quark ($b\bar{b}$) channel.

For the SKA [19] precursor ASKAP [21], its survey project EMU [52] will attain GHz continuum rms sensitivity of 10 µJy over a 30 square degree area with an angular resolution of 10 arcseconds. There are 14 known Milky-Way satellites in this area and many more dwarf galaxy detection could be expected from southern-sky surveys [53–55] (as the Sloan Digital Sky Survey more than doubled the number of known northern-sky dSphs). The authors of [29] argued for an increase in sensitivity over their results of a factor of 5–10, factoring in an increase in the dSph sample observed (as can be seen in Figure 3).

In the case of the SKA precursor MeerKAT [20], the field of view will be smaller than ASKAP but it has a faster survey speed and will be well suited to deep dSph observations. This is because it should potentially obtain $\lesssim 1$ µJy rms sensitivity around the GHz range (with an integration time approaching 1000 h). As can be seen in Figure 3, MeerKAT can probe more of the parameter space than Fermi-LAT for the AVG diffusion scenario.

The SKA itself will achieve up to a two orders of magnitude improvement on its precursors [19] sensitivity for the mid-frequency band ranging 350–1050 MHz. It will also have the advantage of being able to determine magnetic field structures via rotation measures for fields around 1 µG in the dSph environment. In Figure 3, this leads to being able to produce superior limits to Fermi-LAT (for WIMP mass $\geq$ 20 GeV) even in the PES diffusion scenario. These sensitivity improvements come with caveats. The first is that, with the SKA reaching such potentially faint diffuse fluxes, confusion limits will unpredictably affect the source subtraction adjustment to the sensitivity. Secondly, the low level star formation emissions expected within the dSph could be a complicating factor for very faint fluxes. This second caveat can be mitigated through a combination of the use of optical correlations and the fine angular resolution of the SKA itself to identify star formation contributions. However, these caveats do mean that the projection in Figure 3 is likely optimistic.

6. Outlook

The preceding discussion indicates that radio searches for DM annihilation/decay in dwarf galaxies is entering into an age in which it becomes more sensitive to putative DM emissions than gamma-ray telescopes. This means that these kinds of deep radio GHz frequency searches will become a leading candidate for indirect DM hunting as they mature with the arrival of future telescopes such as the SKA and its precursors. Supplementing these GHz projects with lower frequencies via LOFAR or SKA-LOW will make the study of DM via diffuse emission in dwarf galaxies a powerful probe into one of the largest hiatuses in current models of cosmology.

Funding: This research received no external funding.

Acknowledgments: This review is dedicated to the memory of my colleague and mentor Sergio Colafrancesco.

Conflicts of Interest: The author declares no conflict of interest.

Abbreviations

The following abbreviations are used in this manuscript:

DM	Dark Matter
dSph	Dwarf spheroidal galaxy
SKA	Square Kilometre Array
ATCA	Australian Telescope Compact Array
GBT	Green Bank Telescope
JVLA	Jansky Very Large Array
ASKAP	Australian Square Kilometre Array Pathfinder
EMU	Evolutionary Map of the Universe
KAT	Karoo Array Telescope
LOFAR	LOw Frequency ARray

References

1. Mateo, M. Dwarf galaxies of the Local Group. *Ann. Rev. Astron. Astrophys.* **1998**, *36*, 435–506. [CrossRef]
2. Tyler, C. Particle dark matter constraints from the Draco dwarf galaxy. *Phys. Rev. D* **2002**, *66*, 023509. [CrossRef]

3. Profumo, S.; Kamionkowski, M. Dark matter and the cactus gamma-ray excess from draco. *J. Cosmol. Astropart. Phys.* **2006**. [CrossRef]
4. Atwood, W.B.; Abdo, A.A.; Ackermann, M.; Althouse, W.; Anderson, B.; Axelsson, M.; Baldini, L.; Ballet, J.; Band, D.L.; Barbiellini, G.; et al. for the Fermi/LAT collaboration. The Large Area Telescope on the Fermi Gamma-ray Space Telescope Mission. *Astrophys. J.* **2009**, *697*, 1071–1102. [CrossRef]
5. Ackermann, M.; Albert, A.; Baldini, L.; Ballet, J.; Barbiellini, G.; Bastieri, D.; Bechtol, K.; Bellazzini, R.; Blandford, R.D.; Bloom, E.D.; et al. Search for Dark Matter Satellites using the FERMI-LAT. *Astrophys. J.* **2012**, *747*, 121. [CrossRef]
6. Ackermann, M.; Albert, A.; Anderson, B.; Atwood, W.B.; Baldini, L.; Barbiellini, G.; Bastieri, D.; Bechtol, K.; Bellazzini, R.; Bissaldi, E.; et al. Searching for Dark Matter Annihilation from Milky Way Dwarf Spheroidal Galaxies with Six Years of Fermi Large Area Telescope Data. *Phys. Rev. Lett.* **2015**, *115*, 231301. [CrossRef] [PubMed]
7. Geringer-Sameth, A.; Walker, M.G.; Koushiappas, S.M.; Koposov, S.E.; Belokurov, V.; Torrealba, G.; Evans, N.W. Indication of Gamma-Ray Emission from the Newly Discovered Dwarf Galaxy Reticulum II. *Phys. Rev. Lett.* **2015**, *115*, 081101. [CrossRef]
8. Li, S.; Liang, Y.F.; Duan, K.K.; Shen, Z.Q.; Huang, X.; Li, X.; Fan, Y.Z.; Liao, N.H.; Feng, L.; Chang, J. Search for gamma-ray emission from eight dwarf spheroidal galaxy candidates discovered in Year Two of Dark Energy Survey with Fermi-LAT data. *Phys. Rev.* **2016**, *D93*, 043518. [CrossRef]
9. Li, S.; Duan, K.K.; Liang, Y.F.; Xia, Z.Q.; Shen, Z.Q.; Li, X.; Liao, N.H.; Feng, L.; Yuan, Q.; Fan, Y.Z.; et al. Search for gamma-ray emission from the nearby dwarf spheroidal galaxies with 9 years of Fermi-LAT data. *Phys. Rev. D* **2018**, *97*, 122001. [CrossRef]
10. Aharonian, F. Observations of the Sagittarius Dwarf galaxy by the H.E.S.S. experiment and search for a Dark Matter signal. *Astropart. Phys.* **2008**, *29*, 55–62; Erratum in **2010**, *33*, 274–275. [CrossRef]
11. Abramowski, A.; Acero, F.; Aharonian, F.; Akhperjanian, A.G.; Anton, G.; Barnacka, A.; De Almeida, U.B.; Bazer-Bachi, A.R.; Becherini, Y.; Becker, J.; et al. H.E.S.S. constraints on dark matter annihilations towards the sculptor and carina dwarf galaxies. *Astropart. Phys.* **2011**, *34*, 608–616. [CrossRef]
12. Abramowski, A.; Aharonian, F.; Benkhali, F.A.; Akhperjanian, A.G.; Angüner, E.; Backes, M.; Balenderan, S.; Balzer, A.; Barnacka, A.; Becherini, Y.; et al. Search for dark matter annihilation signatures in H.E.S.S. observations of Dwarf Spheroidal Galaxies. *Phys. Rev. D* **2014**, *90*, 112012. [CrossRef]
13. Abdalla, H.; Aharonian, F.; Benkhali, F.A.; Angüner, E.O.; Arakawa, M.; Arcaro, C.; Armand, C.; Arrieta, M.; Backes, M.; Barnard, M.; et al. Searches for gamma-ray lines and 'pure WIMP' spectra from Dark Matter annihilations in dwarf galaxies with H.E.S.S. *J. Cosmol. Astropart. Phys.* **2018**, *11*. [CrossRef]
14. Albert, A.; Alfaro, R.; Alvarez, C.; Álvarez, J.D.; Arceo, R.; Arteaga-Velázquez, J.C.; Rojas, D.A.; Solares, H.A.; Bautista-Elivar, N.; Becerril, A.; et al. Dark Matter Limits from Dwarf Spheroidal Galaxies with The HAWC Gamma-Ray Observatory. *Astrophys. J.* **2018**, *853*, 154. [CrossRef]
15. Klein, U.; Giovanardi, C.; Altschuler, D.R.; Wunderlich, E. A sensitive radio continuum survey of low surface brightness dwarf galaxies. *Astron. Astrophys.* **1992**, *255*, 49–58.
16. Kolb, E.W.; Turner, M.S. The Early Universe. *Front. Phys.* **1990**, *69*, 115–152.
17. Strong, A.W.; Moskalenko, I.V. Propagation of cosmic-ray nucleons in the galaxy. *Astrophys. J.* **1998**, *509*, 212–228. [CrossRef]
18. Perley, R.A.; Chandler, C.J.; Butler, B.J.; Wrobel, J.M. The Expanded Very Large Array: A New Telescope for New Science. *Astrophys. J. Lett.* **2011**, *739*, L1. [CrossRef]
19. Dewdney, P.; Turner, W.; Millenaar, R.; McCool, R.; Lazio, J.; Cornwell, T. SKA Baseline Design Document 2012. Available online: http://www.skatelescope.org/wp-content/uploads/2012/07/SKA-TEL-SKO-DD-001-1_BaselineDesign1.pdf (accessed on 11 November 2018).
20. Booth, R.; de Blok, W.; Jonas, J.; Fanaroff, B. MeerKAT Key Project Science, Specifications, and Proposals. *arXiv* **2009**, arXiv:0910.2935.
21. McConnell, D.; Allison, J.R.; Bannister, K.; Bell, M.E.; Bignall, H.E.; Chippendale, A.P.; Edwards, P.G.; Harvey-Smith, L.; Hegarty, S.; Heywood, I.; et al. The Australian Square Kilometre Array Pathfinder: Performance of the Boolardy Engineering Test Array. *Publ. Astron. Soc. Aust.* **2016**, *33*, e042. [CrossRef]
22. Van Haarlem, M.P.; Wise, M.W.; Gunst, A.W.; Heald, G.; McKean, J.P.; Hessels, J.W.; De Bruyn, A.G.; Nijboer, R.; Swinbank, J.; Fallows, R.; et al. LOFAR: The Low-Frequency Array. *Astron. Astrophys.* **2013**, *556*, A2. [CrossRef]

23. Colafrancesco, S.; Profumo, S.; Ullio, P. Detecting dark matter WIMPs in the Draco dwarf: A multi-wavelength perspective. *Phys. Rev. D* **2007**, *75*, 023513. [CrossRef]
24. Spekkens, K.; Mason, B.S.; Aguirre, J.E.; Nhan, B. A Deep Search for Extended Radio Continuum Emission From Dwarf Spheroidal Galaxies: Implications for Particle Dark Matter. *Astrophys. J.* **2013**, *773*, 61. [CrossRef]
25. Natarajan, A.; Peterson, J.B.; Voytek, T.C.; Spekkens, K.; Mason, B.; Aguirre, J.; Willman, B. Bounds on Dark Matter Properties from Radio Observations of Ursa Major II using the Green Bank Telescope. *Phys. Rev. D* **2013**, *88*, 083535. [CrossRef]
26. Condon, J.J.; Cotton, W.D.; Greisen, E.W.; Yin, Q.F.; Perley, R.A.; Taylor, G.B.; Broderick, J.J. The NRAO VLA Sky survey. *Astron. J.* **1998**, *115*, 1693–1716. [CrossRef]
27. Regis, M.; Richter, L.; Colafrancesco, S.; Massardi, M.; de Blok, W.J.G.; Profumo, S.; Orford, N. Local Group dSph radio survey with ATCA – I: Observations and background sources. *Mon. Not. R. Astron. Soc.* **2015**. *448*, 3731–3746, [CrossRef]
28. Regis, M.; Richter, L.; Colafrancesco, S.; Profumo, S.; de Blok, W.J.G.; Massardi, M. Local Group dSph radio survey with ATCA—II. Non-thermal diffuse emission. *Mon. Not. R. Astron. Soc.* **2015**. *448*, 3747–3765. [CrossRef]
29. Regis, M.; Colafrancesco, S.; Profumo, S.; de Blok, W.; Massardi, M.; Richter, L. Local Group dSph radio survey with ATCA (III): Constraints on particle dark matter. *J. Cosmol. Astropart. Phys.* **2014**, *2014*, 016. [CrossRef]
30. Regis, M.; Richter, L.; Colafrancesco, S. Dark matter in the Reticulum II dSph: A radio search. *J. Cosmol. Astropart. Phys.* **2017**, *2017*, 025.
31. Navarro, J.F.; Frenk, C.S.; White, S.D.M. The Structure of cold dark matter halos. *Astrophys. J.* **1996**, *462*, 563–575. [CrossRef]
32. Burkert, A. The Structure of dark matter halos in dwarf galaxies. *Astrophys. J. Lett.* **1995**, *447*. [CrossRef]
33. Einasto, J. On Galactic Descriptive Functions. *Astron. Nachr.* **1968**, *291*, 97–109. [CrossRef]
34. Walker, M.G.; Mateo, M.; Olszewski, E.W.; Peñarrubia, J.; Evans, N.W.; Gilmore, G. A Universal Mass Profile for Dwarf Spheroidal Galaxies? *Astrophys. J.* **2009**, *704*, 1274. [CrossRef]
35. Adams, J.J.; Simon, J.D.; Fabricius, M.H.; van den Bosch, R.C.; Barentine, J.C.; Bender, R.; Gebhardt, K.; Hill, G.J.; Murphy, J.D.; Swaters, R.A.; et al. Dwarf Galaxy Dark Matter Density Profiles Inferred from Stellar and Gas Kinematics. *Astrophys. J.* **2014**, *789*, 63. [CrossRef]
36. Colafrancesco, S.; Profumo, S.; Ullio, P. Multi-frequency analysis of neutralino dark matter annihilations in the Coma cluster. *Astron. Astrophys.* **2006**, *455*, 21–43. [CrossRef]
37. Colafrancesco, S.; Marchegiani, P.; Beck, G. Evolution of Dark Matter Halos and their Radio Emissions. *J. Cosmol. Astropart. Phys.* **2015**, *2*. [CrossRef]
38. Evoli, C.; Gaggero, D.; Grasso, D.; Maccione, L. Cosmic ray nuclei, antiprotons and gamma rays in the galaxy: A new diffusion model. *J. Cosmol. Astropart. Phys.* **2008**, *2008*, 018. [CrossRef]
39. Baltz, E.A.; Edsjö, J. Positron propagation and fluxes from neutralino annihilation in the halo. *Phys. Rev. D* **1998**, *59*, 023511. . [CrossRef]
40. Baltz, E.A.; Wai, L. Diffuse inverse Compton and synchrotron emission from dark matter annihilations in galactic satellites. *Phys. Rev. D* **2004**, *70*, 023512. [CrossRef]
41. Colafrancesco, S.; Blasi, S. Clusters of Galaxies and the Diffuse Gamma Ray Background. *Astropart. Phys.* **1998**, *9*, 227. [CrossRef]
42. Longair, M.S. *High Energy Astrophysics*; Cambridge University Press: Cambridge, UK, 1994.
43. Jeltema, .E.; Profumo, S. Searching for Dark Matter with X-ray Observations of Local Dwarf Galaxies. *Astrophys. J.* **2008**, *686*, 1045. [CrossRef]
44. Sault, R.J.; Teuben, P.J.; Wright, M.C. A retrospective view of Miriad. In *Astronomical Society of the Pacific Conference Series Vol. 77, Astronomical Data Analysis Software and Systems IV*; Shaw, R.A., Payne, H.E., Hayes, J.J.E., Eds.; Astronomical Society of the Pacific: San Francisco, CA, USA, 1995; p. 433.
45. Bertin, E.; Arnouts, S. SExtractor: Software for source extraction. *Astron. Astrophys. Suppl. Ser.* **1996**, *117*, 393–404. [CrossRef]
46. Prandoni, I.; Gregorini, L.; Parma, P.; De Ruiter, H.R.; Vettolani, G.; Wieringa, M.H.; Ekers, R.D. The ATESP radio survey I. Survey description, observations and data reduction. *Astron. Astrophys. Suppl.* **2000**, *146*, 41–55. [CrossRef]

47. Foley, A.R.; Alberts, T.; Armstrong, R.P.; Barta, A.; Bauermeister, E.F.; Bester, H.; Blose, S.; Booth, R.S.; Botha, D.H.; Buchner, S.J.; et al. Engineering and science highlights of the KAT-7 radio telescope. *Mon. Not. R. Astron. Soc.* **2016**. *460*, 1664–1679, [CrossRef]
48. Bonnivard, V.; Combet, C.; Maurin, D.; Geringer-Sameth, A.; Koushiappas, S.M.; Walker, M.G.; Mateo, M.; Olszewski, E.W.; Bailey, J.I., III. Dark matter annihilation and decay profiles for the Reticulum II dwarf spheroidal galaxy. *Astrophys. J.* **2015**, *808*, L36. [CrossRef]
49. Baring, M.G.; Ghosh, T.; Queiroz, F.S.; Sinha, K. New limits on the dark matter lifetime from dwarf spheroidal galaxies using Fermi-LAT. *Phys. Rev. D* **2016**, *93*, 103009. [CrossRef]
50. Leite, N.; Reuben, R.; Sigl, G.; Tytgat, M.; Vollmann, M. Synchrotron emission from dark matter in galactic subhalos. A look into the Smith cloud. *J. Cosmol. Astropart. Phys.* **2016**, *2016*, 021. [CrossRef]
51. Storm, E.; Jeltema, T.E.; Splettstoesser, M.; Profumo, S. Synchrotron Emission from Dark Matter Annihilation: Predictions for Constraints from Non-detections of Galaxy Clusters with New Radio Surveys. *Astrophys. J.* **2017**, *839*, 33. [CrossRef]
52. Norris, R.P.; Hopkins, A.M.; Afonso, J.; Brown, S.; Condon, J.J.; Dunne, L.; Feain, I.; Hollow, R.; Jarvis, M.; Johnston-Hollitt, M.; et al. EMU: Evolutionary Map of the Universe. *Publ. Astron. Soc. Aust.* **2011**, *28*, 215–248. [CrossRef]
53. Abbott, T.; Abdalla, F.B.; Aleksić, J.; Allam, S.; Amara, A.; Bacon, D.; Balbinot, E.; Banerji, M.; Bechtol, K.; Benoit-Lévy, A.; et al. The Dark Energy Survey: More than dark energy—An overview. *Mon. Not. R. Astron. Soc.* **2016**. *460*, 1270–1299. [CrossRef]
54. Keller, S.C.; Schmidt, B.P.; Bessell, M.S.; Conroy, P.G.; Francis, P.; Granlund, A.; Kowald, E.; Oates, A.P.; Martin-Jones, T.; Preston, T.; et al. SkyMapper and the Southern Sky Survey. *Publ. Astron. Soc. Aust.* **2007**, *24*, 1–12. [CrossRef]
55. Abell, P.A.; Burke, D.L.; Hamuy, M.; Nordby, M.; Axelrod, T.S.; Monet, D.; Vrsnak, B.; Thorman, P.; Ballantyne, D.R.; Simon, J.D.; et al. LSST Science Book, Version 2.0. *arXiv* **2009**, arXiv:0912.0201.

© 2019 by the author. Licensee MDPI, Basel, Switzerland. This article is an open access article distributed under the terms and conditions of the Creative Commons Attribution (CC BY) license (http://creativecommons.org/licenses/by/4.0/).

Article

Remnants of Galactic Subhalos and Their Impact on Indirect Dark-Matter Searches

Martin Stref [1,*], Thomas Lacroix [2] and Julien Lavalle [1]

1   Laboratoire Univers et Particules de Montpellier (LUPM), Université de Montpellier & CNRS, Place Eugène Bataillon, 34095 Montpellier CEDEX 05, France; lavalle@in2p3.fr
2   Instituto de Fisica Teórica, C/ Nicolás Cabrera 13-15, Campus de Cantoblanco UAM, 28049 Madrid, Spain; thomas.lacroix@uam.es
*   Correspondence: martin.stref@umontpellier.fr

Received: 6 May 2019; Accepted: 29 May 2019; Published: 4 June 2019

Abstract: Dark-matter subhalos, predicted in large numbers in the cold-dark-matter scenario, should have an impact on dark-matter-particle searches. Recent results show that tidal disruption of these objects in computer simulations is overefficient due to numerical artifacts and resolution effects. Accounting for these results, we re-estimated the subhalo abundance in the Milky Way using semianalytical techniques. In particular, we showed that the boost factor for gamma rays and cosmic-ray antiprotons is increased by roughly a factor of two.

Keywords: particle dark matter; subhalos; indirect searches

1. Introduction

There is overwhelming evidence that most of the matter in the universe is nonbaryonic [1]. An exciting possibility to account for these puzzling observations is that the universe is filled with exotic particles that interact only very weakly with ordinary matter [2,3]. One of the most elegant and popular dark-matter (DM) particle candidates is the Weakly Interacting Massive Particle (WIMP). These hypothetical particles are being looked for in particle colliders [4–6], in direct detection experiments [7–9], and in cosmic radiation [10,11], so far without success. Although one of the motivations for WIMPs is related to the fact that they emerge naturally in particle theories addressing a hierarchy problem [12–14], WIMPs are also attractive stemming from their very simple thermal-production mechanism in the early universe. Moreover, a large fraction of available parameter space is still unconstrained and currently actively explored [15]. For completeness, it is worth recalling that many alternatives to WIMPs exist that we do not discuss here, like axions [16], sterile neutrinos [17], primordial black holes [18], and extended dark sectors [19]. The cosmological paradigm best supported by current probes is that DM is cold, i.e., collisionless and nonrelativistic. This implies a structuring of matter on scales smaller than typical galaxies, with a model-dependent cutoff [20,21]. Interestingly, subgalactic scales are those where there could be departures from the predictions of the cold DM paradigm because of some observational issues [22]. This might sign new specific properties of the dark matter (e.g., [23]), or it could be due to baryonic effects (e.g., [24]). This motivates a detailed inspection of the impact of DM properties on the smallest scales, irrespective of the underlying scenario.

The small-scale structuring of DM, as treated, for instance, in the WIMP scenario, translates into a large population of subhalos within galactic halos [25–27]. Modeling these subhalos is crucial if one is to make accurate predictions for direct and indirect DM searches. This is a difficult task, as numerical simulations are far from resolving the smallest structures predicted by the cold DM paradigm. To incorporate the smallest structures, one can extrapolate the results of simulations over orders of magnitude in scales (see, e.g., [28]) but this represents a leap of faith. On the other hand, one can employ semianalytical models (see, e.g., [29–32]). The difficulty with the latter is accounting for the tidal effects experienced by subhalos within the host galaxy. These models can be calibrated on cosmological simulations, which are supposed to consistently describe the tidal stripping of subhalos in their host halo. However, it was recently pointed out by van den Bosch and collaborators [33,34] that simulations are plagued with numerical artifacts that lead to a significant overestimate of the tidal stripping efficiency, and therefore to an underestimate of the actual subhalo population even within the numerical resolution limit. An alternative and complementary way to study the tidal stripping of subhalos is to rely on analytical or semianalytical methods, which are based on first principles and allow to deal with subhalo mass scales, down to the free-streaming scale. Here, we review the semianalytical model developed by Stref and Lavalle [35] (SL17 hereafter), which incorporates a realistic and kinematically constrained Milky Way mass model (including baryons) and predicts the galactic subhalo abundance.[1] This model accounts for different sources of tidal effects, and can easily accommodate to different prescriptions for tidal disruption efficiency.

This paper is structured as follows. In Section 2, we briefly review the SL17 model and discuss the resilience of subhalos to tidal effects in light of recent analyses of simulation results [33,34]. In Section 3, we compute the DM mass density within subhalos, as well as the number density of these objects in the Milky Way. Finally, in Section 4, we look at the impact of our results on indirect searches for annihilating DM, focusing on gamma rays and cosmic-ray antiprotons.

2. Semianalytical Model of Galactic Subhalos

In this section, we review the SL17 Galactic subhalo population model and discuss the tidal effects experienced by subhalos. We then propose a way of incorporating the recent results of van den Bosch and collaborators in the model in a consistent calibration procedure.

2.1. Review of the Stref and Lavalle Model

SL17 is a semianalytical model of galactic subhalos that is built upon dynamical constraints and cosmological considerations. The main input of the model is the initial subhalo phase-space density

$$\frac{dN}{dV\,dm\,dc}(\vec{r},m,c) \propto \frac{d\mathcal{P}_v}{dV}(\vec{r}) \times \frac{d\mathcal{P}_m}{dm}(m) \times \frac{d\mathcal{P}_c}{dc}(c,m), \tag{1}$$

where phase space refers to the position–mass–concentration space. Functions $d\mathcal{P}_v/dV$, $d\mathcal{P}_m/dm$ and $d\mathcal{P}_c/dc$ are the spatial, mass, and concentration distributions, respectively. It is assumed that, should subhalos behave as hard spheres (as is the case for single DM "particles" in a cosmological simulation), they would be spatially distributed as $d\mathcal{P}_v/dV \propto \rho_{\rm DM}$ where $\rho_{\rm DM}$ is the total DM density profile of the galaxy. This sets our initial conditions before tidal disruption. The smooth DM mass density is computed through

$$\rho_{\rm sm}(\vec{r}) = \rho_{\rm DM}(\vec{r}) - \langle \rho_{\rm cl} \rangle (\vec{r}), \tag{2}$$

where $\langle \rho_{\rm cl} \rangle$ is the average DM mass density inside clumps (this quantity is explicitly computed in Section 3). In the following, we use the galactic mass models constrained by McMillan [36] on pre-Gaia data for the DM and baryonic mass distributions. In this framework, Equation (2) ensures the compatibility of our subhalo model with the constrained DM profile $\rho_{\rm DM}$. Although this work is devoted to the study of Milky Way subhalos, SL17 can in principle be used to study the substructure population in any virialized DM system. One only needs a mass model for the system in question and a proper calibration of the subhalo mass fraction through the procedure outlined in Section 2.3.

---

[1] We refer to this model as semianalytical because it involves integrals that must be computed numerically. The model does not rely on numerical simulations except at the level of a calibration described in Section 2.3.

Mass m and concentration c refer to cosmological mass m_{200} and concentration c_{200} (defined with respect to the critical density), where we dropped the 200 index for convenience. The subhalo mass function measured in simulations is consistent with a power law [26,27]

$$\frac{d\mathcal{P}_m}{dm}(m) \propto m^{-\alpha_m}\,\Theta(m - m_{\min})\,\Theta(m_{\max} - m), \tag{3}$$

where Θ is the Heaviside step function, and the power-law index is $\alpha_m = 1.9$ or $\alpha_m = 2$. These values of α_m encompass the Press and Schechter [37] mass function and the Sheth and Tormen [38] mass function, as illustrated in Figure 1. These functions can be computed directly from the matter power spectrum in the framework of excursion set theory [39], for spherical collapse (Press–Schechter) and ellipsoidal collapse (Sheth–Tormen). Thus, the two power-law indices we considered bracket the theoretical uncertainties on the small-scale mass function. If the DM is made of WIMPs, mass cutoff $m_{\min}$ can be related to the kinetic decoupling of the DM particle and is found to lie between $10^{-4}\,M_\odot$ and $10^{-10}\,M_\odot$ [29,40–45]. Maximal mass $m_{\max}$ is set to $0.01 \times M_{\rm DM}$, where $M_{\rm DM}$ is the total DM mass in the Milky Way. Concentration distribution $d\mathcal{P}_c/dc$ is generically found to exhibit a log-normal distribution for field halos [46,47], which defines our initial concentration distribution (before tidal stripping). We adopt the peak value and variance fit in Sánchez-Conde and Prada [48], which was shown to provide a good description of cosmological simulations run independently by several groups. Subhalos are assumed to have a Navarro–Frenk–White (NFW) profile [49] with parameters set by m and c (the impact of choosing an Einasto profile [50] instead of NFW was investigated in [35]).

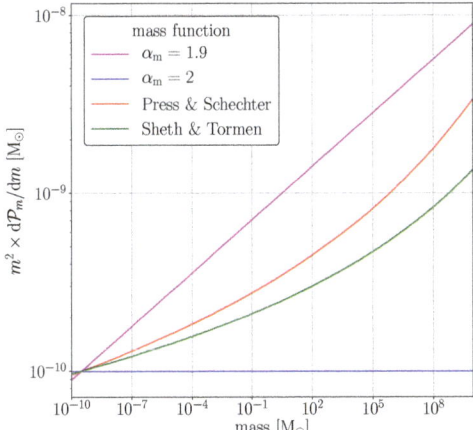

Figure 1. Mass function dn/dm multiplied by m^2. We show the prediction of Press and Schechter [37] (red line), Sheth and Tormen [38] (green line) as well as the power-law mass functions with index $\alpha_m = 1.9$ (magenta line) and $\alpha_m = 2$ (blue line). The Press–Schechter and Sheth–Tormen mass functions were computed for the cosmology of Planck 2018 [1] using the transfer function of Eisenstein and Hu [51] and a sharp-k filter. All mass functions are normalized to unity with $M_{\min} = 10^{-10}\,M_\odot$.

The subhalo population is strongly affected by tidal interactions with the potential of the host galaxy [52]. This is accounted for in the model through the calculation of a tidal radius r_t for each subhalo. The tidal radius should be interpreted as the physical extension of a subhalo, which is in general smaller than the extension it would have on a flat background. The physical mass of a subhalo is then

$$m_t(\vec{r}, m, c) = 4\pi \int_0^{r_t(\vec{r}, m, c)} dx\, x^2 \rho_{\rm sub}(x) \leq m, \tag{4}$$

where ρ_{sub} is the subhalo inner mass density profile. In our modeling, tidal stripping only removes the outer layers of subhalos while leaving the inner parts unchanged. In reality, DM should rearrange itself into a new equilibrium state. Central density however, should be left essentially unchanged [53]. Since we are interested in indirect searches for annihilating DM, and the central density gives the dominant contribution to the annihilation rate, our modeling should lead to a reasonable approximation. Two important contributions are accounted for in SL17: the effect of the smooth galactic potential (including both DM and baryons), and the gravitational shocking induced by the baryonic disk [54,55]. The latter effect turns out to be very efficient at stripping subhalos in the inner 20 kpc of the galaxy, a result also found in numerical studies [56–58]. The strength of SL17 over simulations is that it accounts for the constrained potential of the MW, with a detailed description of baryons.

2.2. Subhalo Disruption?

Whether a subhalo can be completely disrupted by tidal effects is an open question. A number of numerical studies found that a subhalo is completely disrupted when the total energy gained through tidal-stripping or disk-shocking effects is comparable to the binding energy [56,59]. On the other hand, some studies [60–62] found that cuspy subhalos almost always survive mass loss, leaving a small bound remnant behind even after gaining an energy far greater than their binding energy. These contradictory results may have been reconciled in a recent series of papers by van den Bosch and collaborators [33,34,63]. In these studies, it is shown that subhalo disruption in N-body simulations can actually be entirely explained by numerical artifacts. In particular, disruption is shown to be highly sensitive to the value of the force-softening length. If this length is taken sufficiently small, the authors showed that subhalos survive tidal mass loss in the form of a small bound remnant. We aim at quantifying the impact of these results on the whole subhalo population. Tidal disruption is modeled in a very simple way in SL17: given a subhalo with scale radius r_s and tidal radius r_t, we assume

$$\frac{r_t(\vec{r}, m, c)}{r_s(m, c)} < \epsilon_t \Leftrightarrow \text{subhalo is disrupted} \tag{5}$$

In Equation (5), ϵ_t is a dimensionless free parameter assumed universal, i.e., independent of the subhalo's mass, concentration, or position. In SL17, the value of the disruption parameter was set to $\epsilon_t = 1$ in agreement with numerical results (see, e.g., [59]). The results of van den Bosch and collaborators point toward a much lower value for ϵ_t. In this work, we consider two extreme values: $\epsilon_t = 1$ and $\epsilon_t = 0.01$. The latter means a subhalo is disrupted when it has lost around 99.99% of its mass. In the following, we refer to these two configurations as "fragile subhalos" ($\epsilon_t = 1$) and "resilient subhalos" ($\epsilon_t = 0.01$). The final subhalo phase-space density can now be written:

$$\frac{dN}{dV\,dm\,dc}(\vec{r}, m, c) = \frac{N_{tot}}{K_{tot}} \frac{d\mathcal{P}_v}{dV}(\vec{r}) \times \frac{d\mathcal{P}_m}{dm}(m) \times \frac{d\mathcal{P}_c}{dc}(c, m) \Theta\left(\frac{r_t(\vec{r}, m, c)}{r_s(m, c)} - \epsilon_t\right), \tag{6}$$

where N_{tot} is the total number of substructures within the virial radius of the Milky Way, and K_{tot} is a normalization factor:

$$K_{tot} = \int dV \frac{d\mathcal{P}_v}{dV}(\vec{r}) \int dm \frac{d\mathcal{P}_m}{dm}(m) \int dc \frac{d\mathcal{P}_c}{dc}(c, m) \Theta\left(\frac{r_t(\vec{r}, m, c)}{r_s(m, c)} - \epsilon_t\right). \tag{7}$$

2.3. Calibration Procedure

In its current version, the SL17 model requires a calibration of the subhalo abundance in a given mass range (this will change in future versions). To be consistent with results from the highest-resolution simulations available, calibration is done by demanding that the subhalo mass fraction is similar to what is found in the dark matter-only Via Lactea II simulation [26]. This amounts to 11% of the total dark halo mass in the form of subhalos in the virial mass range $[m_1, m_2] = [2.2 \times 10^{-6} M_{DM}, 8.8 \times 10^{-4} M_{DM}]$ where M_{DM} is the total DM mass of the galaxy. We stress that these numbers are expressed in terms of virial masses, not tidal masses (see [35] for further details). To reproduce the (likely overestimated) tidal disruption efficiency in simulations, we have to set $\epsilon_t = 1$ at the calibration stage, and we also neglect the impact of baryons. Disruption efficiency parameter ϵ_t can safely be changed after the calibration has been completed. It is much safer to perform this calibration on dark-matter-only simulations because tidal stripping induced by baryons strongly depends on the details of the stellar distribution, which is acutely constrained in the Milky Way. More formally, the normalization procedure reads:

$$f_{\text{sub}}(m_1, m_2) = \frac{1}{M_{\text{DM}}} \int dV \int_{m_1}^{m_2} dm \int dc \times m \times \left. \frac{dN}{dV \, dm \, dc} \right|_{\text{DMO}, \epsilon_t = 1}. \tag{8}$$

Fixing $f_{\text{sub}}(m_1, m_2) = 0.11$ leads to the total number of clumps $N_{\text{DMO}, \epsilon_t=1}$ in the simulation-like configuration. Note that this value assumes that m is really m_{200} in the equation above, not the tidal mass.

Now that the model is properly calibrated, we incorporate all the effects that are not included in the calibration, i.e., the tidal effects due to the baryons and possibly $\epsilon_t < 1$. This is done by assuming that subhalos in the outskirts of the galaxy are not affected by baryonic tides or the value of ϵ_t. This is motivated by the observation that tidal effects are inefficient far from the center of the galaxy, and subhalos almost behave like isolated halos. The DM mass within clumps per unit of volume can be expressed as

$$\langle \rho_{\text{cl}} \rangle (\vec{r}) = \int_{m_{\min}}^{m_{\max}} dm \int_1^\infty dc \, \frac{dN}{dV \, dm \, dc} \, m_t(\vec{r}, m, c), \tag{9}$$

where m_t is the tidal subhalo mass introduced in Equation (4). Equating $\langle \rho_{\text{cl}} \rangle (r_{200})$, where r_{200} is the virial radius of the galaxy, in the DM-only + $\epsilon_t = 1$ configuration, to the same quantity in the realistic configuration (including baryons and $\epsilon_t \leq 1$) leads to the simple relation

$$\frac{N_{\text{DMO}, \epsilon_t=1}}{K_{\text{DMO}, \epsilon_t=1}} = \frac{N_{\text{tot}}}{K_{\text{tot}}}. \tag{10}$$

The two normalization factors K can be computed using Equation (7), and we obtain the value of N_{tot}. The number of subhalos within the solar radius $r_\odot = 8.21$ kpc is shown in Table 1. This number is highly sensitive to the parameters of mass function α_m and $m_{\min}$, as already shown in Reference [35]. Furthermore, it is quite sensitive to ϵ_t: going from $\epsilon_t = 1$ to $\epsilon_t = 0.01$, the number of subhalos increases by at least an order of magnitude. The impact of ϵ_t on subhalo mass and number density is investigated in the next section.

Table 1. **Top panel:** number of subhalos within $r_\odot = 8.21$ kpc, for different values of mass function parameters α_m and $m_{\min}$, for $\epsilon_t = 1$. **Bottom panel:** same as top panel, for $\epsilon_t = 0.01$.

$\epsilon_t = 1$	$m_{\min} = 10^{-4} M_\odot$	$m_{\min} = 10^{-10} M_\odot$
$\alpha_m = 1.9$	1.90×10^{10}	1.55×10^{16}
$\alpha_m = 2$	2.64×10^{11}	8.40×10^{17}
$\epsilon_t = 0.01$	$m_{\min} = 10^{-4} M_\odot$	$m_{\min} = 10^{-10} M_\odot$
$\alpha_m = 1.9$	6.64×10^{11}	1.68×10^{17}
$\alpha_m = 2$	9.06×10^{12}	9.10×10^{18}

3. Mass and Number Densities of Subhalos

In this section, we computed the mass density within subhalos, as well as the subhalo number density. Subhalo mass density is defined in Equation (9). Once subhalo density is known, it is used to determine the amount of DM smoothly distributed across the galaxy through Equation (2). The DM mass inside subhalos was compared with the total DM density ρ_{DM} in Figure 2. Mass density in the form of subhalos is predicted to be much higher, by orders of magnitude, for resilient subhalos than for fragile subhalos. The former case is also more theoretically justified, although the latter one allows us to compare with very conservative assumptions. At the position of the Solar System, the impact is around one order of magnitude. Although these are large differences, we note the subhalo mass density is still far below the total DM density. This means that most of the DM mass within the orbit of the Sun is smoothly distributed rather than clumpy, irrespective of the efficiency of the tidal disruption set by ϵ_t.

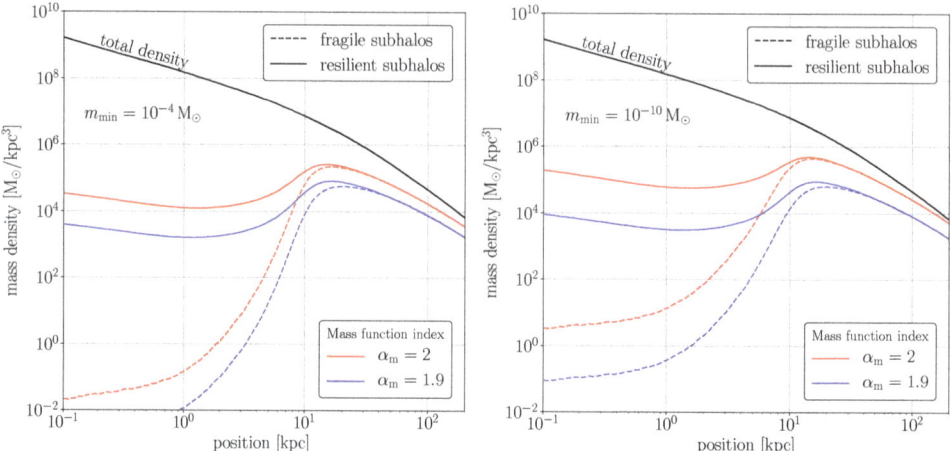

Figure 2. Left panel: Dark matter (DM) mass density inside subhalos $\langle \rho_{cl} \rangle$ for $m_{min} = 10^{-4}\,M_\odot$. The mass function index is $\alpha_m = 2$ (red) or $\alpha_m = 1.9$ (blue). We show the result for $\epsilon_t = 1$ (dashed) and $\epsilon_t = 0.01$ (solid). Total DM density is shown as a black solid curve for comparison. **Right panel:** same as left panel, for $m_{min} = 10^{-10}\,M_\odot$.

The subhalo number density in SL17 can be formally written:

$$\frac{dN}{dV}(\vec{r}) = \int dm \int dc \, \frac{dN}{dV\,dm\,dc} \tag{11}$$

$$= \frac{N_{tot}}{K_{tot}} \frac{d\mathcal{P}_v}{dV}(\vec{r}) \int_{m_{min}}^{m_{max}} dm \, \frac{d\mathcal{P}_m}{dm}(m) \int_1^\infty dc \, \frac{d\mathcal{P}_c}{dc}(c,m) \, \Theta\left(\frac{r_t(\vec{r},m,c)}{r_s} - \epsilon_t \right). \tag{12}$$

The obtained results are shown in Figure 3. Number density, just like mass density, is highly sensitive to α_m and m_{min}, as well as the disruption parameter. Interestingly, the values we get in the Solar neighborhood are comparable to the local number density of stars $n_* \sim 1\,\text{pc}^{-3}$. For a low value of minimal mass m_{min}, the subhalo number density can even be much higher, possibly going as high as $10^5\,\text{pc}^{-3}$. This could have a number of interesting implications for the interactions between subhalos and stars. The tidal heating of subhalos by stars has been investigated in a number of studies [29,60,64–68], with different conclusions. In the next section, we look at the impact of our results on indirect DM searches.

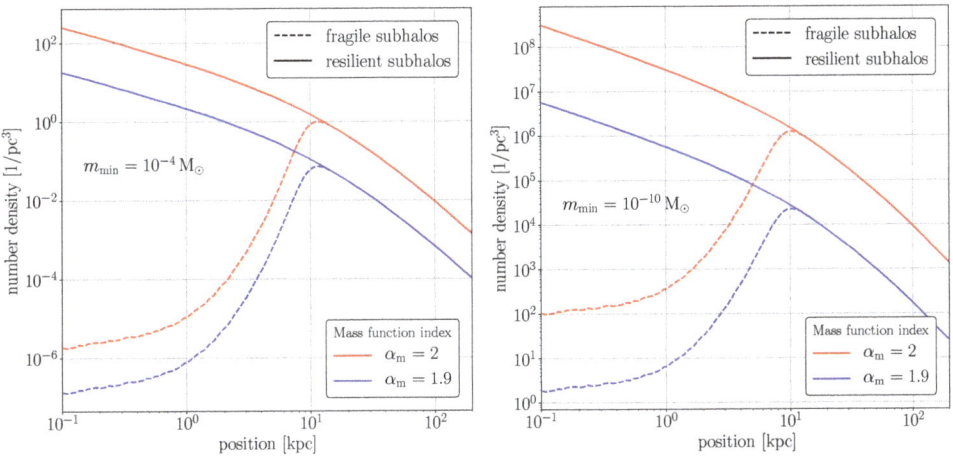

Figure 3. Left panel: subhalo number density for $m_{\min} = 10^{-4}\,\mathrm{M}_\odot$. Mass function index is $\alpha_m = 2$ (red) or $\alpha_m = 1.9$ (blue). We show the result for $\epsilon_t = 1$ (dashed) and $\epsilon_t = 0.01$ (solid). **Right panel:** same as left panel, for $m_{\min} = 10^{-10}\,\mathrm{M}_\odot$.

4. Impact on Indirect Searches for Annihilating Dark Matter

In this section, we quantify the impact of galactic clumps on indirect searches for self-annihilating DM. Inhomogeneities are known to enhance the DM annihilation rate in galactic halos [69]. We computed the local DM self-annihilation rate and evaluated the enhancement due to the survival of clump remnants, referred to as the boost factor. Two complementary channels were then investigated: gamma rays and antiproton cosmic rays.

4.1. Annihilation Profiles and Local Boost Factors

The number of self-annihilation of DM particles at position $\vec{r}$ is proportional to $\rho^2(\vec{r})$, where ρ is DM mass density. If subhalos are discarded, the galactic annihilation profile is

$$\mathcal{L}_0(\vec{r}) = \rho_{\mathrm{DM}}^2(\vec{r})\,. \tag{13}$$

Let us now consistently include the contribution of subhalos. The luminosity of a single clump is

$$L_t(\vec{r}, m, c) = \int_{V_t} \mathrm{d}^3 \vec{r}\, \rho_{\mathrm{sub}}^2(\vec{r})\,, \tag{14}$$

where $V_t(\vec{r}, m, c)$ is the volume of the clump within its tidal radius. The annihilation of the full subhalo population is simply obtained by integrating the luminosity of a single object over the subhalo phase-space number density:

$$\mathcal{L}_{\mathrm{cl}}(\vec{r}) = \int_{m_{\min}}^{m_{\max}} \mathrm{d}m \int_1^{+\infty} \mathrm{d}c\, \frac{\mathrm{d}N}{\mathrm{d}V \mathrm{d}m \mathrm{d}c} L_t(\vec{r}, m, c)\,. \tag{15}$$

The full annihilation profile must also incorporate the annihilation in the smooth halo (different from $\mathcal{L}_0$, which is the density assuming all the DM is smoothly distributed), as well as the annihilation of subhalo particles onto smooth halo particles. The first contribution can be written as:

$$\mathcal{L}_{\mathrm{sm}}(\vec{r}) = \rho_{\mathrm{sm}}^2(\vec{r})\,, \tag{16}$$

where $\rho_{\rm sm}(\vec{r}) = \rho_{\rm DM}(\vec{r}) - \langle \rho_{\rm cl} \rangle (\vec{r})$ is the smooth DM density. The clump-smooth contribution is:

$$\mathcal{L}_{\rm cs}(\vec{r}) = 2\rho_{\rm sm}(\vec{r}) \langle \rho_{\rm cl} \rangle (\vec{r}) \tag{17}$$

$$= 2\rho_{\rm sm}(\vec{r}) \int_{m_{\rm min}}^{m_{\rm max}} dm \int_{1}^{+\infty} dc \, \frac{dN}{dV dm dc} \, m_{\rm t}(\vec{r}, m, c) . \tag{18}$$

The total annihilation profile is simply the sum of all contributions:

$$\mathcal{L}(\vec{r}) = \mathcal{L}_{\rm cl}(\vec{r}) + \mathcal{L}_{\rm sm}(\vec{r}) + \mathcal{L}_{\rm cs}(\vec{r}) . \tag{19}$$

This should be compared to $\mathcal{L}_0(\vec{r})$ to evaluate the impact of clustering on the annihilation rate. This is usually done in terms of a boost factor, which we define as

$$1 + \mathcal{B}(\vec{r}) = \frac{\mathcal{L}(\vec{r})}{\mathcal{L}_0(\vec{r})} . \tag{20}$$

This is not quite the boost factor used in indirect searches, which is defined through a ratio of fluxes (see Equations (23) and 29)). The boost in Equation (20) is rather the local increase in the annihilation rate due to clustering. According to this definition, the boost is zero if $\mathcal{L} = \mathcal{L}_0$, i.e., substructures are not included.[2]

The annihilation profiles are shown on the top panels in Figure 4, and the associated boost factors are shown on the bottom panels. As already shown a long time ago, see e.g., References [29,70], the boost is an increasing function of the galactocentric radius $r = |\vec{r}|$. This is due to the morphology of the annihilation profiles that is modified by the inclusion of clumps: we have $\mathcal{L}_{\rm cl} \propto \rho_{\rm DM}$ while $\mathcal{L}_0 \propto \rho_{\rm DM}^2$. The high sensitivity of the annihilation profile to the mass function index α_m is also noticeable, which is by far the largest source of uncertainty on clump contribution. The value of disruption parameter ϵ_t has almost no impact on the boost above 20 kpc due to the ineffectiveness of tidal effects far from the center of the galaxy. Below 20 kpc, the boost is strongly sensitive to the disruption parameter. In the inner few kiloparsecs, fixing $\epsilon_t = 0.01$ leads to a boost orders of magnitude larger than in the $\epsilon_t = 1$ configuration. We, however, have $\mathcal{B}(\vec{r}) \ll 1$ below 3 kpc regardless of the value of ϵ_t, meaning $\mathcal{L} \simeq \mathcal{L}_0$ and subhalos do not have any impact on the annihilation rate in that region. The region where the impact of ϵ_t on the annihilation profile is the most important is located between 3 and 10 kpc. This region coincidentally includes the Solar System, located at $r \simeq 8$ kpc. This motivates a more detailed investigation of two standard annihilation channels: gamma rays and cosmic-ray antiprotons, which are sensitive to different annihilation regions.

---

[2] This differs by one unit from the definition used in Reference [35].

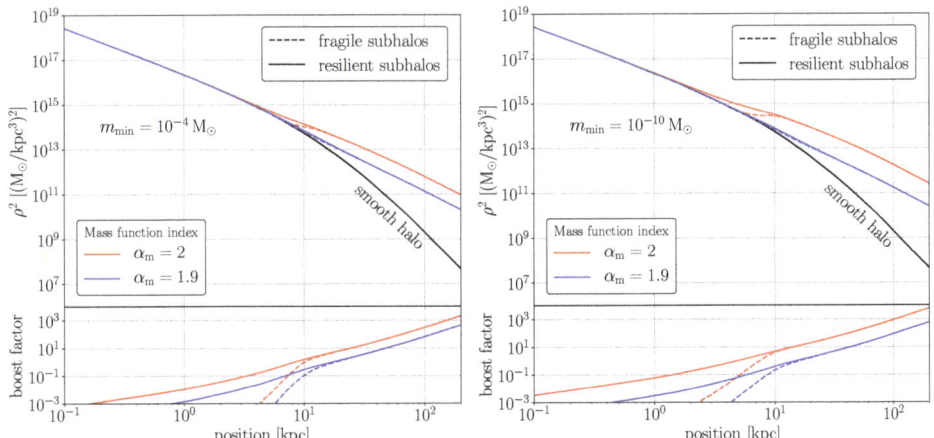

Figure 4. Total luminosity density profiles as defined in Equation (19), for a mass function index $\alpha_m = 2$ (red) and $\alpha_m = 1.9$ (blue). We show the results for efficient tidal disruption ($\epsilon_t = 1$, dashed) and very resilient clumps ($\epsilon_t = 0.01$, solid). (Left panel) Results for $m_{\min} = 10^{-4}\,M_\odot$, (right panel) results for $m_{\min} = 10^{-10}\,M_\odot$. The total luminosity density without clumps is displayed as a solid black line on each panel.

4.2. Application to Gamma Rays

The energy-differential flux of gamma rays originating from DM self-annihilation is, on a given line of sight,

$$\frac{d\Phi_\gamma}{dE\,d\Omega} = \frac{1}{4\pi} \frac{\langle \sigma v \rangle}{2\,m_\chi^2} \frac{dN_\gamma}{dE} \int ds\, \rho^2 \,, \qquad (21)$$

where $\langle \sigma v \rangle$ is the thermally averaged annihilation cross-section, m_χ is the DM mass, dN_γ/dE is the gamma-ray spectrum at annihilation, and s the distance coordinate along the line of sight.[3] If the annihilation cross-section is velocity-independent, astrophysical ingredients only enter through the J-factor, defined as

$$J(\psi) = \int ds\, \rho^2(s, \psi) \,, \qquad (22)$$

where ψ is the angle between the direction of the galactic center and the line of sight (spherical symmetry of the dark halo is assumed). The impact of small-scale clustering on this J-factor has been considered in a number of studies [32,48,70–78]. We define the gamma-ray boost factor as

$$1 + \mathcal{B}_\gamma(\psi) = \frac{J(\psi)}{J_0(\psi)} \,, \qquad (23)$$

where $J_0 = \int ds\, \mathcal{L}_0$ is the J-factor without subhalos. Unlike local boost $\mathcal{B}$ in Equation (20), gamma-ray boost $\mathcal{B}_\gamma$ depends on line of sight rather than the position in the galaxy [70]. The boost is shown as a function of ψ in Figure 5. The growth of local boost $\mathcal{B}(\vec{r})$ as a function of r translates into a growth of $\mathcal{B}_\gamma(\psi)$ as a function of ψ, i.e., the maximal gamma-ray boost is reached at the anticenter. This maximal boost ranges from 0.5 to 9, depending on the values of α_m and $m_{\min}$. The survival of clumps noticeably increases the boost at all latitudes. The gain is greater at small latitudes where substructures are more

---

[3] The expression of $d\Phi_\gamma/dE$ should be multiplied by $1/2$ if the DM particle is not its own antiparticle.

impacted by tidal effects. Below $\psi \simeq 40$ deg, the boost is increased by a factor of at least two in all configurations. This should have important consequences for indirect searches using gamma rays, especially at high latitudes. Interestingly, high latitudes have been shown to be a very sensitive probe of DM annihilation even without the inclusion of clumps [79].

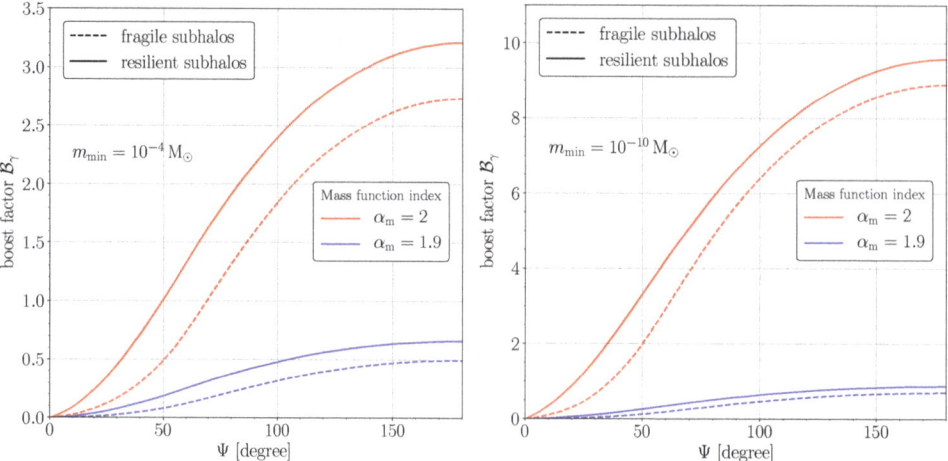

Figure 5. **Left panel:** gamma-ray boost factor as defined in Equation (23), as a function of angle ψ between the direction of the galactic center and the line of sight, for a minimal subhalo mass of $m_{\min} = 10^{-4}\,M_\odot$. We show the results for efficient tidal disruption ($\epsilon_t = 1$, dashed) and very resilient clumps ($\epsilon_t = 0.01$, solid). **Right panel:** same as left panel, for $m_{\min} = 10^{-10}\,M_\odot$.

4.3. Application to Cosmic-Ray Antiprotons

Charged cosmic rays constitute an indirect detection channel, complementary to gamma rays [10,80]. Since their original proposal as a probe of DM annihilation [81], cosmic-ray antiprotons have been shown to be especially sensitive, see e.g., References [82–88]. Antiprotons have been the subject of much scrutiny since the latest measurement of the antiproton flux performed by the AMS-2 collaboration [89]. A number of studies [90–95] have found a discrepancy between the measured flux and a purely secondary origin of antiprotons. This discrepancy could be interpreted as evidence for annihilating DM, although the significance of the excess is debated as it depends on the propagation model used and the modeling of systematic uncertainties. In this context, it is worth evaluating systematic uncertainties coming from small-scale structuring.

Since antiprotons have a random motion due to their diffusion on the inhomogeneities of the magnetic halo, their detection gives little information on their source. This implies that the antiproton boost factor, and the boost for charged cosmic rays in general, is not direction-dependent, unlike for gamma rays. Instead, this boost is energy-dependent [96] and has been shown to be mild, at most a factor of two [31,74]. Although smaller than the gamma-ray boost, this can still be larger than the systematic uncertainties on cosmic-ray propagation. This motivates a new computation of the boost, which we performed here. The antiproton boost factor is defined as:

$$1 + \mathcal{B}_{\bar{p}}(T) = \frac{\Phi_{\bar{p}}(T)}{\Phi_{\bar{p},0}(T)}, \qquad (24)$$

where T is the antiproton kinetic energy, $\Phi_{\bar{p}}$ is the DM-induced antiproton flux including the subhalo contribution, and $\Phi_{\bar{p},0}$ the same flux assuming all the DM is smoothly distributed. To obtain the flux, one must solve the cosmic-ray steady-state propagation equation

$$-K\Delta\Psi + \vec{\nabla}\cdot(\vec{V}_c\Psi) + \partial_E\left[b\Psi - K_{EE}\partial_E\Psi\right] + 2h\delta(z)\Gamma_{\text{ann}}\Psi = Q_{\text{DM}}, \tag{25}$$

which accounts for spatial diffusion, convection, energy losses, diffusive re-acceleration, and spallation processes in the disk (taken as infinitely thin). In Equation (25), Ψ is the antiproton number density per unit energy that is related to the flux through $\Phi_{\bar{p}} = v_{\bar{p}}/(4\pi) \times \Psi$ where $v_{\bar{p}}$ is the antiproton speed. Antiprotons are sourced by DM annihilation:

$$Q_{\text{DM}}(E,\vec{r}) = \frac{\langle\sigma v\rangle}{2}\frac{dN_{\bar{p}}}{dE}\left(\frac{\rho(\vec{r})}{m_\chi}\right)^2, \tag{26}$$

where $dN_{\bar{p}}/dE$ is the antiproton spectrum at annihilation.[4] Several unknown propagation parameters enter Equation (25). These can be constrained using the measured boron-to-carbon ratio (B/C) [97]. We used the best-fit model derived by Reinert and Winkler [92], which includes an energy break in the diffusion coefficient. The B/C ratio can only constrain K_0/L, where K_0 is the normalization of the diffusion coefficient, and L is the half-height of the magnetic halo. As shown in Reference [83], the DM-induced antiproton flux crucially depends on L; hence, we considered two extremal values in this work. A lower bound on L can be obtained from low-energy positron data [98], and the authors of Reference [92] found $L = 4.1$ kpc. For the largest value, we took $L = 15$ kpc. According to the analysis of Reference [92], the B/C data are consistent with negligible re-acceleration. Furthermore, we neglected energy losses that are unimportant for high-energy antiprotons. The resulting transport equation can be solved semianalytically using Green's function formalism (see Reference [31] for the solution), and the differential flux can be written as:

$$\frac{d\Phi_{\bar{p}}}{dT\,d\Omega} = \frac{v_{\bar{p}}}{4\pi}\frac{\langle\sigma v\rangle}{2\,m_\chi^2}\int dE_s\int d^3\vec{r}_s\, G(E \leftarrow E_s; \vec{r}_\odot \leftarrow \vec{r}_s)\frac{dN_{\bar{p}}}{dE}(E_s)\rho^2(\vec{r}_s). \tag{27}$$

Since all energy-dependent terms have been neglected in Equation (25), the energy part of Green's function is trivial

$$G(E \leftarrow E_s; \vec{r}_\odot \leftarrow \vec{r}_s) = \delta(E - E_s) \times \overline{G}(\vec{r}_\odot \leftarrow \vec{r}_s), \tag{28}$$

and the boost factor can be simply written:

$$1 + \mathcal{B}_{\bar{p}}(T) = \frac{\int d^3\vec{r}_s\,\overline{G}(\vec{r}_\odot \leftarrow \vec{r}_s)\,\mathcal{L}(\vec{r}_s)}{\int d^3\vec{r}_s\,\overline{G}(\vec{r}_\odot \leftarrow \vec{r}_s)\,\mathcal{L}_0(\vec{r}_s)}. \tag{29}$$

The boost factor is shown as a function of the antiproton kinetic energy in Figure 6. We first note that the boost is roughly energy-independent. This is because antiprotons probe the entire volume of the magnetic halo during their lifetime, independently of their energy. This is not true at low energies, below a few GeVs, where energy losses become relevant. The half-height of the magnetic halo has a small impact on the boost, with $L = 15$ kpc leading to a slightly larger $\mathcal{B}_{\bar{p}}$ than $L = 4.1$ kpc. The main source of uncertainties are coming from subhalo parameters α_m, $m_{\min}$, and ϵ_t. Survival parameter ϵ_t had significant impact on the result, with a small value $\epsilon_t = 0.01$ leading to a boost roughly twice as large as in the $\epsilon_t = 1$ case. As for gamma rays, the most critical parameter is α_m. For a low-value $\alpha_m = 1.9$, the boost never exceeded 10%, while it was always higher for $\alpha_m = 2$.

---

[4] If DM is not its own antiparticle, Q should be divided by 2.

Overall, playing with the propagation and subhalo parameters, we found that the antiproton boost can conservatively range from 2% to 140%. These values are in agreement with earlier results. Although it is conservative to ignore small-scale clustering when deriving limits on the annihilation cross-section using data, the boost should be included when interpreting an excess as a signature of DM annihilation. Indeed, a factor of two in the DM contribution would change the inferred mass and cross-section of the hypothetical DM particle.

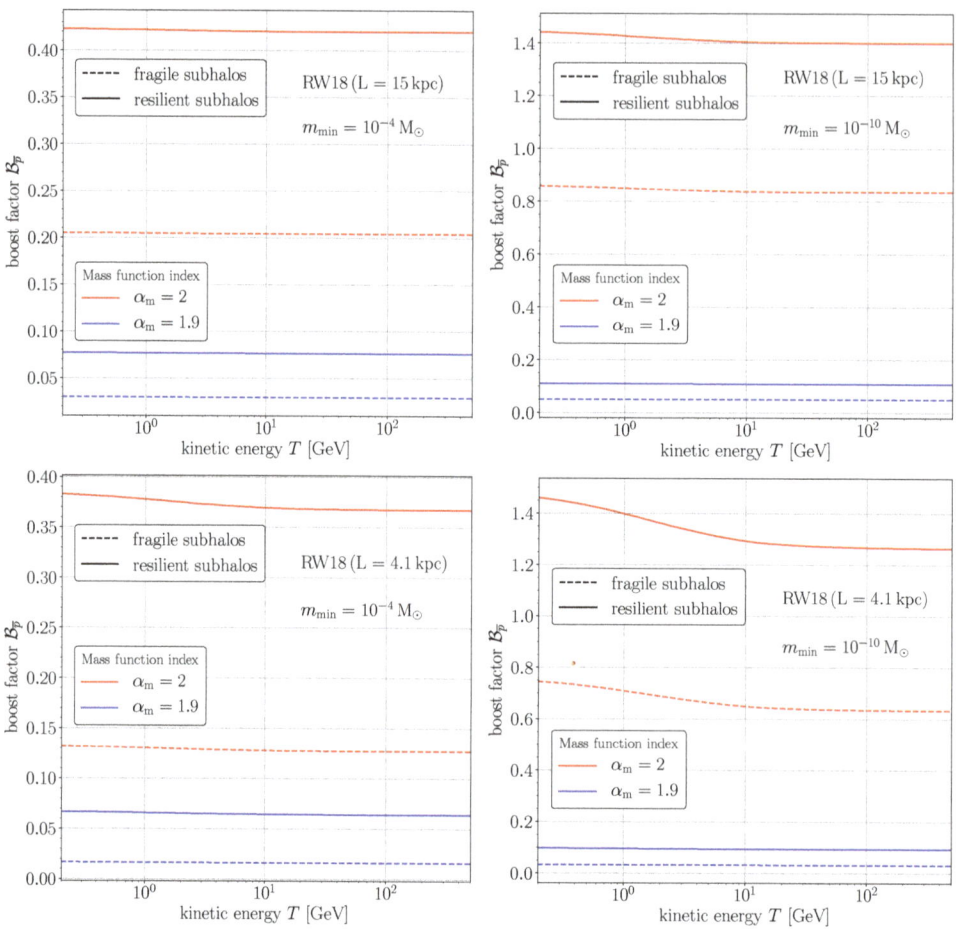

Figure 6. Top panel: antiproton boost factor as a function of kinetic energy, as defined in Equation (29). We show the result for a half-height of the magnetic halo of $L = 15$ kpc, and a minimal subhalo mass of $m_{\min} = 10^{-4}\,M_\odot$ (left) or $m_{\min} = 10^{-10}\,M_\odot$ (right). **Bottom panel**: same as top panel, for half-height $L = 4.1$ kpc.

5. Conclusions

Subhalos suffer mass loss due to their interaction with the tidal field of the galaxy, which makes their modeling very challenging. Consequently, most subhalo models rely at least partly on numerical simulations to calibrate their predictions. However, it was recently shown that numerical simulations might not properly account for the tidal disruption of subhalos, as artificial effects lead to a serious overestimation of the efficiency of these processes [33,34]. We note that the resistance of subhalos to tidal stripping is further supported by theoretical arguments, like adiabatic invariance that should prevail in their inner parts [55], as already emphasized in Reference [35]. We derived some of the consequences of these results using the semianalytical galactic subhalo population model of Stref and Lavalle [35], assuming tidal disruption efficiency $\epsilon_t = 0.01$. We predicted the spatial dependence of the subhalo properties due to tides induced both by the global gravitational potential and baryonic disk shocking. We remind the reader that this model is built from constrained mass models for the Milky Way and is therefore consistent with current kinematic constraints, which is usually not the case in extrapolations from "Milky Way-like" simulations. We found that the local mass density is still dominated by the smooth component of the dark halo. The local number density of subhalos is increased by roughly one order of magnitude with respect to estimates based on a tidal disruption efficiency similar to that inferred from simulations ($\epsilon_t = 1$). This makes the subhalo number density comparable, for a broad range of minimal subhalo mass m_{min}, to the local star number density. Since our description of the subhalo population relies solely on gravitational principles, it should be valid for a wide range of cold DM candidate such as WIMPs, axions, or primordial black holes. One only needs to modify the mass function, in particular, m_{min}, to explore alternatives to the WIMP scenario. The resilience of subhalos increases the local WIMP annihilation rate, which, in turn, affects predictions for indirect searches. For gamma rays, we found that the boost factor was increased by at least a factor of two for $\psi < 40$ deg, and slightly less for higher values of ψ. The boost factor for antiprotons is also increased by a rough factor of two if subhalos are resilient to tidal disruption. For a complementary study comparing the SL17 model to simulation results regarding indirect searches in gamma rays and neutrinos, we refer the reader to Reference [99].

For future work, we plan on including a more detailed mass function, directly deriving from the primordial power spectrum, as well as the tidal heating of subhalos due to individual stars, and studying the consequences of having a large population of small objects in the Solar neighborhood.

Author Contributions: conceptualization, all authors; software, M.S.; investigation, M.S.; writing–original-draft preparation, M.S.; writing–review and editing, T.L. and J.L.; supervision, J.L.

Acknowledgments: We wish to thank M. A. Sánchez-Conde and M. Doro for inviting us to contribute to this topical review.

Funding: J.L. and M.S. are partly supported by the Agence Nationale pour la Recherche (ANR) Project No. ANR-18-CE31-0006, the Origines, Constituants, et EVolution de l'Univers (OCEVU) Labex (No. ANR-11-LABX-0060), the CNRS IN2P3-Theory/INSU-PNHE-PNCG project "Galactic Dark Matter," and the European Union's Horizon 2020 Research and Innovation Program under Marie Skłodowska-Curie Grant Agreements No. 690575 and No. 674896, in addition to recurrent funding by the Centre National de la Recherche Scientifique (CNRS) and the University of Montpellier. T.L. is supported by the European Union's Horizon 2020 Research and Innovation Program under the Marie Skłodowska-Curie grant agreement No. 713366. The work of TL was also supported by the Spanish Agencia Estatal de Investigación through grants PGC2018-095161-B-I00, IFT Centro de Excelencia Severo Ochoa SEV-2016-0597, and Red Consolider MultiDark FPA2017-90566-REDC.

Conflicts of Interest: The authors declare no conflict of interest. The funders had no role in the design of the study; in the collection, analyses, or interpretation of data; in the writing of the manuscript; or in the decision to publish the results.

References

1. Aghanim, N.; Akrami, Y.; Ashdown, M.; Aumont, J.; Baccigalupi, C.; Ballardini, M.; Banday, A.J.; Barreiro, R.B.; Bartolo, N.; Basak, S.; et al. Planck 2018 results. VI. Cosmological parameters. *arXiv* **2018**, arXiv:1807.06209.
2. Bertone, G.; Hooper, D.; Silk, J. Particle dark matter: Evidence, candidates and constraints. *Phys. Rep.* **2005**, *405*, 279–390. [CrossRef]
3. Feng, J.L. Dark Matter Candidates from Particle Physics and Methods of Detection. *Ann. Rev. Astron. Astrophys.* **2010**, *48*, 495–545. [CrossRef]
4. Fairbairn, M.; Kraan, A.C.; Milstead, D.A.; Sjöstrand, T.; Skands, P.; Sloan, T. Stable massive particles at colliders. *Phys. Rep.* **2007**, *438*, 1–63. [CrossRef]
5. Goodman, J.; Ibe, M.; Rajaraman, A.; Shepherd, W.; Tait, T.M.P.; Yu, H.B. Constraints on Dark Matter from Colliders. *Phys. Rev.* **2010**, *D82*, 116010. [CrossRef]

6. Kahlhoefer, F. Review of LHC Dark Matter Searches. *Int. J. Mod. Phys.* **2017**, *A32*, 1730006. [CrossRef]
7. Cerdeno, D.G.; Green, A.M. Direct detection of WIMPs. In *Particle Dark Matter: Observations, Models and Searches*; Cambridge University Press: Cambridge, UK, 2010; pp. 347–369.
8. Baudis, L. Direct dark matter detection: The next decade. *Phys. Dark Univ.* **2012**, *1*, 94–108. [CrossRef]
9. Freese, K.; Lisanti, M.; Savage, C. Colloquium: Annual modulation of dark matter. *Rev. Mod. Phys.* **2013**, *85*, 1561–1581. [CrossRef]
10. Lavalle, J.; Salati, P. Dark matter indirect signatures. *C. R. Phys.* **2012**, *13*, 740–782. [CrossRef]
11. Gaskins, J.M. A review of indirect searches for particle dark matter. *Contemp. Phys.* **2016**, *57*, 496–525. [CrossRef]
12. Jungman, G.; Kamionkowski, M.; Griest, K. Supersymmetric dark matter. *Phys. Rep.* **1996**, *267*, 195–373. [CrossRef]
13. Servant, G.; Tait, T.M.P. Is the lightest Kaluza-Klein particle a viable dark matter candidate? *Nuclear Phys. B* **2003**, *650*, 391–419. [CrossRef]
14. Agashe, K.; Servant, G. Warped Unification, Proton Stability, and Dark Matter. *Phys. Rev. Lett.* **2004**, *93*, 231805. [CrossRef]
15. Leane, R.K.; Slatyer, T.R.; Beacom, J.F.; Ng, K.C.Y. GeV-scale thermal WIMPs: Not even slightly ruled out. *Phys. Rev.* **2018**, *D98*, 023016. [CrossRef]
16. Marsh, D.J.E. Axion Cosmology. *Phys. Rep.* **2016**, *643*, 1–79. [CrossRef]
17. Boyarsky, A.; Ruchayskiy, O.; Shaposhnikov, M. The Role of sterile neutrinos in cosmology and astrophysics. *Ann. Rev. Nucl. Part. Sci.* **2009**, *59*, 191–214. [CrossRef]
18. Carr, B.; Kuhnel, F.; Sandstad, M. Primordial Black Holes as Dark Matter. *Phys. Rev.* **2016**, *D94*, 083504. [CrossRef]
19. Alexander, J.; Battaglieri, M.; Echenard, B.; Essig, R.; Graham, M.; Izaguirre, G.; Jaros, J.; Krnjaic, G.; Mardon, J.; Morrissey, D.; et al. Dark Sectors 2016 Workshop: Community Report. *arXiv* **2016**, arXiv:1608.08632.
20. Blumenthal, G.R.; Faber, S.M.; Primack, J.R.; Rees, M.J. Formation of galaxies and large-scale structure with cold dark matter. *Nature* **1984**, *311*, 517–525. [CrossRef]
21. Davis, M.; Efstathiou, G.; Frenk, C.S.; White, S.D.M. The evolution of large-scale structure in a universe dominated by cold dark matter. *Astrophys. J.* **1985**, *292*, 371–394. [CrossRef]
22. Bullock, J.S.; Boylan-Kolchin, M. Small-Scale Challenges to the ΛCDM Paradigm. *Ann. Rev. Astron. Astrophys.* **2017**, *55*, 343–387. [CrossRef]
23. Tulin, S.; Yu, H.B. Dark Matter Self-interactions and Small Scale Structure. *Phys. Rep.* **2018**, *730*, 1–57. [CrossRef]
24. Pontzen, A.; Governato, F. How supernova feedback turns dark matter cusps into cores. *Mon. Not. R. Astron. Soc.* **2012**, *421*, 3464. [CrossRef]
25. Moore, B.; Ghigna, S.; Governato, F.; Lake, G.; Quinn, T.R.; Stadel, J.; Tozzi, P. Dark matter substructure within galactic halos. *Astrophys. J.* **1999**, *524*, L19–L22. [CrossRef]
26. Diemand, J.; Kuhlen, M.; Madau, P.; Zemp, M.; Moore, B.; Potter, D.; Stadel, J. Clumps and streams in the local dark matter distribution. *Nature* **2008**, *454*, 735–738. [CrossRef]
27. Springel, V.; Wang, J.; Vogelsberger, M.; Ludlow, A.; Jenkins, A.; Helmi, A.; Navarro, J.F.; Frenk, C.S.; White, S.D.M. The Aquarius Project: The subhalos of galactic halos. *Mon. Not. R. Astron. Soc.* **2008**, *391*, 1685–1711. [CrossRef]
28. Springel, V.; White, S.D.M.; Frenk, C.S.; Navarro, J.F.; Jenkins, A.; Vogelsberger, M.; Wang, J.; Ludlow, A.; Helmi, A. Prospects for detecting supersymmetric dark matter in the Galactic halo. *Nature* **2008**, *456*, 73–76. [CrossRef]
29. Berezinsky, V.; Dokuchaev, V.; Eroshenko, Y. Small-scale clumps in the galactic halo and dark matter annihilation. *Phys. Rev.* **2003**, *D68*, 103003. [CrossRef]
30. Penarrubia, J.; Benson, A.J. Effects of dynamical evolution on the distribution of substructures. *Mon. Not. R. Astron. Soc.* **2005**, *364*, 977–989. [CrossRef]
31. Lavalle, J.; Yuan, Q.; Maurin, D.; Bi, X.J. Full calculation of clumpiness boost factors for antimatter cosmic rays in the light of ΛCDM N-body simulation results. Abandoning hope in clumpiness enhancement? *Astron. Astroph.* **2008**, *479*, 427–452. [CrossRef]

32. Bartels, R.; Ando, S. Boosting the annihilation boost: Tidal effects on dark matter subhalos and consistent luminosity modeling. *Phys. Rev.* **2015**, *D92*, 123508. [CrossRef]
33. van den Bosch, F.C.; Ogiya, G.; Hahn, O.; Burkert, A. Disruption of Dark Matter Substructure: Fact or Fiction? *Mon. Not. R. Astron. Soc.* **2018**, *474*, 3043–3066. [CrossRef]
34. van den Bosch, F.C.; Ogiya, G. Dark Matter Substructure in Numerical Simulations: A Tale of Discreteness Noise, Runaway Instabilities, and Artificial Disruption. *Mon. Not. R. Astron. Soc.* **2018**, *475*, 4066–4087. [CrossRef]
35. Stref, M.; Lavalle, J. Modeling dark matter subhalos in a constrained galaxy: Global mass and boosted annihilation profiles. *Phys. Rev.* **2017**, *D95*, 063003. [CrossRef]
36. McMillan, P.J. The mass distribution and gravitational potential of the Milky Way. *Mon. Not. R. Astron. Soc.* **2017**, *465*, 76–94. [CrossRef]
37. Press, W.H.; Schechter, P. Formation of Galaxies and Clusters of Galaxies by Self-Similar Gravitational Condensation. *Astrophys. J.* **1974**, *187*, 425–438. [CrossRef]
38. Sheth, R.K.; Tormen, G. Large-scale bias and the peak background split. *Mon. Not. R. Astron. Soc.* **1999**, *308*, 119–126. [CrossRef]
39. Bond, J.R.; Cole, S.; Efstathiou, G.; Kaiser, N. Excursion set mass functions for hierarchical Gaussian fluctuations. *Astrophys. J.* **1991**, *379*, 440–460. [CrossRef]
40. Hofmann, S.; Schwarz, D.J.; Stoecker, H. Damping scales of neutralino cold dark matter. *Phys. Rev.* **2001**, *D64*, 083507. [CrossRef]
41. Loeb, A.; Zaldarriaga, M. The Small-scale power spectrum of cold dark matter. *Phys. Rev.* **2005**, *D71*, 103520. [CrossRef]
42. Green, A.M.; Hofmann, S.; Schwarz, D.J. The First wimpy halos. *J. Cosmol. Astropart. Phys.* **2005**, *0508*, 003. [CrossRef]
43. Bertschinger, E. The Effects of Cold Dark Matter Decoupling and Pair Annihilation on Cosmological Perturbations. *Phys. Rev.* **2006**, *D74*, 063509. [CrossRef]
44. Bringmann, T.; Hofmann, S. Thermal decoupling of WIMPs from first principles. *J. Cosmol. Astropart. Phys.* **2007**, *0704*, 016. [CrossRef]
45. Bringmann, T. Particle Models and the Small-Scale Structure of Dark Matter. *New J. Phys.* **2009**, *11*, 105027. [CrossRef]
46. Bullock, J.S.; Kolatt, T.S.; Sigad, Y.; Somerville, R.S.; Kravtsov, A.V.; Klypin, A.A.; Primack, J.R.; Dekel, A. Profiles of dark haloes: Evolution, scatter and environment. *Mon. Not. R. Astron. Soc.* **2001**, *321*, 559–575. [CrossRef]
47. Macciò, A.V.; Dutton, A.A.; van den Bosch, F.C. Concentration, spin and shape of dark matter haloes as a function of the cosmological model: WMAP1, WMAP3 and WMAP5 results. *Mon. Not. R. Astron. Soc.* **2008**, *391*, 1940–1954. [CrossRef]
48. Sánchez-Conde, M.A.; Prada, F. The flattening of the concentration mass relation towards low halo masses and its implications for the annihilation signal boost. *Mon. Not. R. Astron. Soc.* **2014**, *442*, 2271–2277. [CrossRef]
49. Navarro, J.F.; Frenk, C.S.; White, S.D.M. A Universal Density Profile from Hierarchical Clustering. *Astrophys. J.* **1997**, *490*, 493–508. [CrossRef]
50. Einasto, J. On the Construction of a Composite Model for the Galaxy and on the Determination of the System of Galactic Parameters. *Trudy Astrofizicheskogo Instituta Alma-Ata* **1965**, *5*, 87–100.
51. Eisenstein, D.J.; Hu, W. Baryonic features in the matter transfer function. *Astrophys. J.* **1998**, *496*, 605. [CrossRef]
52. Binney, J.; Tremaine, S. *Galactic Dynamics*; Princeton University Press: Princeton, NJ, USA, 1987.
53. Drakos, N.E.; Taylor, J.E.; Benson, A.J. The phase-space structure of tidally stripped haloes. *Mon. Not. R. Astron. Soc.* **2017**, *468*, 2345–2358. [CrossRef]
54. Ostriker, J.P.; Spitzer, L., Jr.; Chevalier, R.A. On the Evolution of Globular Clusters. *Astrophys. J. Lett.* **1972**, *176*, L51. [CrossRef]
55. Gnedin, O.Y.; Ostriker, J.P. On the selfconsistent response of stellar systems to gravitational shocks. *Astrophys. J.* **1999**, *513*, 626. [CrossRef]
56. D'Onghia, E.; Springel, V.; Hernquist, L.; Keres, D. Substructure Depletion in the Milky Way Halo by the Disk. *Astrophys. J.* **2010**, *709*, 1138–1147. [CrossRef]

57. Errani, R.; Peñarrubia, J.; Laporte, C.F.P.; Gómez, F.A. The effect of a disc on the population of cuspy and cored dark matter substructures in Milky Way-like galaxies. *Mon. Not. R. Astron. Soc.* **2017**, *465*, L59–L63. [CrossRef]
58. Kelley, T.; Bullock, J.S.; Garrison-Kimmel, S.; Boylan-Kolchin, M.; Pawlowski, M.S.; Graus, A.S. Phat ELVIS: The inevitable effect of the Milky Way's disk on its dark matter subhaloes. *arXiv* **2018**, arXiv:1811.12413.
59. Hayashi, E.; Navarro, J.F.; Taylor, J.E.; Stadel, J.; Quinn, T.R. The Structural evolution of substructure. *Astrophys. J.* **2003**, *584*, 541–558. [CrossRef]
60. Goerdt, T.; Gnedin, O.Y.; Moore, B.; Diemand, J.; Stadel, J. The survival and disruption of CDM micro-haloes: Implications for direct and indirect detection experiments. *Mon. Not. R. Astron. Soc.* **2007**, *375*, 191–198. [CrossRef]
61. Berezinsky, V.; Dokuchaev, V.; Eroshenko, Y. Remnants of dark matter clumps. *Phys. Rev.* **2008**, *D77*, 083519. [CrossRef]
62. Peñarrubia, J.; Benson, A.J.; Walker, M.G.; Gilmore, G.; McConnachie, A.W.; Mayer, L. The impact of dark matter cusps and cores on the satellite galaxy population around spiral galaxies. *Mon. Not. R. Astron. Soc.* **2010**, *406*, 1290–1305. [CrossRef]
63. van den Bosch, F.C. Dissecting the Evolution of Dark Matter Subhaloes in the Bolshoi Simulation. *Mon. Not. R. Astron. Soc.* **2017**, *468*, 885–909. [CrossRef]
64. Berezinsky, V.; Dokuchaev, V.; Eroshenko, Y. Destruction of small-scale dark matter clumps in the hierarchical structures and galaxies. *Phys. Rev.* **2006**, *D73*, 063504. [CrossRef]
65. Zhao, H.; Taylor, J.E.; Silk, J.; Hooper, D. Tidal Disruption of the First Dark Microhalos. *Astrophys. J.* **2007**, *654*, 697–701. [CrossRef]
66. Angus, G.W.; Zhao, H. Analysis of galactic tides and stars on CDM microhalos. *Mon. Not. R. Astron. Soc.* **2007**, *375*, 1146–1156. [CrossRef]
67. Green, A.M.; Goodwin, S.P. On mini-halo encounters with stars. *Mon. Not. R. Astron. Soc.* **2007**, *375*, 1111–1120. [CrossRef]
68. Schneider, A.; Krauss, L.; Moore, B. Impact of Dark Matter Microhalos on Signatures for Direct and Indirect Detection. *Phys. Rev.* **2010**, *D82*, 063525. [CrossRef]
69. Silk, J.; Stebbins, A. Clumpy cold dark matter. *Astrophys. J.* **1993**, *411*, 439–449. [CrossRef]
70. Bergstrom, L.; Edsjo, J.; Gondolo, P.; Ullio, P. Clumpy neutralino dark matter. *Phys. Rev.* **1999**, *D59*, 043506. [CrossRef]
71. Diemand, J.; Kuhlen, M.; Madau, P. Dark matter substructure and gamma-ray annihilation in the Milky Way halo. *Astrophys. J.* **2007**, *657*, 262–270. [CrossRef]
72. Strigari, L.E.; Koushiappas, S.M.; Bullock, J.S.; Kaplinghat, M. Precise constraints on the dark matter content of Milky Way dwarf galaxies for gamma-ray experiments. *Phys. Rev.* **2007**, *D75*, 083526. [CrossRef]
73. Pieri, L.; Bertone, G.; Branchini, E. Dark Matter Annihilation in Substructures Revised. *Mon. Not. R. Astron. Soc.* **2008**, *384*, 1627. [CrossRef]
74. Pieri, L.; Lavalle, J.; Bertone, G.; Branchini, E. Implications of high-resolution simulations on indirect dark matter searches. *Phys. Rev. D* **2011**, *83*, 023518. [CrossRef]
75. Blanchet, S.; Lavalle, J. Diffuse gamma-ray constraints on dark matter revisited. I: The impact of subhalos. *J. Cosmol. Astropart. Phys.* **2012**, *1211*, 021. [CrossRef]
76. Ishiyama, T. Hierarchical Formation of Dark Matter Halos and the Free Streaming Scale. *Astrophys. J.* **2014**, *788*, 27. [CrossRef]
77. Moliné, Á.; Sánchez-Conde, M.A.; Palomares-Ruiz, S.; Prada, F. Characterization of subhalo structural properties and implications for dark matter annihilation signals. *Mon. Not. R. Astron. Soc.* **2017**, *466*, 4974–4990. [CrossRef]
78. Hiroshima, N.; Ando, S.; Ishiyama, T. Modeling evolution of dark matter substructure and annihilation boost. *Phys. Rev.* **2018**, *D97*, 123002. [CrossRef]
79. Chang, L.J.; Lisanti, M.; Mishra-Sharma, S. Search for dark matter annihilation in the Milky Way halo. *Phys. Rev.* **2018**, *D98*, 123004. [CrossRef]
80. Maurin, D.; Taillet, R.; Donato, F.; Salati, P.; Barrau, A.; Boudoul, G. Galactic cosmic ray nuclei as a tool for astroparticle physics. *arXiv* **2002**, arXiv:astro-ph/0212111.
81. Silk, J.; Srednicki, M. Cosmic-ray antiprotons as a probe of a photino-dominated universe. *Phys. Rev. Lett.* **1984**, *53*, 624–627. [CrossRef]

82. Bergström, L.; Edsjö, J.; Ullio, P. Cosmic Antiprotons as a Probe for Supersymmetric Dark Matter? *Astrophys. J.* **1999**, *526*, 215–235. [CrossRef]
83. Donato, F.; Fornengo, N.; Maurin, D.; Salati, P. Antiprotons in cosmic rays from neutralino annihilation. *Phys. Rev.* **2004**, *D69*, 063501. [CrossRef]
84. Barrau, A.; Salati, P.; Servant, G.; Donato, F.; Grain, J.; Maurin, D.; Taillet, R. Kaluza-Klein dark matter and Galactic antiprotons. *Phys. Rev.* **2005**, *D72*, 063507. [CrossRef]
85. Bringmann, T.; Salati, P. The galactic antiproton spectrum at high energies: Background expectation vs. exotic contributions. *Phys. Rev.* **2007**, *D75*, 083006. [CrossRef]
86. Cirelli, M.; Kadastik, M.; Raidal, M.; Strumia, A. Model-independent implications of the e+-, anti-proton cosmic ray spectra on properties of Dark Matter. *Nucl. Phys.* **2009**, *B813*, 1–21. [CrossRef]
87. Boudaud, M.; Cirelli, M.; Giesen, G.; Salati, P. A fussy revisitation of antiprotons as a tool for Dark Matter searches. *J. Cosmol. Astropart. Phys.* **2015**, *5*, 13. [CrossRef]
88. Giesen, G.; Boudaud, M.; Génolini, Y.; Poulin, V.; Cirelli, M.; Salati, P.; Serpico, P.D. AMS-02 antiprotons, at last! Secondary astrophysical component and immediate implications for Dark Matter. *J. Cosmol. Astropart. Phys.* **2015**, *1509*, 023. [CrossRef]
89. Aguilar, M.; Ali Cavasonza, L.; Alpat, B.; Ambrosi, G.; Arruda, L.; Attig, N.; Aupetit, S.; Azzarello, P.; Bachlechner, A.; Barao, F.; et al. Antiproton Flux, Antiproton-to-Proton Flux Ratio, and Properties of Elementary Particle Fluxes in Primary Cosmic Rays Measured with the Alpha Magnetic Spectrometer on the International Space Station. *Phys. Rev. Lett.* **2016**, *117*, 091103. [CrossRef]
90. Cuoco, A.; Krämer, M.; Korsmeier, M. Novel Dark Matter Constraints from Antiprotons in Light of AMS-02. *Phys. Rev. Lett.* **2017**, *118*, 191102. [CrossRef]
91. Cui, M.Y.; Yuan, Q.; Tsai, Y.L.S.; Fan, Y.Z. Possible dark matter annihilation signal in the AMS-02 antiproton data. *Phys. Rev. Lett.* **2017**, *118*, 191101. [CrossRef]
92. Reinert, A.; Winkler, M.W. A Precision Search for WIMPs with Charged Cosmic Rays. *J. Cosmol. Astropart. Phys.* **2018**, *1801*, 055. [CrossRef]
93. Cui, M.Y.; Pan, X.; Yuan, Q.; Fan, Y.Z.; Zong, H.S. Revisit of cosmic ray antiprotons from dark matter annihilation with updated constraints on the background model from AMS-02 and collider data. *J. Cosmol. Astropart. Phys.* **2018**, *1806*, 024. [CrossRef]
94. Cuoco, A.; Heisig, J.; Klamt, L.; Korsmeier, M.; Krämer, M. Scrutinizing the evidence for dark matter in cosmic-ray antiprotons. *arXiv* **2019**, arXiv:1903.01472.
95. Cholis, I.; Linden, T.; Hooper, D. A Robust Excess in the Cosmic-Ray Antiproton Spectrum: Implications for Annihilating Dark Matter. *arXiv* **2019**, arXiv:1903.02549.
96. Lavalle, J.; Pochon, J.; Salati, P.; Taillet, R. Clumpiness of dark matter and positron annihilation signal: Computing the odds of the galactic lottery. *Astron. Astrophys.* **2007**, *462*, 827–848. [CrossRef]
97. Strong, A.W.; Moskalenko, I.V. Propagation of cosmic-ray nucleons in the galaxy. *Astrophys. J.* **1998**, *509*, 212–228. [CrossRef]
98. Lavalle, J.; Maurin, D.; Putze, A. Direct constraints on diffusion models from cosmic-ray positron data: Excluding the minimal model for dark matter searches. *Phys. Rev.* **2014**, *D90*, 081301. [CrossRef]
99. Hütten, M.; Stref, M.; Combet, C.; Lavalle, J.; Maurin, D. γ-ray and ν searches for dark matter subhalos in the Milky Way with a baryonic potential. *arXiv* **2019**, arXiv:1904.10935.

© 2019 by the authors. Licensee MDPI, Basel, Switzerland. This article is an open access article distributed under the terms and conditions of the Creative Commons Attribution (CC BY) license (http://creativecommons.org/licenses/by/4.0/).

Article

γ-ray and ν Searches for Dark-Matter Subhalos in the Milky Way with a Baryonic Potential

Moritz Hütten [1],*, Martin Stref [2],*, Céline Combet [3], Julien Lavalle [2] and David Maurin [3]

1 Max-Planck-Institut für Physik, Föhringer Ring 6, D-80805 München, Germany
2 Laboratoire Univers & Particules de Montpellier (LUPM), CNRS & Université de Montpellier (UMR-5299), Place Eugène Bataillon, F-34095 Montpellier CEDEX 05, France; lavalle@in2p3.fr
3 Laboratoire de Physique Subatomique et de Cosmologie, Université Grenoble-Alpes, CNRS/IN2P3, 53 Avenue des Martyrs, 38026 Grenoble, France; celine.combet@lpsc.in2p3.fr (C.C.); dmaurin@lpsc.in2p3.fr (D.M.)
* Correspondence: mhuetten@mpp.mpg.de (M.H.); martin.stref@umontpellier.fr (M.S.)

Received: 25 April 2019; Accepted: 18 May 2019; Published: 28 May 2019

Abstract: The distribution of dark-matter (DM) subhalos in our galaxy remains disputed, leading to varying γ-ray and ν flux predictions from their annihilation or decay. In this work, we study how, in the inner galaxy, subhalo tidal disruption from the galactic baryonic potential impacts these signals. Based on state-of-the art modeling of this effect from numerical simulations and semi-analytical results, updated subhalo spatial distributions are derived and included in the CLUMPY code. The latter is used to produce a thousand realizations of the γ-ray and ν sky. Compared to predictions based on DM only, we conclude a decrease of the flux of the brightest subhalo by a factor of 2 to 7 for annihilating DM and no impact on decaying DM: the discovery prospects or limits subhalos can set on DM candidates are affected by the same factor. This study also provides probability density functions for the distance, mass, and angular distribution of the brightest subhalo, among which the mass may hint at its nature: it is most likely a dwarf spheroidal galaxy in the case of strong tidal effects from the baryonic potential, whereas it is lighter and possibly a dark halo for DM only or less pronounced tidal effects.

Keywords: dark matter; galactic subhalos; semi-analytic modeling; gamma-rays and neutrinos

1. Introduction

In this contribution to *The Role of Halo Substructure in Gamma-Ray Dark-Matter Searches*, we revisit a previous study on the detectability of galactic dark clumps in γ-rays [1]. The latter relied on the best knowledge we had a few years ago of the properties of dark-matter (DM) clumps in the Milky Way. These properties (e.g., the mass and spatial distributions of galactic subhalos) were inferred from numerical simulations with a typical mass resolution of a few $10^5 \, M_\odot$, and extrapolated down ten orders of magnitude to the model-dependent minimal masses of subhalos [2,3]. Functional parametrizations were incorporated in the CLUMPY code [4–6] to generate γ-ray skymaps, accounting for the whole population of subhalos. For each combination of subhalo properties we explored, hundreds of skymap realizations were drawn, allowing us to calculate the average properties of the brightest clump. In the context of the future CTA γ-ray observatory [7] and its foreseen extragalactic survey, we concluded that the limits on DM set from this brightest clump should be "competitive and complementary to those based on long observation of a single bright dwarf spheroidal galaxy".

In the recent years, numerical simulations [8–10] and semi-analytical studies [11–15] have investigated the impact of the baryonic components of disk galaxies on their subhalo population by tidal stripping and disruption. These works reached the generic conclusion of a strong depletion of subhalos in the disk regions (i.e., also the Solar neighborhood), though with different quantitative estimates. Such a difference is expected due to the diversity of assumptions and methods used by different groups. In any case, this immediately questions the conclusion reached in [1], where the brightest subhalo was found, on average, at ~10 kpc from the Galactic center and at similar distance from Earth. Experimental limits on DM from galactic subhalos, derived from *Fermi*-LAT [16–24] or expected from prolonged operation [25], from HAWC [26,27], or future instruments such as GAMMA 400 [28] and CTA [1,29], should also be impacted by this result.

The paper is organized as follows: Section 2 presents the overall methodology and recalls how all relevant subhalos are efficiently accounted for with CLUMPY. Section 3 lists the subhalo critical parameters, highlighting the very different spatial distributions considered in this analysis. Section 4 presents updated statistics of the subhalo population and provides probability density functions (PDFs) of the brightest subhalo's properties (distance to the observer, mass, brightness, etc.). The analysis is performed for both annihilating DM [30–32] via the so-called *J*-factors, or decaying DM [33,34] (*D*-factors). We also show one realization of a subhalo skymap for all configurations considered. We conclude and briefly comment on the consequences for DM indirect detection limits in Section 5.

2. Important Quantities and Methodology

2.1. γ-ray and ν Fluxes from Dark Matter

The γ-ray or ν flux from annihilating/decaying DM particles, at energy E, along the line-of-sight (l.o.s.) in the direction (ψ, θ), and integrated over the solid angle $\Delta\Omega = 2\pi(1 - \cos \alpha_{\text{int}})$, is given by

$$\frac{d\Phi_{\gamma,\nu}}{dE}\left(E, \psi, \theta, \Delta\Omega\right)^{\text{Annihil.}}_{\text{Decay}} = \underbrace{\frac{1}{4\pi} \sum_f \frac{dN^f_{\gamma,\nu}}{dE} B_f \times \begin{Bmatrix} \frac{\langle \sigma v \rangle}{m^2_{\text{DM}}} \delta \\ \frac{1}{\tau_{\text{DM}} \, m_{\text{DM}}} \end{Bmatrix}}_{\text{Particle physics term: } \frac{d\Phi^{pp}_{\gamma,\nu}}{dE}(E)} \times \underbrace{\int_0^{\Delta\Omega} \int_{\text{l.o.s.}} dl \, d\Omega \times \begin{Bmatrix} \rho^2(\vec{r}) \\ \rho(\vec{r}) \end{Bmatrix}}_{\text{Astrophysics term: } J\text{- or } D\text{-factor}}, \quad (1)$$

where we made the distinction between the cases of DM self-annihilation (top) and decay (bottom). In both cases, m_χ is the mass of the DM particle, dN^f_γ/dE and B_f correspond to the spectrum per interaction and branching ratio of annihilation or decay channel f, and l is the distance from the observer. In case of annihilation, $\langle \sigma v \rangle$ is the velocity-averaged cross-section[1], while τ_{DM} is the DM particle lifetime in the decay scenario. Finally, $\rho(\vec{r})$ is the overall Galactic DM density distribution. The latter can be cast formally as the sum of a smooth distribution ρ_{sm} of unclustered DM particles, and a collection of $i = 1 \ldots N_{\text{subs}}$ subhalos, each with a density $\rho_i(\vec{r})$. The astrophysical term for annihilating DM[2] then reads [4,5,35],

$$\frac{dJ_{\text{tot}}}{d\Omega} = \int_{\text{l.o.s.}} \left(\rho_{\text{sm}} + \sum_i^{N_{\text{subs}}} \rho_i\right)^2 dl = \underbrace{\iint \rho^2_{\text{sm}} \, dl}_{dJ_{\text{sm}}/d\Omega} + \underbrace{\iint \left(\sum_i^{N_{\text{subs}}} \rho_i\right)^2 dl}_{dJ_{\text{subs}}/d\Omega} + \underbrace{2 \iint \rho_{\text{sm}} \sum_i^{N_{\text{subs}}} \rho_i \, dl}_{dJ_{\text{cross-prod}}/d\Omega}. \quad (2)$$

The above formula corresponds to a single realization of the underlying distribution of subhalos in the galaxy. The statistical properties of this distribution can be partly obtained from the formalism of hierarchical structure formation (e.g., [36]) or extracted from numerical simulations, as discussed below.

---

[1] We assume here the DM particle to be a Majorana particle, so that $\delta = 2$ (for a Dirac, $\delta = 4$).
[2] For conciseness, we present in Equations (2) and (3) formulae for annihilating DM only. Analogous formulae without a cross-product term and linear in the DM density can also be written for the *D*-factor of decaying DM.

2.2. Generating Skymaps with CLUMPY v3.0

For all our calculations, we rely on the public CLUMPY code described in [4–6]. It is a flexible code that efficiently emulates the end-product of numerical simulations in terms of γ-ray and neutrino signals for DM annihilation or decay. It allows easy exploration of how results are affected when changing the properties of the DM halos. CLUMPY v2.0 [5] was used for this purpose, to estimate the sensitivity of the CTA [1] and HAWC [27] γ-ray telescopes to galactic DM subhalos. Aside from galactic subhalo studies, CLUMPY has also been used by several teams to model DM annihilation or decay in γ-rays or ν in many targets: dwarf spheroidal galaxies [37–46], the galactic halo [47–49], the Smith HI cloud [50], nearby galaxies [51], galaxy clusters [52–54], and also for the extragalactic diffuse emission [55].

The present analysis is performed with CLUMPY v3.0 [6].[3] For completeness, we recap below the main steps of the CLUMPY calculation used for this work:

- CLUMPY *and the particle physics term:* Equation (1) shows that the particle physics term and the astrophysical terms are decoupled.[4] As the flux depends on the specific DM candidate chosen, we provide results in terms of J- and D-factors only; CLUMPY can easily be used to transform those into γ-ray or ν fluxes for any user-defined DM candidate (see CLUMPY's online documentation[5]).
- CLUMPY *and the astrophysics term:* to calculate skymaps of $dJ/d\Omega$, one should rely in principle on Equation (2). However, this is impractical in terms of computing time, as $\sim 10^{14}$ subhalos are expected in a Milky Way-sized DM halo. This problem can be overcome by formally separating Equation (2) in an average and "resolved" component,

$$\frac{dJ_{tot}}{d\Omega} = \frac{dJ_{sm}}{d\Omega} + \left\langle \frac{dJ_{subs}}{d\Omega} \right\rangle + \left\langle \frac{dJ_{cross-prod}}{d\Omega} \right\rangle + \sum_{k=1}^{N_{subs}^{drawn}} \frac{dJ_{drawn}^k}{d\Omega}. \quad (3)$$

With this ansatz, only a limited number N_{subs}^{drawn} of subhalos need to have their J-factor profiles calculated individually, while an average description is sought for the remaining "unresolved" DM. The criterion to discriminate between resolved and unresolved components often relies on a simple subhalo mass threshold, e.g., as done in works directly relying on numerical simulations [57] or their subhalo catalogs [58]. CLUMPY has been developed to treat this problem in a more efficient way, acknowledging the fact that rather light, but close-by subhalos may show J-factors comparable to heavier, more distant objects. The CLUMPY approach relies on the notion that the overall DM signal fluctuates around an average description, $\langle J_{tot} \rangle \pm \sigma_{J_{subs}}$, and we refer to [4] for a detailed description of our criterion to accordingly discriminate between unresolved and resolved halos. For the purpose of this work (and also the previous [1]), this approach allows us to preselect halos likely to shine bright at Earth and to consider all decades down to the smallest subhalo masses in the calculation.

3. Modeling the Galactic Subhalo Distribution

In this study, we focus on the impact of tidal disruption of subhalos in interaction with the baryonic components of the Milky Way, and compare four parametric models of the resulting galactic subhalo

---

[3] When releasing CLUMPY v3.0, we corrected a misprint that was present in v2.0, related to our implementation of the virial overdensity from [56]. This issue was responsible for obtaining in [1] about a factor 3 more subhalos than expected per flux decade (see full details in the CLUMPY documentation). We recall that in [1], we found that galactic variance is responsible for a factor $\lesssim 10$ uncertainty on the value of the brightest subhalo, and that other subhalo properties were responsible for another factor $\sim$6. Given these very large uncertainties, the conclusions on DM limits set from dark clumps with CLUMPY v2.0 are not qualitatively changed, but we urge users to rely on CLUMPY v3.0 for future studies.

[4] Strictly speaking, this factorization holds true only for DM candidates for which $\langle \sigma v \rangle$ is independent of the velocity and consideration of small redshift cells, $\Delta z/z \ll 1$.

[5] https://lpsc.in2p3.fr/clumpy.

abundance and their J/D-factors. We consider three quantities to be most sensitive to the J/D-factor distribution: (i) the spatial PDF of subhalos in the Milky Way, $d\mathcal{P}/dV$, (ii) the mass-concentration[6] relationship, $c(m,r)$, and (iii) the calibration of the total number of subhalos in the Milky Way, N_{calib}. The latter number is determined from numerical simulations, in a range where subhalos are resolved. Here, N_{calib} is defined for the mass range 10^8–10^{10} $M_\odot$, and it typically falls in the range 100–300.[7]

For modeling the subhalo distribution with these parameters, we start from an "unevolved" distribution, where we assume the position and mass of a subhalo to be uncorrelated,

$$\frac{d^3\mathcal{P}}{dV\,dm\,dc} = \frac{d\mathcal{P}}{dV}(\vec{r}) \times \frac{d\mathcal{P}}{dm}(m) \times \frac{d\mathcal{P}}{dc}(c,m). \tag{4}$$

Here, $d\mathcal{P}/dm$ and $d\mathcal{P}/dc$ describe the PDFs for a subhalo to have a given mass and a given concentration c. In reality, the factorization in Equation (4) may break down when subhalos gravitationally interact with the DM and baryonic potentials of their host halo [15,62], entangling their mass and positional distributions. We will consider this effect by "evolving" the distribution of Equation (4) in the model presented in Section 3.2.3.

3.1. Fixed Subhalo-Related Quantities

This work focuses on the impact of a baryonic disk potential on the subhalo population, mainly through the spatial PDF $d\mathcal{P}/dV$ (see Section 3.2). We keep several other subhalo-related quantities fixed to isolate this effect. For details on these parameters and how they affect the subhalo emission, we refer to our earlier work in [1] and only provide a brief summary below:

- *Index α_m of the power-law subhalo mass PDF $d\mathcal{P}/dm \propto m^{-\alpha_m}$ and subhalo mass range:* We choose $\alpha_m = 1.9$, $m_{\text{min}} = 10^{-6}\,M_\odot$, and $m_{\text{max}} = 1.3 \times 10^{10}\,M_\odot$. The maximum clump mass for all models is set to $10^{-2} \times M_{200}$ of the NFW halo from Section 3.2.3. This is motivated by the fact that we do not consider the possibility of any subhalos heavier than the Magellanic clouds, the heaviest satellites of our galaxy. The minimal clump mass and α_m mostly affect the diffuse emission boost from unresolved halos. For a fixed normalization N_{calib}, a steeper mass function ($\alpha_m = 2.0$) decreases the number of bright halos ($J \gtrsim 10^{20}\,\text{GeV}^2\,\text{cm}^{-3}$) by not more than $\sim$30%.
- *Width of $d\mathcal{P}/dc$:* We set $\sigma_c = 0.14$ [63]. Using a larger scatter $\sigma_c = 0.24$ [64] increases only by a few percent the number of subhalos per flux decade. Reducing the scatter or no scatter has the opposite effect.
- *Subhalo density profile:* We model all subhalos with a spherically symmetric NFW profile [65]. Using an Einasto profile [65,66] instead amounts to a global increase $\sim$2 of the number of subhalos per flux decade within the considered integration regions $\Delta\Omega$. Please note that micro-halos with $m \ll M_\odot$ may show steeper inner slopes [67–69]; however, we have found that these micro-halos do not provide new bright, resolved subhalo candidates [1].
- *Level of sub-substructures:* We do not consider an emission boost from substructure within subhalos. Such a boost from additional levels of substructure[8] increases the number of subhalos per flux decade, with the largest increase of almost a factor 2 for the largest luminosities. Sub-substructures actually increase the signal in the outskirts of halos (see Figure 4 of [1]), the impact of which depends on the instrument angular resolution or containment angle used in the analysis. For instance, in [1], no impact was found for dark clumps within the angular resolution of CTA.

---

[6] The concentration is defined to be $c = r_\Delta/r_{-2}$, with r_Δ, taken to be the subhalo boundary, is the radius at which the mean subhalo density is Δ times the critical density (see, e.g., [6]), and r_{-2} is the position in the subhalo for which the slope of the density is -2. We use $\Delta = 200$ in this work.

[7] This range was recently shown to be in agreement with the observed number of dwarf spheroidal galaxies SDSS corrected by the detection efficiency [59], alleviating the tension caused by the so-called missing satellite problem in CDM scenarios [60,61]. Given the minimal mass of subhalos, m_{min}, N_{calib} can be used to calculate the mass fraction, f_{DM}, of DM in subhalos.

[8] As shown in [5] (see their Figure 1), only the first level of substructure significantly boosts the halo luminosity, the next levels bringing a few extra percent at most.

Please note that our choice for these constant parameters will lead to a rather conservative number of detectable subhalos and J-factor of the brightest 'resolved' subhalo—hence conservative limits for DM indirect detection—compared to other choices. Pushing all the parameters to get the most optimistic case would lead to a factor $\sim$2–3 increase of the J-factor of the brightest subhalo [1].

3.2. The Spatial Distribution $d\mathcal{P}/dV$ of Subhalos

We consider four configurations of $d\mathcal{P}/dV$ in this work, which are described below and summarized in Table 1. The first configuration (model #1) is close to one of our 2016 study [1], i.e., based on results from DM-only simulations. It is used as a reference to which the other configurations describing interaction with the baryonic disk are compared to. We consider a spherically symmetric Galactic DM halo and correspondingly, also $d\mathcal{P}/dV$ distribution. The maximum distance of any subhalo from the Galactic center is set to $R_{200} = 231.7$ kpc for all configurations, inspired by the NFW halo from Section 3.2.3. We show later that the brightest subhalo is found only with negligible probability at larger Galactocentric radii for all models. Please note that despite the common value for R_{200}, the total Galactic DM profile is different from one configuration to another. We however focus here on the clumpy part of the halo only and do not consider the smoothly distributed DM.[9] While this article is dedicated to the derivation and comparison of statistical properties of the subhalo population brightness for these different models, we still emphasize that when going to firm predictions and limits, the overall DM profile should matter. Indeed, the Milky Way is a strongly constrained system [70,71], which must be taken into account when extrapolating simulation results. This aspect goes beyond the scope of this work though, but is worth mentioning as the Gaia mission is currently boosting our handle on Milky Way dynamics [72,73].

Table 1. Subhalo parameters for the models investigated in this study, with model #1 based on results of DM-only numerical simulations, while models #2 to #4 are different implementations of DM subhalos post-processed in the Milky Way halo and baryonic disk potential. For models #2, #3, and #4, we also show the number of surviving subhalos with tidal masses between 10^8 M$_\odot$ and 10^{10} M$_\odot$. See Section 3 for details and parameters common to all subhalo configurations.

	Model #1	Model #2	Model #3	Model #4
	Aquarius [74]	Phat-ELVIS [10]	SL17 [15] with $\epsilon_t = 10^{-2}$	SL 17 [15] with $\epsilon_t = 1$
	Einasto	Sigmoid-Einasto Equation (5)	$\propto \rho_{sm}$	$\propto \rho_{sm}$
$\dfrac{d\mathcal{P}}{dV}$	$\alpha_E = 0.68$	$\alpha_E = 0.68$	NFW *	NFW *
	$r_{-2} = 199$ kpc	$r_{-2} = 128$ kpc	$r_{-2} = r_s = 19.6$ kpc *	$r_{-2} = r_s = 19.6$ kpc *
	-	$r_0 = 29.2$ kpc	-	-
	-	$r_c = 4.24$ kpc	-	-
$N_{\rm calib}$	300	-	276 *	276 *
$N_{\rm surviving}$	-	90	114 ± 11	112 ± 10
$c(m)$	Moliné et al. [75]	Moliné et al. [75]	Sánchez-Conde & Prada [63]	Sánchez-Conde & Prada [63]

* Properties of the initial subhalos from which the surviving ones are obtained after interaction with the baryonic potential.

3.2.1. Model #1: DM only (as Implemented in Hütten et al., 2016)

This first configuration uses the position-dependent concentration parametrization of Moliné et al. [75]. The latter is based on the analysis of the DM-only simulations VL-II [76] and ELVIS [77], and it predicts that subhalos of a given mass are more concentrated close to the galactic center than in the outer parts of their host halos. This effect is related to tidal disruption in the DM potential of the host halo. In the outer parts, when tidal disruption in the DM potential becomes

---

[9] We still provide later the total DM profile for the semi-analytical configurations (model #3 and model #4) because it is one of the building blocks of the model.

negligible, the concentration is very close to the concentration found for field halos [63].[10] DM-only tidal effects also affect the spatial PDF of subhalos [62], and all recent DM-only simulations found it to be flatter than the DM distribution in the smooth halo. We use an Einasto profile for $\mathrm{d}\mathcal{P}/\mathrm{d}V$ with $\alpha_E = 0.68$ and $r_{-2} = 199$ kpc, following the results of the Aquarius A-1 halo [74], and we fix the number of subhalos above 10^8 M$_\odot$ to $N_{\mathrm{calib}} = 300$, as an upper bound to what was found in the Aquarius simulations [74]. Subhalo outer radii and corresponding masses in this model are kept to be *cosmological* radii and masses, and subhalos are truncated where their mean density reaches 200 times the critical density of the Universe (see, e.g., [6] for details). This model is contained in the parameter space already explored in our previous study [1].

3.2.2. Model #2: DM + Galactic Disk Potential (Numerical, Phat-ELVIS)

In the recent Phat-ELVIS simulations, Kelley et al. [10] have accounted for the effect of the Milky Way baryon potential (including stellar and gas disk, and bulge). They found that subhalos were strongly destroyed in the inner part of the halo, leaving basically no subhalo with mass $m \lesssim 5 \times 10^6$ M$_\odot$ within the inner ~ 30 kpc of the host DM halo. This is at odds with the predicted $\mathrm{d}\mathcal{P}/\mathrm{d}V$ of previous simulation sets (e.g., Aquarius [74] that was previously used as reference in [1]).

Using the subhalo catalog from the simulations provided by the authors, we compute the normalized subhalo PDF per unit volume, $\mathrm{d}\mathcal{P}/\mathrm{d}V$, at $z = 0$, averaged over the 12 Milky-Way-like halos available. The dispersion in each bin provides the error bar. The data are fitted using the following parametrization:

$$\frac{\mathrm{d}\mathcal{P}}{\mathrm{d}V}(r) = \frac{A}{1 + e^{-(r-r_0)/r_c}} \times \exp\left\{-\frac{2}{\alpha_E}\left[\left(\frac{r}{r_{-2}}\right)^\alpha - 1\right]\right\}, \qquad (5)$$

and results are shown in the left panel of Figure 1. The first term in Equation (5) is a Sigmoid function, centered on r_0 and increasing from 0 to A given r_c, that allows us to capture the sharp decrease of the number of subhalos in the inner regions of the parent halo. The Sigmoid then transitions to an Einasto profile with characteristic scale r_{-2} and index α_E. In order to have an Einasto profile close to the DM-only case in the outer parts (see previous paragraph), we fix $\alpha_E = 0.68$, and the best-fit parameters, obtained on all subhalos in the catalog are $r_0 = 29.2$ kpc, $r_c = 4.24$, and $r_{-2} = 128$ kpc.[11] The constant A is set to ensure $\mathrm{d}\mathcal{P}/\mathrm{d}V$ is a PDF.

Finally, among the 12 Milky-Way-like host halos of the Phat-ELVIS simulations, we select halos with masses similar to the NFW halo introduced in the following Section 3.2.3, in the range of $1.1 \times 10^{12} - 1.4 \times 10^{12}$ M$_\odot$; averaging the number of subhalos between 10^8 and 10^{10} M$_\odot$ that have survived interaction with the baryonic potential, we fix in CLUMPY the normalization of the number of subhalos to $N_{\mathrm{calib}} = N_{\mathrm{surviving}} = 90$. In the same way as in the DM-only model # 1, we define and calculate subhalo masses based on their cosmological radii.

3.2.3. Models #3 and #4: DM + Disk Potential (Semi-Analytical, SL17)

Complementary to numerical approaches, semi-analytical models have also been considered by several authors, e.g., [11–15]. In this work, we use the study by Stref and Lavalle [15] (SL17 hereafter) to capture the effects of tidal stripping from the smooth galactic potential of DM and baryons and disk shocking by the baryonic disk. The model in SL17 is built on the dynamically constrained mass models from McMillan [71], where the latter used kinematic data (including maser observations, the Solar

---

[10] The difference between using the space-dependent or field halo concentration was found to be a factor ~ 2 larger on the brightest subhalo in [1].

[11] The Milky-Way-like halos in the Phat-ELVIS simulations have masses ranging from 7×10^{11} M$_\odot$ to 1.9×10^{12} M$_\odot$, and virial radii from 235 kpc to 329 kpc respectively. The fit above was performed on the radial range common to all host halos, namely from 0 to 235 kpc. The value of the parameters remain compatible at the one-sigma level when increasing the radial range to 329 kpc, or when cutting on masses above 5×10^6 M$_\odot$.

velocity, terminal velocity curves, the vertical force and the mass within large radii) to determine the Milky Way's DM distribution following the parametrization of Zhao [78]

$$\langle \rho_{\text{tot}}(r) \rangle = \rho_s \left(\frac{r}{r_s}\right)^{-\gamma} \left[1 + \left(\frac{r}{r_s}\right)^\alpha\right]^{(\gamma-\beta)/\alpha}. \quad (6)$$

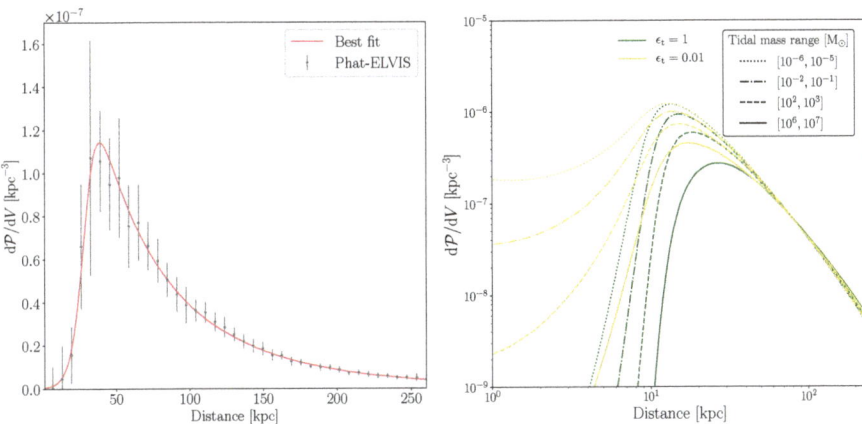

Figure 1. Spatial PDFs of subhalos surviving interaction with the baryonic disk potential. (**Left panel**): directly computed from the catalogs of the Phat-ELVIS simulations [10]. Dots correspond to the average over the 12 Milky-Way-like halos in the simulations, with error bars obtained from the dispersion over the 12 halos. The best-fit model (red curve) has been computed using the parametrization given by Equation (5). (**Right panel**): SL17 model for various mass ranges (line styles) and values of ϵ_t (colors). See Figure 2 for a comparison between all $d\mathcal{P}/dV$ models used in the analysis.

In this work, we only consider the results based on the NFW parametrization ($\alpha, \beta, \gamma = 1, 3, 1$), for which the best-fit parameters are $r_s = 19.6$ kpc, $\rho_s = 8.517 \times 10^6$ $M_\odot$ kpc^{-3}, resulting in $R_{200} = 231.7$ kpc and $M_{200} = 1.31 \times 10^{12}$ $M_\odot$. We checked that our conclusions are left unchanged if considering a cored profile ($\alpha, \beta, \gamma = 1, 3, 0$) instead. In SL17, the initial population of subhalos traces the above Galactic DM halo mass PDF and there is *initially* no correlation between a subhalo's mass and its position,

$$\frac{d\mathcal{P}}{dV}(r, \text{initial}) \propto \langle \rho_{\text{tot}}(r) \rangle. \quad (7)$$

The addition of tidal interactions modifies this picture because tidal effects select subhalos based on their mass and concentration to produce an 'evolved' subhalo population. This evolution manifests itself in two aspects: (i) subhalos with a given mass at a given position with too small a concentration are disrupted, while (ii) subhalos that survive are stripped from a large fraction of their mass. Mass stripping is encoded into the subhalo tidal radius r_t, which is computed as described in [15]. Please note that tidal effects in SL17 are computed assuming circular orbits for the clumps. Cosmological simulations show that subhalos are efficiently disrupted by tidal interactions. For instance, Hayashi et al. [79] find that a subhalo with tidal radius r_t and scale radius r_s is disrupted if $r_t \lesssim 0.77\, r_s$. It has however been recently pointed out by van den Bosch et al. [80,81] that disruption within simulations might be largely explained due to a lack of numerical resolution, Poisson noise, or runaway instabilities. According to these authors, subhalos are far more resilient to tidal disruption than numerical simulations tend to show, implying that a subhalo could survive even if $r_t \ll r_s$ (this result is expected from theoretical grounds [82,83] and in agreement with earlier findings by

Peñarrubia et al. [84]). The SL17 model includes a free parameter ϵ_t that allows us to simply investigate both possibilities. The ϵ_t parameter is defined such that

$$\frac{r_t}{r_s} < \epsilon_t \Leftrightarrow \text{subhalo is disrupted.} \tag{8}$$

This disruption criterion can also be expressed in terms of the concentration: the subhalo is disrupted if $c < c_{\min}$ where $c_{\min}$ is referred to as the minimal concentration (which depends a priori on the subhalo's position and mass). A "cosmological-simulation-like" configuration, where subhalos are efficiently disrupted, corresponds to $\epsilon_t \sim 1$. Conversely, a model of very resilient subhalos is obtained by setting $\epsilon_t \ll 1$. In the following, we will consider two extreme values : $\epsilon_t = 0.01$ (model #3) and $\epsilon_t = 1$ (model #4). The former value implies a subhalo is disrupted when it has lost around 99.99% of its cosmological mass. Please note that the value of r_t does depend on the choice of ϵ_t.

The behavior of the SL17 model is illustrated in Figure 1 (right panel) where we show the spatial distributions of surviving subhalos for different mass decades. We see that the distribution of lighter objects extends to lower radii than the distribution of heavier ones. This is because, as already pointed out in [15], smaller objects are more concentrated on average and therefore more resilient to tidal disruption. Interestingly, if simulations overestimate tidal disruption as pointed out in van den Bosch et al. [80,81], i.e., $\epsilon_t = 0.01$ with our parametrization, SL17 predicts a large population of very light subhalos in the innermost regions of the Milky Way.

The number of subhalos in SL17 is calibrated onto the Via Lactea II cosmological simulations [76], by the mass fraction in resolved subhalos in these simulations. Since Via Lactea II is a DM-only simulation, the calibration is performed without the baryonic tides and setting $\epsilon_t = 1$ for consistency. This leads to $N_{\text{calib}} = 276$ for initial cosmological subhalo masses, m_{200}, between 10^8 M$_\odot$ and 10^{10} M$_\odot$ as quoted in Table 1. Adding a baryonic potential and considering tidally stripped masses leads to ~110 surviving subhalos in the same mass range for both choices of ϵ_t (see also Table 1).

3.2.4. d$\mathcal{P}$/dV Model Comparison

The four subhalo PDFs considered in this study are compared in Figure 2. The configurations based on the Phat-ELVIS and the Aquarius simulations are shown in red and blue, respectively, while the SL17 models are shown in yellow ($\epsilon_t = 0.01$) and green ($\epsilon_t = 1$). In order to make a meaningful comparison with the simulation results, we only show the SL17 prediction for large subhalo tidal masses ($m > 10^6$ M$_\odot$), comparable to the mass of the smallest objects identified in Phat-ELVIS.

The effect of a galactic disk as implemented in the Phat-ELVIS simulation is to disrupt most subhalos in the inner 30 kpc of the galactic halo, as opposed to a DM-only simulation such as Aquarius where subhalos can survive down to much lower radii. The subhalo distribution predicted by SL17, which accounts for the effect of the disk, resembles the one of Phat-ELVIS in that massive subhalos are disrupted at the center. We note that d$\mathcal{P}$/dV peaks at a lower radius in SL17 compared to Phat-ELVIS. Setting $\epsilon_t = 0.01$ pushes the peak radius to an even lower value with respect to the $\epsilon_t = 1$ case, as expected since subhalos are then more resilient to disruption.

A more detailed understanding of the difference between these models is beyond the scope of this paper, since the SL17 models are semi-analytically constructed from the kinematically constrained mass model of the Milky Way from [71] taking into account the mass dependence of the subhalo spatial distribution, while the DM-only and Phat-ELVIS configurations are based on approximate fits to Milky-Way-like halos from numerical simulations. We still note a similar trend between the SL17 ($\epsilon_t = 1$) and Phat-ELVIS configurations, which may indicate that the semi-analytical method associated with the former (pending some simplifying assumptions) somewhat capture the main features of the latter (pending numerical effects likely to dominate in the central regions).

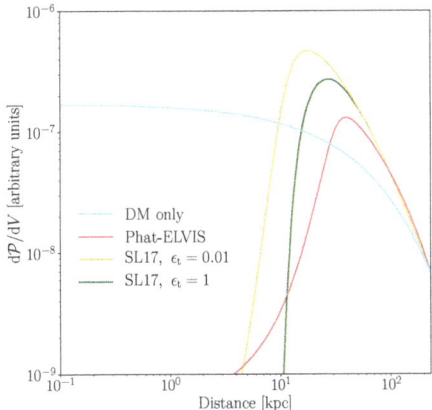

Figure 2. The four spatial PDFs of subhalos considered in this work: SL17 models for subhalos with $m > 10^6\,\mathrm{M}_\odot$ (yellow and green), PDF based on the Phat-ELVIS simulation (red) [10], and on the Aquarius simulation (blue) [74]. To highlight the behavior at low radii where tidal effects are the most relevant, the curves are shifted to match the value of $d\mathcal{P}/dV(231.7\,\mathrm{kpc})$ in the Phat-ELVIS configuration.

4. Results

Using CLUMPY, we generate fullsky subhalo populations and corresponding J- and D-factors according to the models from Table 1. For all configurations, the distance between the Sun and the Galactic center is set to $R_\odot = 8.21$ kpc [71]. We consider two estimations of the J-/D-factors, integrating either over the full angular extent of a subhalo or up to a radius of $\alpha_{\mathrm{int}} = 0.5°$. Averages and PDFs are then obtained from a statistical sample of 1000 realizations of the subhalo population for each model.

4.1. Subhalo Source Count Distributions

Figure 3 presents the source count distribution for all models and J-factors $J > 2 \times 10^{16}\,\mathrm{GeV}^2\,\mathrm{cm}^{-5}$ ($D > 2 \times 10^{15}\,\mathrm{GeV\,cm}^{-2}$),[12] similar to Figure 3 of our earlier work [1]. Solid lines show J-/D-factors within $\alpha_{\mathrm{int}} = 0.5°$ ($\Delta\Omega = 2.5 \times 10^{-4}$ sr), dashed lines the full signal. For two configurations, we also show the variance bands of the distributions. For the first time, we also show in this work the D-factor distribution for decaying DM. Comparing the solid and dashed curves, it can be seen that in the case of annihilation, most of the fainter halos' emission is contained within the innermost $0.5°$ except for the $\sim$100 brightest halos (left panel). This is different in the case of DM decay, for which the emission profile is much more extended (right panel). Comparing the models of this work, we reach the following conclusions for *annihilating* DM (left panel):

- Model #2 (Phat-ELVIS, red lines) predicts about a factor 5 less halos per flux decade than the Aquarius-like DM-only reference model #1. The average brightest halo (within $0.5°$) is about a factor 4 fainter than expected for the DM-only case. This drastic decrease of bright objects is both attributed to the fact that the Phat-ELVIS simulations [10] find (i) overall less subhalos in Milky Way-like galactic halos ($N_{\mathrm{calib}} = 90$ vs. $N_{\mathrm{calib}} = 300$) and (ii), no subhalos are found close to Earth in the innermost 30 kpc of the galactic halo.
- Model #3 (SL17, yellow lines) predicts almost the same abundance of bright halos as the DM-only model #1. Model #3 starts from an initial subhalo distribution biased towards the Galactic center

---

[12] We checked for convergence in order to obtain a complete ensemble of objects above these thresholds.

following the overall DM distribution, and accounts for tidal subhalo disruption and stripping afterwards according to the semi-analytical model of [15]. With $\epsilon_t = 0.01$ few subhalos are affected. In turn, the DM-only model #1 already includes a subhalo distribution anti-biased towards the Galactic center in an evolved galactic halo according to the Aquarius simulations (although the considered fitting to the Aquarius simulations [74] does not account for a mass dependence of the halo depletion).

- Model #4 (SL17, green lines) applies a much stronger condition on tidal stripping and total depletion than the model #3 configuration within the semi-analytical approach of [15]. Illustratively, we calculate a total of 1.41×10^6 initial subhalos (in the full range between $m_{\min} = 10^{-6} \, M_\odot$ and $m_{\max} = 1.3 \times 10^{10} \, M_\odot$) for the subhalo models #3 and #4, out of which 20,000 are completely disrupted for the model #3 ($\epsilon_t = 10^{-2}$). In contrast, 530,000 halos are disrupted for $\epsilon_t = 1$ in the model #4.[13] In result, a factor 2 less halos are present above the lower end of the displayed brightness distributions, the ratio increasing for the brightest decades. Recall that surviving halos are truncated at the same tidal radius in models #3 and #4.

For unstable, *decaying* DM, the flux is proportional to the distance-scaled DM column density (Figure 3, right panel). Contrary to the case of annihilation, we find:

- Changing the signal integration region $\Delta\Omega$ drastically impacts the collected signal, as the emission shows a much broader profile than for annihilation. This loss is most drastic for the brightest halos.
- For an integration angle of $\alpha_{\rm int} = 0.5°$, all models are in remarkable agreement at the brightest end. For fainter flux decades and considering the signal over the full halos extent, models differ by a factor $\sim$5 (however, with a rather large spread in the D-factor PDF of the brightest halo in the individual models, see the later Figure 5). This suggests that predictions for the largest subhalo flux from decaying DM should be rather model-independent.

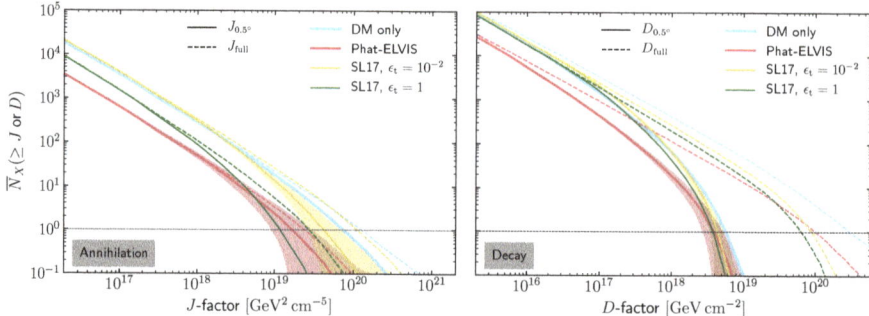

Figure 3. Cumulative source count distribution of galactic subhalos (full sky, averaged over 1000 simulations) for all configurations gathered in Table 1. (**Left panel**): annihilating DM. (**Right panel**): decaying DM. In both panels, the solid lines show the distribution of the J-factors within $\alpha_{\rm int} = 0.5°$, whereas the dashed lines for integrating over the full halo extents up to r_{200} or r_t.

For illustrative purposes, Figure 4 displays subhalo skymaps of a random realization of each of the four models under scrutiny. For each model and to ease comparison, the same DM subhalo sky is used in case of annihilation (*J*-factors, left) or decay (*D*-factors, right). The sky realization varies of course from one model to the other. We do not include the average and smooth DM distributions here. The maps include all subhalos with masses above $10^4 \, M_\odot$ (cosmological masses in the models

---

[13] Please note that most drawn subhalos are disrupted at masses below a tidal mass of $10^4 \, M_\odot$, the scale above which subhalos are shown in the later Figure 4.

#1 & #2, tidal masses in the models #3 & #4). For example, in the DM-only case, this corresponds to 1,214,313 halos included in the map, and for a HEALPix resolution of $N_{\text{side}} = 1024$, CLUMPY requires ~30 CPUh for its computation in case of annihilation (~20 CPUh in case of decay). Please note that we did not select the shown random sky realizations to reflect some particular average or extreme case. In Appendix A, we list some properties of the brightest objects in these maps, which can be compared to the average properties derived in the remainder of this section.

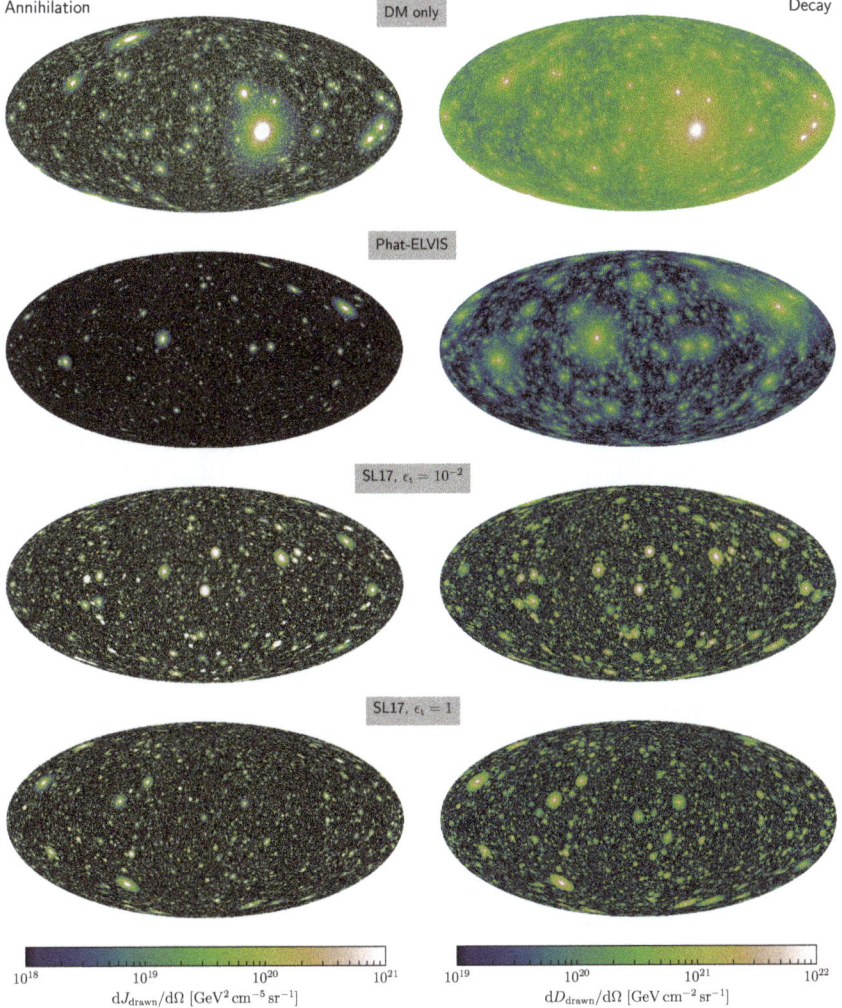

Figure 4. One random realization of the Galactic DM subhalo sky (all subhalos above 10^4 M$_\odot$, ignoring the smooth contribution) in case of annihilation (**left**) or decay (**right**), derived from the models gathered in Table 1. Maps are drawn in galactic coordinates (Mollweide projection) with $(l, b) = (0, 0)$ at their centers. (**From top to bottom**): Model #1 emulating numerical DM-only simulations (1,214,313 drawn halos); model #2 emulating the Phat-ELVIS simulations [10] (364,064 drawn halos); and the semi-analytical models #3 (subhalos more resilient against tidal disruption, 549,572 surviving halos) and #4 (less subhalos surviving tidal destruction, 546,096 surviving halos) according to SL17 [15]. The displayed maps (fits format, 50 MB in file size) can, along with their subhalo catalogs, be provided upon request. In Appendix A, we list some properties of the brightest objects in these maps.

4.2. Statistical Properties of the Brightest Halo

Finally, we focus on the statistics linked to the halo with the largest J or D-factor (its properties are marked with a "$\star$" symbol in the following). For cold DM particles structuring on small scales, fully dark subhalos represent interesting targets for indirect searches. Conversely, not detecting the brightest of them can be used to set limits. We remark that focusing on the brightest halo alone for setting limits is a simplistic assumption in some circumstances. For example, there are numerous yet unidentified sources detected by the *Fermi*-LAT which could include a (i.e., the brightest) DM subhalo [16–21,85,86], so constraints set should be correspondingly weaker in this scenario.

The bottom row of Figure 5 shows the PDFs of $J^\star$ and $D^\star$, distilled from 1000 realizations of a DM subhalo sky for each model.[14] From this follows that the brightest expected signal from subhalos has in median a J-factor of $\widetilde{J}^\star_{0.5°} = 8.8^{+11}_{-4.0} \times 10^{19}$ GeV^2 cm^{-5} ($\widetilde{J}^\star_{\text{full}} = 1.6^{+2.9}_{-0.9} \times 10^{20}$ GeV^2 cm^{-5}) for the optimistic DM-only model. The signal is expected a factor 7 lower, at $\widetilde{J}^\star_{0.5°} = 1.3^{+0.8}_{-0.4} \times 10^{19}$ GeV^2 cm^{-5} ($\widetilde{J}^\star_{\text{full}} = 3.2^{+2.5}_{-1.3} \times 10^{19}$ GeV^2 cm^{-5}; factor 5 lower) in the case of a largely depleted inner galactic halo (model #4). For decaying DM, all models produce remarkably similar fluxes within $\alpha_{\text{int}} = 0.5°$ (lower left panel of Figure 5) of $D^\star_{0.5°} \sim 4^{+3}_{-1} \times 10^{18}$ GeV cm^{-2}. Over the full extent of the halo, however, the width of the D-factor distributions is much larger, and D-factors between $D^\star_{\text{full}} \gtrsim 10^{20}$ GeV cm^{-2} up to $D^\star_{\text{full}} \lesssim 10^{22}$ GeV cm^{-2} are obtained.

The PDFs of the brightest subhalo's properties may also be derived. The top row of Figure 5 (left) shows that for annihilating DM, the brightest halo is found at a distance of ~10 kpc from Earth for the models #1 (see also [1]) and #3, and also on similar Galactocentric radii (left panel of second row). For the models #2 and #4, which reflect strong tidal disruption of halos in the inner galactic region, bright halos are found at about ~30–40 kpc distance from the Galactic center, and equivalent distances from Earth. For decaying DM, models #2 and #4 give similar predictions, while models #1 and #3 tend to predict the brightest halo at larger Galactocentric and observer distances (right panels of the top rows).

More importantly, the third row of Figure 5 sheds light on the question whether the brightest DM halo is likely to be a dark halo or a satellite galaxy. For annihilating DM, models #2 and #4 predict subhalos with masses $m^\star \gtrsim 10^8$ M$_\odot$ to shine brightest in γ-rays or ν; these objects are most probably associated with a (dwarf) satellite hosted in their center [88]. For the DM-only case #1 this is not anymore so obvious, as discussed in [1], as lighter objects become probable candidates to provide the highest fluxes. Finally, model #3 predicts very light and close-by halos to shine the brightest: If this model reflects the true nature of DM in our galaxy, the highest γ-rays or ν signal from Galactic DM substructure will likely arise from a dark spot in the sky. For decaying DM, the brightest Galactic DM subhalo is likely a satellite galaxy with mass $m^\star \gtrsim 10^8$ M$_\odot$.

For blind searches of this brightest halo, it is finally useful to check whether there is a preferred direction in the sky to search for the brightest halo. As all our configurations are symmetrical around the direction towards the Galactic center, we present in the fourth row of Figure 5 the probability per unit area, $d\mathcal{P}/d\cos(\theta^\star)$, to find the brightest halo at the angular distance $\theta^\star$ from the Galactic center (GC). As found in [1], in the DM-only case #1, the brightest halo is more probably found in a direction close to the GC. In contrast, model #3 suggests to preferentially search in a ring-like region at ~90° distance from the GC; while models #2 and #4 predict the highest probability to find the brightest subhalo towards the galactic anticenter. While these trends also apply for searches for DM decay in subhalos (right panel), they are much more pronounced for signals from annihilating DM.

---

[14] Probability distributions were derived using a kernel density estimation (KDE) with an adaptive Gaussian kernel according to [87] (except $d\mathcal{P}/d(\cos\theta^\star)$, which was obtained from a histogram). To handle the boundary conditions of $d\mathcal{P}/dm(m > m_{\text{max}} = 1.3 \times 10^{10}$ M$_\odot) = 0$ and $d\mathcal{P}/dV(R > R_{200} = 231.7$ kpc$) = 0$, we use the PyQt-Fit KDE implementation by P. Barbier de Reuille, https://pyqt-fit.readthedocs.io (not anymore maintained as of submission of the manuscript), which accounts for a renormalization algorithm at the boundary. Please note that for a precise power-law source count distribution, the PDF of $J^\star/D^\star$ follows a Fréchet distribution, see App. B of [1].

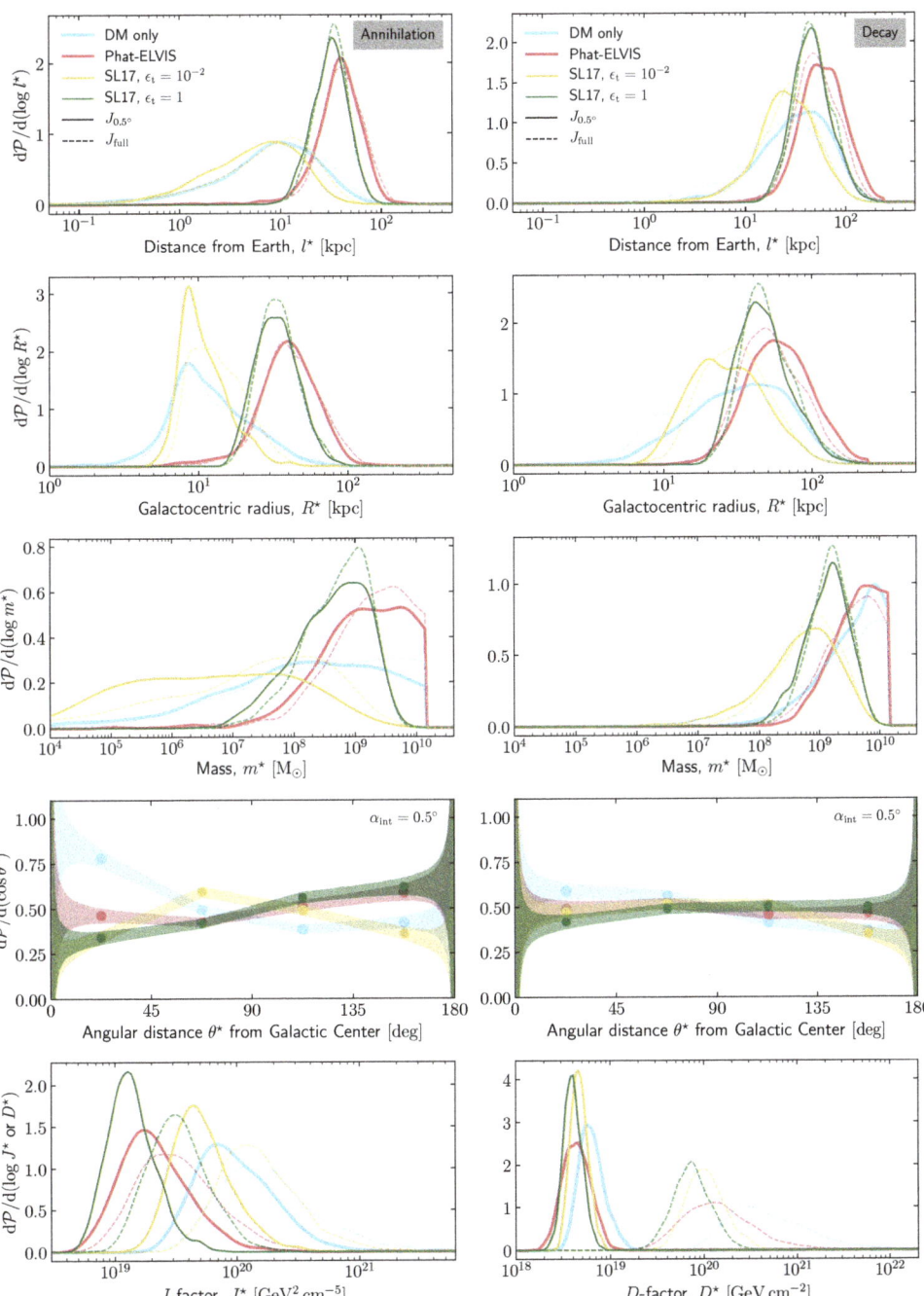

Figure 5. PDFs of the γ-ray (or ν) brightest galactic subhalo properties for the four investigated models. (**Left panel**): annihilating DM. (**Right panel**): decaying DM. Solid lines show the statistics for only the emission from the innermost $\alpha_{int} = 0.5°$ of a subhalo, dashed lines the emission over the full extent.

5. Conclusions

In this work, we have studied the impact of tidal disruption of subhalos by Milky Way's baryonic potential on the properties of the γ-ray and ν signals from subhalos. This effect is mostly encoded in the spatial distribution of subhalos, and four models have been considered. The first model serves as reference and is based on 'DM-only' simulations, not including tidal disruption in the baryonic potential. Similar models based on this assumption have been applied by many authors in the past, e.g., [1,16,18,29,58]. A second model was obtained by implementing the recent results from the Phat-ELVIS numerical simulation (resolution-limited to a few 10^6 M$_\odot$) where the inner 30 kpc of the Milky Way are depleted of subhalos. The last two models relied on semi-analytical calculations applicable to the whole subhalo mass range: these calculations find a Galactocentric radius $\lesssim 10$ kpc, slightly dependent on the subhalo mass, below which subhalos are stripped of their outer parts or even disrupted, depending on a disruption criterion ϵ_t, taken to be to 1 or 10^{-2} in this study.

These models lead to significantly different brightness populations of DM annihilation and decay signals. To quantify the difference, we have simulated 1000 realizations of the subhalo population for each model. Focusing on the brightest subhalo, whose properties can be used to study DM detectability [1], we find (for an integration angle of $0.5°$) a factor 2 to 7 less signal compared to the case of subhalos in the DM-only configuration, but no significant difference for the decay signal. Our large statistical sample also allowed us to reconstruct the PDF of several properties of this brightest subhalo (mass, distance to the observer, angular distribution in the sky). In particular, the mass information indicates that in models without or little disruption in the disk potential, the brightest subhalo can be close by, and with a mass range below that of known dwarf spheroidal galaxies, i.e., a dark clump. On the other hand, in models with strong tidal disruption, the brightest subhalo is farther away, and its preferred mass is shifted to values similar to those of satellite galaxies, i.e., it could be a known dSph. The latter situation would worsen the prospects of blind galactic dark clump searches with background-dominated instruments that were discussed in [1].

In any case, our results highlight the importance of better characterizing the spatial PDF of the subhalo population, in particular by constraining further the level of tidal disruption. Although both numerical and semi-analytical approaches show the same trend in reducing the number and brightness of subhalos, there remain serious quantitative differences. We recall that it is still debated whether or not tidal disruption could be significantly amplified by numerical artifacts in simulations [80,81]. On the other hand, present-day semi-analytical methods currently rely on simplifying assumptions, some of which should be relaxed, e.g., include a more realistic distribution of orbits—see complementary studies in [89,90]. A further question is related to the possible redistribution of DM inside stripped subhalos, see, e.g., [12,91,92]. Until a clearer picture surfaces, all possibilities from weak to strong tidal effects must be equally considered for indirect DM searches. Finally, we stress again that any complete DM halo model comprising a subhalo population should be checked against kinematic constraints, which are increasingly stringent for the Milky Way in the context of the Gaia results [72,73]. This is particularly important for tidal disruption as it strongly depends on the detailed description of the baryonic components of the galaxy. Predictions or limits based on galactic halo models which do not account for these constraints should be taken with caution.

All our calculations were performed with the public code CLUMPY, and our results illustrate how this code can quickly be used to incorporate and exploit any progress made by numerical simulations and/or semi-analytical calculations. All computations and drawing of random realizations of the discussed models can be repeated at one's own account. Also, the subhalo skymaps and catalogs shown here as illustration for the various models are available upon request.

Author Contributions: Conceptualization, all authors; Formal analysis, M.H., M.S., C.C.; Software, all authors; Data curation, M.H.; Writing—original draft preparation, D.M. and C.C.; Writing—review & editing, all authors.

Funding: This work was supported by the "Investissements d'avenir, Labex ENIGMASS" and by the Max Planck society (MPG). We also acknowledge support from the ANR project GaDaMa (ANR-18-CE31-0006), the OCEVU Labex (ANR-11-LABX-0060), the CNRS IN2P3-Theory/INSU-PNHE-PNCG project "Galactic Dark Matter", and European Union's Horizon 2020 research and innovation program under the Marie Skłodowska-Curie grant agreements N° 690575 and N° 674896.

Acknowledgments: We thank T. Kelley and J. S. Bullock for providing us, prior to the public release, with the Phat-ELVIS subhalo catalog. We also thank M. Doro and M. A. Sánchez-Conde for inviting us on this special issue, and for organizing the *Halo Substructure and Dark-Matter Searches* workshop held in Madrid in 2018, where discussions prompted this work. Calculations were performed at the Max Planck Computing & Data Facility at Forschungszentrum Garching. We finally thank the anonymous referees for their reviews and the valuable comments which helped to improve the quality of the manuscript.

Conflicts of Interest: The authors declare no conflict of interest. The founding sponsors had no role in the design of the study; in the collection, analyses, or interpretation of data; in the writing of the manuscript, or in the decision to publish the results.

Appendix A. Properties of the Brightest Subhalos in the Example Maps

In Table A1, we list the properties of the brightest subhalos in the random realizations displayed in Figure 4. Please note that different objects may provide the largest flux for annihilation or decay and for different integration regions $\Delta\Omega$, and the largest values are marked in boldface (except for the DM-only halo, where the very same object is the brightest in all considered scenarios). Further properties of the objects (structural parameters, brightness of lower ranked subhalos, etc.) can be retrieved from the full subhalo catalogs which can be provided to the reader upon request.

Table A1. Properties of the brightest subhalos in the random realizations displayed in Figure 4.

	Position in Map (l, b)	Distance $l^\star$ [kpc]	Mass $m^\star$ [$M_\odot$]	$J_{0.5°}$ [GeV^2 cm^{-5}]	J_{tot} [GeV^2 cm^{-5}]	$D_{0.5°}$ [GeV cm^{-2}]	D_{tot} [GeV cm^{-2}]
DM only	$(-54°, -13°)$	16.3	4.0×10^9	1.8×10^{20}	5.5×10^{20}	1.1×10^{19}	1.8×10^{21}
Phat-ELVIS	$(-141°, +31°)$	42.2	2.0×10^9	$\mathbf{1.4 \times 10^{19}}$	$\mathbf{2.3 \times 10^{19}}$	3.6×10^{18}	$\mathbf{1.3 \times 10^{20}}$
	$(+37°, +8°)$	69.5	3.8×10^9	1.4×10^{19}	2.0×10^{19}	$\mathbf{4.1 \times 10^{18}}$	9.2×10^{19}
SL17, $\epsilon_t = 10^{-2}$	$(-19°, -73°)$	7.8	3.7×10^6	$\mathbf{3.3 \times 10^{19}}$	$\mathbf{6.1 \times 10^{19}}$	2.3×10^{18}	7.1×10^{18}
	$(-73°, +20°)$	54.8	2.3×10^9	1.0×10^{19}	2.6×10^{19}	$\mathbf{4.1 \times 10^{18}}$	$\mathbf{9.0 \times 10^{19}}$
SL17, $\epsilon_t = 1$	$(+71°, -25°)$	30.1	1.3×10^8	$\mathbf{9.0 \times 10^{18}}$	$\mathbf{1.5 \times 10^{19}}$	2.3×10^{18}	1.6×10^{19}
	$(+93°, -50°)$	68.5	3.4×10^9	5.2×10^{18}	1.4×10^{19}	$\mathbf{3.4 \times 10^{18}}$	$\mathbf{8.4 \times 10^{19}}$

References

1. Hütten, M.; Combet, C.; Maier, G.; Maurin, D. Dark matter substructure modelling and sensitivity of the Cherenkov Telescope Array to Galactic dark halos. *J. Cosmol. Astropart. Phys.* **2016**, *9*, 047. [CrossRef]
2. Profumo, S.; Sigurdson, K.; Kamionkowski, M. What Mass Are the Smallest Protohalos? *Phys. Rev. Lett.* **2006**, *97*, 031301. [CrossRef] [PubMed]
3. Bringmann, T. Particle models and the small-scale structure of dark matter. *New J. Phys.* **2009**, *11*, 105027. [CrossRef]
4. Charbonnier, A.; Combet, C.; Maurin, D. CLUMPY: A code for γ-ray signals from dark matter structures. *Comput. Phys. Commun.* **2012**, *183*, 656–668. [CrossRef]
5. Bonnivard, V.; Hütten, M.; Nezri, E.; Charbonnier, A.; Combet, C.; Maurin, D. CLUMPY: Jeans analysis, γ-ray and ν fluxes from dark matter (sub-)structures. *Comput. Phys. Commun.* **2016**, *200*, 336–349. [CrossRef]
6. Hütten, M.; Combet, C.; Maurin, D. CLUMPY v3: γ-ray and ν signals from dark matter at all scales. *Comput. Phys. Commun.* **2019**, *235*, 336–345. [CrossRef]
7. Acharya, B.S.; Actis, M.; Aghajani, T.; Agnetta, G.; Aguilar, J.; Aharonian, F.; Ajello, M.; Akhperjanian, A.; Alcubierre, M.; Aleksić, J.; et al. Introducing the CTA concept. *Astropart. Phys.* **2013**, *43*, 3–18. [CrossRef]
8. D'Onghia, E.; Springel, V.; Hernquist, L.; Keres, D. Substructure Depletion in the Milky Way Halo by the Disk. *Astrophys. J.* **2010**, *709*, 1138–1147. [CrossRef]

9. Zhu, Q.; Marinacci, F.; Maji, M.; Li, Y.; Springel, V.; Hernquist, L. Baryonic impact on the dark matter distribution in Milky Way-sized galaxies and their satellites. *Mon. Not. R. Astron. Soc.* **2016**, *458*, 1559–1580. [CrossRef]
10. Kelley, T.; Bullock, J.S.; Garrison-Kimmel, S.; Boylan-Kolchin, M.; Pawlowski, M.S.; Graus, A.S. Phat ELVIS: The inevitable effect of the Milky Way's disk on its dark matter subhaloes. *arXiv* **2018**, arXiv:1811.12413.
11. Green, A.M.; Goodwin, S.P. On mini-halo encounters with stars. *Mon. Not. R. Astron. Soc.* **2007**, *375*, 1111–1120. [CrossRef]
12. Goerdt, T.; Gnedin, O.Y.; Moore, B.; Diemand, J.; Stadel, J. The survival and disruption of cold dark matter microhaloes: Implications for direct and indirect detection experiments. *Mon. Not. R. Astron. Soc.* **2007**, *375*, 191–198. [CrossRef]
13. Berezinsky, V.; Dokuchaev, V.; Eroshenko, Y. Remnants of dark matter clumps. *Phys. Rev. D* **2008**, *77*, 083519. [CrossRef]
14. Berezinsky, V.S.; Dokuchaev, V.I.; Eroshenko, Y.N. Small-scale clumps of dark matter. *Phys. Uspekhi* **2014**, *57*, 1–36. [CrossRef]
15. Stref, M.; Lavalle, J. Modeling dark matter subhalos in a constrained galaxy: Global mass and boosted annihilation profiles. *Phys. Rev. D* **2017**, *95*, 063003. [CrossRef]
16. Berlin, A.; Hooper, D. Stringent constraints on the dark matter annihilation cross section from subhalo searches with the Fermi Gamma-Ray Space Telescope. *Phys. Rev. D* **2014**, *89*, 016014. [CrossRef]
17. Bertoni, B.; Hooper, D.; Linden, T. Examining The Fermi-LAT Third Source Catalog in search of dark matter subhalos. *J. Cosmol. Astropart. Phys.* **2015**, *12*, 035. [CrossRef]
18. Schoonenberg, D.; Gaskins, J.; Bertone, G.; Diemand, J. Dark matter subhalos and unidentified sources in the Fermi 3FGL source catalog. *J. Cosmol. Astropart. Phys.* **2016**, *5*, 028. [CrossRef]
19. Mirabal, N.; Charles, E.; Ferrara, E.C.; Gonthier, P.L.; Harding, A.K.; Sánchez-Conde, M.A.; Thompson, D.J. 3FGL Demographics Outside the Galactic Plane using Supervised Machine Learning: Pulsar and Dark Matter Subhalo Interpretations. *Astrophys. J.* **2016**, *825*, 69. [CrossRef]
20. Bertoni, B.; Hooper, D.; Linden, T. Is the gamma-ray source 3FGL J2212.5+0703 a dark matter subhalo? *J. Cosmol. Astropart. Phys.* **2016**, *5*, 049. [CrossRef]
21. Wang, Y.P.; Duan, K.K.; Ma, P.X.; Liang, Y.F.; Shen, Z.Q.; Li, S.; Yue, C.; Yuan, Q.; Zang, J.J.; Fan, Y.Z.; et al. Testing the dark matter subhalo hypothesis of the gamma-ray source 3FGL J 2212.5 +0703. *Phys. Rev. D* **2016**, *94*, 123002. [CrossRef]
22. Liang, Y.F.; Xia, Z.Q.; Duan, K.K.; Shen, Z.Q.; Li, X.; Fan, Y.Z. Limits on dark matter annihilation cross sections to gamma-ray lines with subhalo distributions in N-body simulations and Fermi LAT data. *Phys. Rev. D* **2017**, *95*, 063531. [CrossRef]
23. Calore, F.; De Romeri, V.; Di Mauro, M.; Donato, F.; Marinacci, F. Realistic estimation for the detectability of dark matter subhalos using Fermi-LAT catalogs. *Phys. Rev. D* **2017**, *96*, 063009. [CrossRef]
24. Hooper, D.; Witte, S.J. Gamma rays from dark matter subhalos revisited: Refining the predictions and constraints. *J. Cosmol. Astropart. Phys.* **2017**, *4*, 018. [CrossRef]
25. Charles, E.; Sánchez-Conde, M.; Anderson, B.; Caputo, R.; Cuoco, A.; Di Mauro, M.; Drlica-Wagner, A.; Gomez-Vargas, G.A.; Meyer, M.; Tibaldo, L.; et al. Sensitivity projections for dark matter searches with the Fermi large area telescope. *Phys. Rep.* **2016**, *636*, 1–46. [CrossRef]
26. Harding, J.P.; Dingus, B. Dark Matter Annihilation and Decay Searches with the High Altitude Water Cherenkov (HAWC) Observatory. In Proceedings of the 34th International Cosmic Ray Conference (ICRC2015), The Hague, The Netherlands, 30 July–6 August 2015; Volume 34, p. 1227.
27. Abeysekara, A.U.; Albert, A.; Alfaro, R.; Alvarez, C.; Arceo, R.; Arteaga-Velázquez, J.C.; Rojas, D.A.; Solares, H.A.; Belmont-Moreno, E.; BenZvi, S.Y.; et al. Searching for Dark Matter Sub-structure with HAWC. *arXiv* **2018**, arXiv:1811.11732.
28. Egorov, A.E.; Galper, A.M.; Topchiev, N.P.; Leonov, A.A.; Suchkov, S.I.; Kheymits, M.D.; Yurkin, Y.T. Detactability of Dark Matter Subhalos by Means of the GAMMA-400 Telescope. *Phys. Atom. Nuclei* **2018**, *81*, 373–378. [CrossRef]
29. Brun, P.; Moulin, E.; Diemand, J.; Glicenstein, J.F. Searches for dark matter subhaloes with wide-field Cherenkov telescope surveys. *Phys. Rev. D* **2011**, *83*, 015003. [CrossRef]
30. Gunn, J.E.; Lee, B.W.; Lerche, I.; Schramm, D.N.; Steigman, G. Some astrophysical consequences of the existence of a heavy stable neutral lepton. *Astrophys. J.* **1978**, *223*, 1015–1031. [CrossRef]

31. Lake, G. Detectability of gamma-rays from clumps of dark matter. *Nature* **1990**, *346*, 39–40. [CrossRef]
32. Silk, J.; Stebbins, A. Clumpy cold dark matter. *Astrophys. J.* **1993**, *411*, 439–449. [CrossRef]
33. Berezinsky, V.; Masiero, A.; Valle, J.W.F. Cosmological signatures of supersymmetry with spontaneously broken R parity. *Phys. Lett. B* **1991**, *266*, 382–388. [CrossRef]
34. Bertone, G.; Buchmüller, W.; Covi, L.; Ibarra, A. Gamma-rays from decaying dark matter. *J. Cosmol. Astropart. Phys.* **2007**, *11*, 3. [CrossRef]
35. Bergström, L.; Ullio, P.; Buckley, J.H. Observability of γ rays from dark matter neutralino annihilations in the Milky Way halo. *Astropart. Phys.* **1998**, *9*, 137–162. [CrossRef]
36. Press, W.H.; Schechter, P. Formation of Galaxies and Clusters of Galaxies by Self-Similar Gravitational Condensation. *Astrophys. J.* **1974**, *187*, 425–438. [CrossRef]
37. Walker, M.G.; Combet, C.; Hinton, J.A.; Maurin, D.; Wilkinson, M.I. Dark Matter in the Classical Dwarf Spheroidal Galaxies: A Robust Constraint on the Astrophysical Factor for γ-Ray Flux Calculations. *Astrophys. J. Lett.* **2011**, *733*, L46. [CrossRef]
38. Charbonnier, A.; Combet, C.; Daniel, M.; Funk, S.; Hinton, J.A.; Maurin, D.; Power, C.; Read, J.I.; Sarkar, S.; Walker, M.G.; et al. Dark matter profiles and annihilation in dwarf spheroidal galaxies: Prospectives for present and future γ-ray observatories—I. The classical dwarf spheroidal galaxies. *Mon. Not. R. Astron. Soc.* **2011**, *418*, 1526–1556. [CrossRef]
39. Bonnivard, V.; Combet, C.; Maurin, D.; Walker, M.G. Spherical Jeans analysis for dark matter indirect detection in dwarf spheroidal galaxies—Impact of physical parameters and triaxiality. *Mon. Not. R. Astron. Soc.* **2015**, *446*, 3002–3021. [CrossRef]
40. Bonnivard, V.; Combet, C.; Daniel, M.; Funk, S.; Geringer-Sameth, A.; Hinton, J.A.; Maurin, D.; Read, J.I.; Sarkar, S.; Walker, M.G.; et al. Dark matter annihilation and decay in dwarf spheroidal galaxies: The classical and ultrafaint dSphs. *Mon. Not. R. Astron. Soc.* **2015**, *453*, 849–867. [CrossRef]
41. Bonnivard, V.; Combet, C.; Maurin, D.; Geringer-Sameth, A.; Koushiappas, S.M.; Walker, M.G.; Mateo, M.; Olszewski, E.W.; Bailey, J.I., III. Dark Matter Annihilation and Decay Profiles for the Reticulum II Dwarf Spheroidal Galaxy. *Astrophys. J. Lett.* **2015**, *808*, L36. [CrossRef]
42. Bonnivard, V.; Maurin, D.; Walker, M.G. Contamination of stellar-kinematic samples and uncertainty about dark matter annihilation profiles in ultrafaint dwarf galaxies: The example of Segue I. *Mon. Not. R. Astron. Soc.* **2016**, *462*, 223–234. [CrossRef]
43. Genina, A.; Fairbairn, M. The potential of the dwarf galaxy Triangulum II for dark matter indirect detection. *Mon. Not. R. Astron. Soc.* **2016**, *463*, 3630–3636. [CrossRef]
44. Walker, M.G.; Mateo, M.; Olszewski, E.W.; Koposov, S.; Belokurov, V.; Jethwa, P.; Nidever, D.L.; Bonnivard, V.; Bailey, J.I., III; Bell, E.F.; et al. Magellan/M2FS Spectroscopy of Tucana 2 and Grus 1. *Astrophys. J.* **2016**, *819*, 53. [CrossRef]
45. Campos, M.D.; Queiroz, F.S.; Yaguna, C.E.; Weniger, C. Search for right-handed neutrinos from dark matter annihilation with gamma-rays. *J. Cosmol. Astropart. Phys.* **2017**, *7*, 016. [CrossRef]
46. Albert, A.; Alfaro, R.; Alvarez, C.; Álvarez, J.D.; Arceo, R.; Arteaga-Velázquez, J.C.; Rojas, D.A.; Solares, H.A.; Bautista-Elivar, N.; Becerril, A.; et al. Dark Matter Limits from Dwarf Spheroidal Galaxies with the HAWC Gamma-Ray Observatory. *Astrophys. J.* **2018**, *853*, 154. [CrossRef]
47. The ANTARES Collaboration. Search of dark matter annihilation in the galactic centre using the ANTARES neutrino telescope. *J. Cosmol. Astropart. Phys.* **2015**, *10*, 068.
48. Albert, A.; André, M.; Anghinolfi, M.; Anton, G.; Ardid, M.; Aubert, J.J.; Avgitas, T.; Baret, B.; Barrios-Martí, J.; Basa, S.; et al. Results from the search for dark matter in the Milky Way with 9 years of data of the ANTARES neutrino telescope. *Phys. Lett. B* **2017**, *769*, 249–254. [CrossRef]
49. Balázs, C.; Conrad, J.; Farmer, B.; Jacques, T.; Li, T.; Meyer, M.; Queiroz, F.S.; Sánchez-Conde, M.A. Sensitivity of the Cherenkov Telescope Array to the detection of a dark matter signal in comparison to direct detection and collider experiments. *Phys. Rev. D* **2017**, *96*, 083002. [CrossRef]
50. Nichols, M.; Mirabal, N.; Agertz, O.; Lockman, F.J.; Bland-Hawthorn, J. The Smith Cloud and its dark matter halo: Survival of a Galactic disc passage. *Mon. Not. R. Astron. Soc.* **2014**, *442*, 2883–2891. [CrossRef]
51. Albert, A.; Alfaro, R.; Alvarez, C.; Álvarez, J.D.; Arceo, R.; Arteaga-Velázquez, J.C.; Rojas, D.A.; Solares, H.A.; Becerril, A.; Belmont-Moreno, E.; et al. Search for dark matter gamma-ray emission from the Andromeda Galaxy with the High-Altitude Water Cherenkov Observatory. *J. Cosmol. Astropart. Phys.* **2018**, *6*, 043. [CrossRef]

52. Combet, C.; Maurin, D.; Nezri, E.; Pointecouteau, E.; Hinton, J.A.; White, R. Decaying dark matter: Stacking analysis of galaxy clusters to improve on current limits. *Phys. Rev. D* **2012**, *85*, 063517. [CrossRef]
53. Nezri, E.; White, R.; Combet, C.; Hinton, J.A.; Maurin, D.; Pointecouteau, E. γ-rays from annihilating dark matter in galaxy clusters: Stacking versus single source analysis. *Mon. Not. R. Astron. Soc.* **2012**, *425*, 477–489. [CrossRef]
54. Maurin, D.; Combet, C.; Nezri, E.; Pointecouteau, E. Disentangling cosmic-ray and dark-matter induced γ-rays in galaxy clusters. *Astron. Astrophys.* **2012**, *547*, A16. [CrossRef]
55. Hütten, M.; Combet, C.; Maurin, D. Extragalactic diffuse γ-rays from dark matter annihilation: Revised prediction and full modelling uncertainties. *J. Cosmol. Astropart. Phys.* **2018**, *2*, 005. [CrossRef]
56. Bryan, G.L.; Norman, M.L. Statistical Properties of X-Ray Clusters: Analytic and Numerical Comparisons. *Astrophys. J.* **1998**, *495*, 80–99. [CrossRef]
57. Springel, V.; White, S.D.M.; Frenk, C.S.; Navarro, J.F.; Jenkins, A.; Vogelsberger, M.; Wang, J.; Ludlow, A.; Helmi, A. Prospects for detecting supersymmetric dark matter in the Galactic halo. *Nature* **2008**, *456*, 73–76. [CrossRef] [PubMed]
58. Lange, J.U.; Chu, M.C. Can galactic dark matter substructure contribute to the cosmic gamma-ray anisotropy? *Mon. Not. R. Astron. Soc.* **2015**, *447*, 939–947. [CrossRef]
59. Kim, S.Y.; Peter, A.H.G.; Hargis, J.R. Missing Satellites Problem: Completeness Corrections to the Number of Satellite Galaxies in the Milky Way are Consistent with Cold Dark Matter Predictions. *Phys. Rev. Lett.* **2018**, *121*, 211302. [CrossRef]
60. Klypin, A.; Kravtsov, A.V.; Valenzuela, O.; Prada, F. Where Are the Missing Galactic Satellites? *Astrophys. J.* **1999**, *522*, 82–92. [CrossRef]
61. Strigari, L.E.; Bullock, J.S.; Kaplinghat, M.; Diemand, J.; Kuhlen, M.; Madau, P. Redefining the Missing Satellites Problem. *Astrophys. J.* **2007**, *669*, 676–683. [CrossRef]
62. Han, J.; Cole, S.; Frenk, C.S.; Jing, Y. A unified model for the spatial and mass distribution of subhaloes. *Mon. Not. R. Astron. Soc.* **2016**, *457*, 1208–1223. [CrossRef]
63. Sánchez-Conde, M.A.; Prada, F. The flattening of the concentration-mass relation towards low halo masses and its implications for the annihilation signal boost. *Mon. Not. R. Astron. Soc.* **2014**, *442*, 2271–2277. [CrossRef]
64. Bullock, J.S.; Kolatt, T.S.; Sigad, Y.; Somerville, R.S.; Kravtsov, A.V.; Klypin, A.A.; Primack, J.R.; Dekel, A. Profiles of dark haloes: Evolution, scatter and environment. *Mon. Not. R. Astron. Soc.* **2001**, *321*, 559–575. [CrossRef]
65. Navarro, J.F.; Frenk, C.S.; White, S.D.M. The Structure of Cold Dark Matter Halos. *Astrophys. J.* **1996**, *462*, 563. [CrossRef]
66. Einasto, J.; Haud, U. Galactic models with massive corona. I—Method. II—Galaxy. *Astron. Astrophys.* **1989**, *223*, 89–106.
67. Ishiyama, T.; Makino, J.; Ebisuzaki, T. Gamma-ray Signal from Earth-mass Dark Matter Microhalos. *Astrophys. J. Lett.* **2010**, *723*, L195–L200. [CrossRef]
68. Ishiyama, T. Hierarchical Formation of Dark Matter Halos and the Free Streaming Scale. *Astrophys. J.* **2014**, *788*, 27. [CrossRef]
69. Angulo, R.E.; Hahn, O.; Ludlow, A.D.; Bonoli, S. Earth-mass haloes and the emergence of NFW density profiles. *Mon. Not. R. Astron. Soc.* **2017**, *471*, 4687–4701. [CrossRef]
70. Piffl, T.; Binney, J.; McMillan, P.J.; Steinmetz, M.; Helmi, A.; Wyse, R.F.G.; Bienayme, O.; Bland-Hawthorn, J.; Freeman, K.; Gibson, B.; et al. Constraining the Galaxy's dark halo with RAVE stars. *Mon. Not. R. Astron. Soc.* **2014**, *445*, 3133–3151. [CrossRef]
71. McMillan, P.J. The mass distribution and gravitational potential of the Milky Way. *Mon. Not. R. Astron. Soc.* **2017**, *465*, 76–94. [CrossRef]
72. Binney, J. Self-consistent modelling of our Galaxy with Gaia data. *arXiv* **2017**, arxiv:1706.01374.
73. Brown, A.G.A.; Vallenari, A.; Prusti, T.; de Bruijne, J.H.J.; Babusiaux, C.; Bailer-Jones, C.A.L.; Biermann, M.; Evans, D.W.; Eyer, L.; Jansen, F.; et al. Gaia Data Release 2. Summary of the contents and survey properties. *Astron. Astrophys.* **2018**, *616*, A1.
74. Springel, V.; Wang, J.; Vogelsberger, M.; Ludlow, A.; Jenkins, A.; Helmi, A.; Navarro, J.F.; Frenk, C.S.; White, S.D. The Aquarius Project: The subhaloes of galactic haloes. *Mon. Not. R. Astron. Soc.* **2008**, *391*, 1685–1711. [CrossRef]

75. Moliné, Á.; Sánchez-Conde, M.A.; Palomares-Ruiz, S.; Prada, F. Characterization of subhalo structural properties and implications for dark matter annihilation signals. *Mon. Not. R. Astron. Soc.* **2017**, *466*, 4974–4990. [CrossRef]
76. Diemand, J.; Kuhlen, M.; Madau, P.; Zemp, M.; Moore, B.; Potter, D.; Stadel, J. Clumps and streams in the local dark matter distribution. *Nature* **2008**, *454*, 735–738. [CrossRef]
77. Garrison-Kimmel, S.; Boylan-Kolchin, M.; Bullock, J.S.; Lee, K. ELVIS: Exploring the Local Volume in Simulations. *Mon. Not. R. Astron. Soc.* **2014**, *438*, 2578–2596. [CrossRef]
78. Zhao, H. Analytical models for galactic nuclei. *Mon. Not. R. Astron. Soc.* **1996**, *278*, 488–496. [CrossRef]
79. Hayashi, E.; Navarro, J.F.; Taylor, J.E.; Stadel, J.; Quinn, T.R. The Structural evolution of substructure. *Astrophys. J.* **2003**, *584*, 541–558. [CrossRef]
80. van den Bosch, F.C.; Ogiya, G.; Hahn, O.; Burkert, A. Disruption of dark matter substructure: Fact or fiction? *Mon. Not. R. Astron. Soc.* **2018**, *474*, 3043–3066. [CrossRef]
81. van den Bosch, F.C.; Ogiya, G. Dark matter substructure in numerical simulations: A tale of discreteness noise, runaway instabilities, and artificial disruption. *Mon. Not. R. Astron. Soc.* **2018**, *475*, 4066–4087. [CrossRef]
82. Weinberg, M.D. Adiabatic invariants in stellar dynamics. 1: Basic concepts. *Astrophys. J.* **1994**, *108*, 1398–1402. [CrossRef]
83. Gnedin, O.Y.; Lee, H.M.; Ostriker, J.P. Effects of Tidal Shocks on the Evolution of Globular Clusters. *Astrophys. J.* **1999**, *522*, 935–949. [CrossRef]
84. Peñarrubia, J.; Benson, A.J.; Walker, M.G.; Gilmore, G.; McConnachie, A.W.; Mayer, L. The impact of dark matter cusps and cores on the satellite galaxy population around spiral galaxies. *Mon. Not. R. Astron. Soc.* **2010**, *406*, 1290–1305. [CrossRef]
85. Ackermann, M.; Ajello, M.; Baldini, L.; Ballet, J.; Barbiellini, G.; Bastieri, D.; Bellazzini, R.; Bissaldi, E.; Blandford, R.D.; Bloom, E.D.; et al. The Search for Spatial Extension in High-latitude Sources Detected by the Fermi Large Area Telescope. *Astrophys. J. Suppl. Ser.* **2018**, *237*, 32. [CrossRef]
86. The Fermi-LAT collaboration. Fermi Large Area Telescope Fourth Source Catalog. *arXiv* **2019**, arxiv:1902.10045.
87. Scott, D.W. *Multivariate Density Estimation: Theory, Practice, and Visualization*; John Wiley and Sons: Hoboken, NJ, USA, 2015.
88. Strigari, L.E. Dark matter in dwarf spheroidal galaxies and indirect detection: A review. *Rep. Prog. Phys.* **2018**, *81*, 056901. [CrossRef] [PubMed]
89. Hiroshima, N.; Ando, S.; Ishiyama, T. Modeling evolution of dark matter substructure and annihilation boost. *Phys. Rev. D* **2018**, *97*, 123002. [CrossRef]
90. Ando, S.; Ishiyama, T.; Hiroshima, N. Halo substructure boosts to the signatures of dark matter annihilation. *arXiv* **2019**, arxiv:1903.11427.
91. Peñarrubia, J.; Navarro, J.F.; McConnachie, A.W. The Tidal Evolution of Local Group Dwarf Spheroidals. *Astrophys. J.* **2008**, *673*, 226–240. [CrossRef]
92. Drakos, N.E.; Taylor, J.E.; Benson, A.J. The phase-space structure of tidally stripped haloes. *Mon. Not. R. Astron. Soc.* **2017**, *468*, 2345–2358. [CrossRef]

© 2019 by the authors. Licensee MDPI, Basel, Switzerland. This article is an open access article distributed under the terms and conditions of the Creative Commons Attribution (CC BY) license (http://creativecommons.org/licenses/by/4.0/).

Article

Gamma-Ray Sensitivity to Dark Matter Subhalo Modelling at High Latitudes

Francesca Calore [1,*], Moritz Hütten [2] and Martin Stref [1,3]

1   Laboratoire d'Annecy-le-Vieux de Physique Théorique (LAPTh), Université Grenoble Alpes, USMB, CNRS, F-74000 Annecy, France; martin.stref@lapth.cnrs.fr
2   Max-Planck-Institut für Physik, Föhringer Ring 6, D-80805 München, Germany; mhuetten@mpp.mpg.de
3   Laboratoire Univers & Particules de Montpellier (LUPM), CNRS & Université de Montpellier, Place Eugène Bataillon, CEDEX 05, F-34095 Montpellier, France
*   Correspondence: calore@lapth.cnrs.fr

Received: 28 October 2019; Accepted: 22 November 2019; Published: 26 November 2019

Abstract: Searches for "dark" subhaloes in gamma-ray point-like source catalogues are among promising strategies for indirect dark matter detection. Such a search is nevertheless affected by uncertainties related, on the one hand, to the modelling of the dark matter subhalo distribution in Milky-Way-like galaxies, and, on the other hand, to the sensitivity of gamma-ray instruments to the dark matter subhalo signals. In the present work, we assess the detectability of dark matter subhaloes in Fermi-LAT catalogues, taking into accounts uncertainties associated with the modelling of the galactic subhalo population. We use four different halo models bracketing a large set of uncertainties. For each model, adopting an accurate detection threshold of the LAT to dark matter subhalo signals and comparing model predictions with the number of unassociated point-sources in Fermi-LAT catalogues, we derive upper limits on the annihilation cross section as a function of dark matter mass. Our results show that, even in the best-case scenario (i.e., DMonly subhalo model), which does not include tidal disruption from baryons, the limits on the dark matter parameter space are less stringent than current gamma-ray limits from dwarf spheroidal galaxies. Comparing the results obtained with the different subhalo models, we find that baryonic effects on the subhalo population are significant and lead to dark matter constraints that are less stringent by a factor of ~2 to ~5. This uncertainty comes from the unknown resilience of dark matter subhaloes to tidal disruption.

Keywords: dark matter; galactic sub-halos; gamma rays

1. Introduction

The identification of dark matter (DM) is one of the major endeavours of particle physics and cosmology of the 21st century. Despite theoretical and experimental efforts deployed to detect DM particles, the nature of this elusive form of matter remains mostly unknown. We know cold DM to be successful in describing the universe on large scales [1]. However, null outcomes of weakly-interacting massive particle (WIMP, [2]) searches in direct, indirect, and collider experiments, together with deviations from cold DM predictions on small scales [3], challenge this paradigm and feed the interest for alternative DM scenarios.

However, deviations from cold DM predictions on small scales tantalise this paradigm and cast serious doubts on the weakly-interacting massive particle hypothesis, the most scrutinised model for cold DM so far [2]. Additionally, searches for DM particle candidates at the weak scale have been until now unsuccessful with current instruments, on ground and in space. In particular, attempts of indirect detection of high-energy photons from WIMP self-annihilation provide some among the strongest limits on WIMP DM [4,5]. At this stage, it is unclear if the WIMP (and cold DM) paradigm has to be revised in favour of other, still viable, DM particle models (warm and ultra-light DM models), or if it is instead kinematically outside of the main explored range and can be discovered with the next generation of gamma-ray telescopes, e.g., the Cherenkov telescope array (CTA, [6]).

Indirect detection constraints on the WIMP parameter space are unavoidably affected by background model systematics. This is particularly severe in the inner region of the galaxy, where the gamma-ray emission is dominated by the interactions of cosmic rays with the interstellar matter and fields (i.e., galactic diffuse emission). "Cleaner" and, in this respect, more promising targets for DM identification are dwarf spheroidal galaxies, optically faint galaxies whose dynamics has been proved to be dominated by large haloes of DM [7]. Those faintest detectable galaxies can probe the WIMP paradigm with multi wavelength observations, from optical to gamma rays (see for example [8,9]). Moreover, the DM haloes hosting dwarf spheroidal galaxies are thought to be the most massive of a vast population of DM subhaloes, overdensities in the DM host halo surrounding our galaxy [10,11]. While the majority of these DM subhaloes lacks an optical counterpart, a steady gamma-ray signal from directions where no object can be associated in other wavelengths would be a hint for WIMP annihilation.

Searches for DM subhaloes are typically performed in point-source catalogues of the Large Area Telescope (LAT), aboard the Fermi satellite. Point-source catalogues like the Third Fermi-LAT Source Catalogue (3FGL) [12], and the Second Catalogue of Hard Fermi-LAT Sources (2FHL) [13] contain a number of gamma-ray sources which are not associated with any known astrophysical object. Classification algorithms, utilizing in particular spectral information, are applied on these unassociated sources in order to single out potential DM subhalo candidates [14–16].

Limits on the DM parameter space (annihilation cross-section vs. mass) are derived by comparing the number of expected DM subhalo candidates in the catalogues with predictions of the number of subhaloes above the Fermi-LAT detection threshold expected from theoretical models of subhaloes [16–22]. To this end, one needs to know how many subhaloes are expected to be bright enough in gamma rays to be seen above the standard astrophysical background. This requires, on the one hand, a detailed description of the galactic subhalo population. The complicated physics of subhalo evolution inside the potential of their host leads to different quantitative pictures depending on the models. To bracket these uncertainties, various models, either analytical or based on numerical simulations, are considered in this study, see Section 2. On the other hand, the number of expected detectable subhaloes is obtained by convolving the DM subhalo signal with the Fermi-LAT detection threshold. The LAT detection threshold depends on the spectral signal that is looked for. Reference [22] showed that computing the sensitivity of the LAT to DM subhalo signals, adopting the specific spectral energy distribution determined by the particle physics DM model (see also Section 3), provides more accurate predictions on the number of expected detectable subhalo and that important differences with respect to assuming a fixed sensitivity threshold arise. We will therefore use the Fermi-LAT detection threshold as derived in [22].

The goal of the present paper is to assess the detectability of DM subhaloes as predicted by state-of-the-art DM subhalo models [23]. We will do so by using the more accurate Fermi-LAT sensitivity threshold to DM subhalo signals [22]. In Section 2 we describe the galactic subhalo models, in Section 3 we remind the reader the main ingredients to compute the gamma-ray DM signal from dark subhaloes, and in Section 4 how the LAT sensitivity is computed. We present the results in Section 5, and conclude in Section 6.

2. Galactic Subhalo Modelling

Subhaloes are subject to a variety of phenomena, including tidal stripping, gravitational shocking, and dynamical friction, which make their modelling challenging. Subhaloes can be studied by the means of fully-numerical cosmological simulations or simplified analytical models. These different approaches lead to similar qualitative pictures regarding the galactic subhalo population but often differ on a quantitative level. To get a handle on the modelling uncertainties, four models are considered in this study. These models share some common features: spherical symmetry is assumed for the galactic halo, subhaloes all have a Navarro–Frenk–White (NFW) density profile and their mass function is a power law with index $\alpha_m = 1.9$.[1] These assumptions are all verified on the scales resolved by numerical simulations, see e.g., [10,26]. Four configurations are considered, which are identical to those used in Hütten et al. [23], which the reader is referred to for further details.

Our first model is based on the Aquarius DM-only N-body simulation [26] and as such is called DMonly. The subhalo spatial distribution in Aquarius is found to be well fitted by an Einasto profile with parameters $\alpha_E = 0.68$ and $r_{-2} = 199$ kpc. The core in the distribution is created by tidal interactions which tend to disrupt subhaloes at the center of the host. The total number of subhaloes is fixed by assuming 300 high-mass clumps with masses larger than 10^8 $M_\odot$, as an upper bound to the Aquarius findings [26]. Subhaloes are further assumed to follow the mass-concentration relation given by Moliné et al. [27]. While a well-known effect of tides is to remove matter from the outskirts of subhaloes, this is not accounted for in DMonly and all the subhaloes have their cosmological extension (defined with respect to the critical density of the universe).

The Phat-ELVIS model is based on a suite of DM-only simulations which incorporate a static disc potential [28]. Through gravitational shocking, the disc is very efficient at disrupting most subhaloes in the inner 30 kpc of the host galaxy. The spatial distribution of the remaining population is well fitted by the following function:

$$\frac{dP}{dV}(r) = \frac{A}{1 + e^{-(r-r_0)/r_c}} \times \exp\left\{-\frac{2}{\alpha}\left[\left(\frac{r}{r_{-2}}\right)^\alpha - 1\right]\right\}, \quad (1)$$

with $\alpha = 0.68$, $r_0 = 29.2$ kpc, $r_c = 4.24$ kpc and $r_{-2} = 128$ kpc. Similar to the DMonly model, the mass-concentration relation is taken from [27] and the density profile of subhaloes extends to their cosmological extension.

Our next configurations are based on the semi-analytical model of Stref and Lavalle [29] (SL17 from now on). This model relies on a realistic description of the Milky Way and incorporates the stripping effect due to the gravitational potential of the galaxy as well as the shocking effect from the disc. It is not clear yet whether the efficient disruption of DM subhaloes as observed in simulations is realistic or not [30–32]. This can be of importance because it has been shown to impact predictions for indirect searches [33]. To account for this uncertainty, we consider two scenarios. In the first one, called SL17-fragile, subhaloes are disrupted when their tidal radius r_t is equal to their scale radius r_s. In the second one, called SL17-resilient, subhaloes are more robust and survive unless $r_t < 10^{-2} r_s$. Unlike the DMonly and Phat-ELVIS models, subhaloes in the SL17 configurations are stripped down to their tidal radius.

Knowing the DM subhalo spatial density ρ_{DM}, it is possible to compute the so-called astrophysical or $\mathcal{J}$-factor towards the direction—line of sight (l.o.s.)—of the subhalo of interest:

$$\int_0^{\Delta\Omega}\int_{\text{l.o.s.}} d\ell\, d\Omega\, \rho_{DM}^2(\ell), \quad (2)$$

where the integral along the l.o.s. is further integrated over the solid angle $\Delta\Omega = 2\pi(1 - \cos\theta_{\text{int}})$. In what follows, we will set $\theta_{\text{int}} = 0.1°$ effectively considering subhaloes as point-like sources. We note that previous works have overestimated the $\mathcal{J}$-factor—and thus got too stringent limits on DM—by integrating up to 0.5° [22] or, up to the DM profile scale radius, e.g., [16]. Indeed, as we will explain below, the way in which the LAT sensitivity to DM spectra is computed strictly applies to point-like sources having an angular extension of 0.1–0.3°. Cutting the integration radius up-to 0.1° worsens the final limits on the DM annihilation cross section by a factor of 2, over all the DM mass range.

---

[1] The mass function is sharply cut at $m_{\min} = 10^{-6}$ $M_\odot$. This mass cut-off can be related to the kinetic decoupling of the DM particle in the early universe, see e.g., [24,25].

Again, we believe this choice to be truly conservative. The $\mathcal{J}$-factor is one of the crucial ingredients to compute gamma-ray DM fluxes, as we will see below.

Having incorporated these models in the CLUMPY code [34–36], we consider 1000 Monte Carlo realisations for each configuration, and we select all subhaloes with $\mathcal{J}$ (<0.1°) >10^{17} GeV^{2}cm^{-5}. The choice of this cut guarantees that the flux from DM annihilation (for e.g., cross-section values ~10^{26}–10^{-23} cm^3/s and masses ~100 GeV is well below the Fermi-LAT catalogues threshold, and therefore that we do not miss any detectable subhalo. As we highlight below, this cut also allows us to study what is the role, if any, of low-mass subhaloes. We note that relying on the simulations done in [23] guarantees that the subhalo population is complete in brightness.

In Figure 1, we show the scatter plots of $\mathcal{J}$-factor values, $\mathcal{J}$ (<0.1°), as a function of subhalo mass, M_{SH}, in one realisation of the Monte Carlo simulations for each subhalo model.

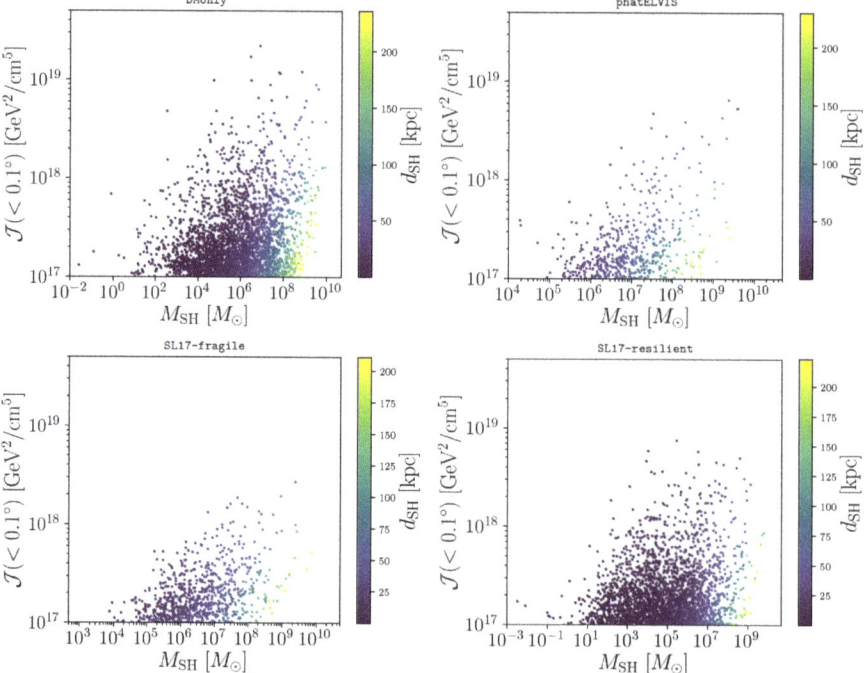

Figure 1. Scatter plot of $\mathcal{J}$-factor values, $\mathcal{J}$ within 0.1°, as a function of subhalo mass, M_{SH}, in one realisation of the Monte Carlo simulations for each subhalo model—**top left**: DMonly, **top right**: Phat-ELVIS, **bottom left**: SL17-fragile, **bottom right**: SL17-resilient. The colour-bar represents the subhalo distance from Earth, hereafter d_{SH}. The realisation shown is the one containing the lowest mass subhalo. We remind that we have applied a cut of $\mathcal{J}$ (<0.1°) > 10^{17} GeV^{2}cm^{-5}.

3. Gamma Rays from Subhaloes

The $\mathcal{J}$-factor is proportional to the predicted gamma-ray flux from WIMP DM annihilation. We therefore expect that the most-likely detectable subhaloes will be also the ones with the highest $\mathcal{J}$-factor. However, the sensitivity of a gamma-ray telescope to a DM (or astrophysical) signal does also depend on the gamma-ray spectrum that is looked for—in general harder spectra (e.g., BL Lacertae objects, spectral index ~2.2) are detected more easily—as we will see below.

To compute the predicted flux from DM annihilation, we have to specify the particle physics content of the underlying DM particle model we consider. In what follows, we provide equations for Majorana DM candidates (such as the neutralino in supersymmetric extensions of the Standard

Model)—predictions for Dirac DM particles can be obtained by multiplying the flux for an additional factor of $1/2$.

The flux of photons expected in a given energy range from annihilation of DM particles of mass $m_{\rm DM}$, distributed spatially following the DM distribution $\rho_{\rm DM}$, writes generally as:

$$\mathcal{F}(E_{\min}, E_{\max}) = \frac{\langle \sigma v \rangle}{8\pi m_{\rm DM}^2} \, \mathcal{J} \int_{E_{\min}}^{E_{\max}} \frac{{\rm d}N_{\rm DM}^i}{{\rm d}E} \, dE, \qquad (3)$$

where $\langle \sigma v \rangle$ is the thermally averaged annihilation cross section, and ${\rm d}N_{\rm DM}^i/{\rm d}E$ is the energy spectrum providing the number of gamma rays per annihilation of DM in a given final state i (e.g., $b\bar{b}$, $\tau^+\tau^-$, etc.). We use tabulated DM spectra from [37].

4. Fermi-LAT Sensitivity to DM Subhalos

We adopt the flux sensitivity calculation of Calore et al. [22], where the authors provided an accurate calculation of the LAT sensitivity to DM annihilation signals from subhaloes and showed that such a determination of the detection threshold leads to significant differences with respect to adopting a fixed flux threshold. The Fermi-LAT source detection simulation of DM subhaloes was performed for the third Fermi-LAT catalog of point sources (3FGL) [12], and the second catalog of hard Fermi-LAT sources (2FHL) [13]. In both catalogues, unassociated sources represent a significant fraction of all detected sources: About 15% in the 2FHL and 30% in the 3FGL.

Interestingly, some gamma-ray emitting DM subhaloes can hide among unassociated sources in the Fermi-LAT catalogues. In the present work, we assess what is the sensitivity to DM subhalo modelling of the Fermi-LAT 3FGL and 2FHL catalogues.

As can be seen from Figures 5–8 in [22], the flux sensitivity threshold of Fermi-LAT for the 3FGL and 2FHL set-ups depends both on latitude and mass of the DM candidate: Regardless of the annihilation channel, the flux sensitivity threshold decreases by a factor of about 2 between $20°$ and $80°$ in latitude for all masses, for both the 3FGL and 2FHL set-up. Also, higher (lower) DM masses are more easily detected in the 3FGL (2FHL) set-up, as thoroughly explained in [22]. We note that our sensitivity threshold for the $b\bar{b}$ and $\tau^+\tau^-$ channels is very similar to the one more recently derived by Coronado-Blazquez et al. [16]. Given the contamination of the galactic diffuse foreground, the sensitivity calculation of [22] is truly accurate for $|b| > 20°$. We will therefore consider only subhlaoes at high latitudes.

5. Results

To derive the number of detectable subhaloes for a given mass and final state annihilation channel, we computed the corresponding gamma-ray flux (Equation (3)) in the same energy range of the catalogue we want to compare with ($E > 0.1$ GeV for the 3FGL and $E > 50$ GeV for the 2FHL). For each subhalo in the Monte Carlo simulations, we then compared the DM gamma-ray flux with the Fermi-LAT sensitivity threshold at the position of the subhalo: A subhalo was detected if its gamma-ray flux was larger than the flux threshold at its position.

In general, the number of detectable subhaloes was almost linearly proportional to the annihilation cross section. As found in [22], the number of detectable subhaloes did not strongly depend on the DM mass for annihilation into bottom quarks, while, because of the harder spectrum, the DM mass was more relevant in the case of annihilation into τ leptons. In Table 1, for each model and catalogue configuration, we provide the annihilation cross section required to have at least one subhalo detectable for annihilation into b-quarks and τ leptons. We note that the minimal cross section needed to detect at least one subhalo was about a few 10^{-25} for annihilation into $b\bar{b}$ in the 3FGL, in the case of the DMonly model. The minimal cross section for SL17-resilient was found to be a factor of $\sim$2 higher, while it was $\sim$4–5 higher for the Phat-ELVIS and SL17-fragile models. The hierarchy between the models was similar for the 2FHL catalogue and for annihilation into $\tau^+\tau^-$. These minimal cross sections exceed current bounds from Fermi-LAT observations towards dwarf spheroidal galaxies, see e.g., [38]. Dwarf spheroidal galaxies are traditionally believed to give the strongest and most robust limits of the DM parameter space—although several, independent, works addressed the robustness of such a bound showing that it is prone to uncertainties of a factor of a few mainly because on the uncertainty in the modelling of the foreground at the dwarf position [9] and of the dwarf DM distribution [39–41].

Table 1. Cross-section required to have at least one subhalo detectable in the 3FGL (2FHL) catalogue set-up for a 100 GeV (1.5 TeV) DM particle mass.

	One Detectable Subhalo Cross-section (cm^3/s)			
	3FGL, $b\bar{b}$	3FGL, $\tau^+\tau^-$	2FHL, $b\bar{b}$	2FHL, $\tau^+\tau^-$
DMonly	8.80×10^{-25}	17.25×10^{-25}	3.81×10^{-23}	10.52×10^{-23}
Phat-ELVIS	34.64×10^{-25}	76.96×10^{-25}	18.99×10^{-23}	50.91×10^{-23}
SL17-fragile	44.50×10^{-25}	100.02×10^{-25}	28.91×10^{-23}	63.82×10^{-23}
SL17-resilient	19.32×10^{-25}	34.23×10^{-25}	9.70×10^{-23}	19.30×10^{-23}

It is of interest to have a look at the distribution of the $\mathcal{J}$-factor of detectable subhaloes versus their mass. This is shown in Figure 2 for the 3FGL catalogue set-up. In contrast to Figure 1, all subhaloes are here represented by grey dots, while the ones detectable in the 3FGL catalogue are shown by coloured points. Note that these subhaloes, represented by their $\mathcal{J}$-factors, are detectable for fluxes from a DM particle with mass of 100 GeV and annihilation cross section into $b\bar{b}$ of 5×10^{-24} cm^3/s. Figure 3 shows the same for the 2FHL catalogue set-up, DM mass of 1.5 TeV and annihilation cross section into $b\bar{b}$ of 5×10^{-22} cm^3/s.

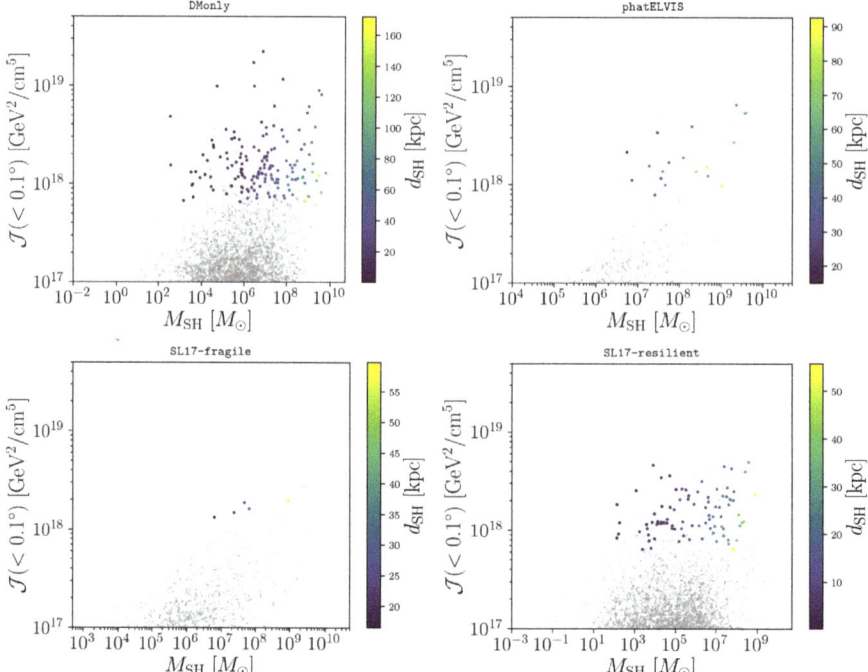

Figure 2. Same as in Figure 1 displaying all subhaloes selected as grey dots and those which would be detectable in the 2FHL catalogue as coloured points. The results are shown for a DM mass of 100 GeV, b-quark annihilation, and a cross section of 5×10^{-24} cm^3/s.

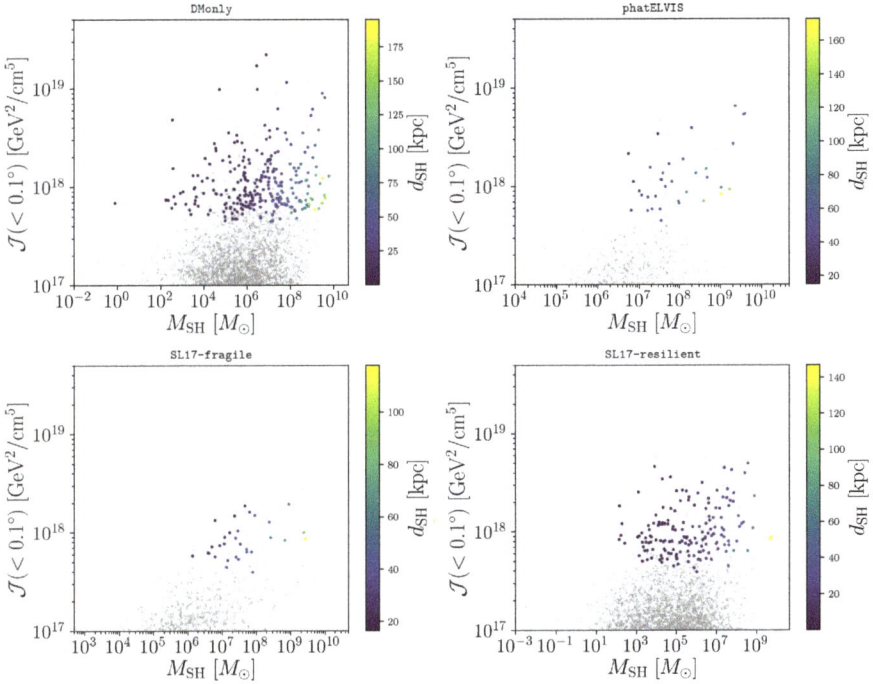

Figure 3. Same as in Figure 1 displaying all subhaloes selected as grey dots and those which would be detectable in the 2FHL catalogue as coloured points. The results are shown for a DM mass of 1.5 TeV, b-quark annihilation, and a cross section of 5×10^{-22} cm^3/s.

A few considerations are in order. First, for the set of particle physics parameter chosen, for the 3FGL (2FHL) set-up the $\mathcal{J}$-factor threshold for subhalo detection was about 7×10^{17} (4×10^{17}) GeV^2/cm^5 for all four models. However, not all subhaloes with $\mathcal{J}$-factor above this threshold were detectable in the Fermi-LAT catalogues; indeed, the DM mass and latitude dependence of the flux sensitivity threshold implied that the highest $\mathcal{J}$-factor subhaloes were sometimes not the most likely detectable ones with the LAT. This could be clearly seen, for example, in the bottom right panel of Figure 2: While the brightest gamma-ray subhalo had $\mathcal{J} \sim 8 \times 10^{18}$ GeV^2/cm^5, the detectable subhalo with the highest $\mathcal{J}$-factor had $\mathcal{J} \sim 5 \times 10^{18}$ GeV^2/cm^5. The same occurred for the 2FHL set-up. Secondly, the mass of detectable subhaloes could span up to seven orders in magnitude (from $\sim 10^2\,M_\odot$ to $\sim 10^{10}\,M_\odot$) depending on the configuration, as was the case for DMonly and SL17-resilient. We concluded that among detectable sources there were both dwarf galaxies and dark subhaloes. Indeed, galaxy formation models agreed that DM haloes with mass $> 10^8\,M_\odot$ are massive enough to systematically host galaxies. Although the exact threshold for star formation is quite debated and dark subhaloes can even coexist with luminous ones above that star formation threshold [42], low-mass subhaloes (below $10^7\,M_\odot$) are almost surely optically dark objects. However, those can still have large $\mathcal{J}$-factor and be among detectable subhaloes. This occurred for our DMonly, but also in a model where the effect of baryons in the galaxy was fully modelled (SL17-resilient). Finally, we note that in subhalo models where tidal disruption was less efficient (DMonly and SL17-resilient), most detectable subhaloes were located at a distance less than 20 kpc from us. Some of them were even closer than 10 kpc. On the other hand, when tidal disruption was efficient (as in Phat-ELVIS and SL17-fragile), a larger fraction of detectable subhaloes was located farther away (see also [23] for details). This was because the stellar disc disrupted most objects orbiting within the inner $\sim$20 kpc of the galaxy.

In Figure 4, we display the all-sky gamma-ray maps of selected haloes corresponding to the realisations shown in Figure 1. Fluxes were computed again assuming a DM mass of 100 GeV and an annihilation cross section into $b\bar{b}$ of 5×10^{-24} cm^3/s, for the 3FGL catalogue set-up. Subhaloes whose flux exceeded the sensitivity threshold are highlighted by light orange circles on the skymaps. Besides the latitude cut $|b| > 20°$, we could see that bright clumps at high latitude remained undetectable because of the latitude (and DM mass dependence) of the LAT detection threshold. We also note that subhaloes could have a very small angular extension on the sky and still be detectable, as could be seen in particular on the SL17-resilient skymap (bottom right). This was due to the cuspy density profile of DM haloes: Even if the structure was stripped off its outer layers by tidal effects, the $\mathcal{J}$-factor was only mildly affected and could remain quite high.

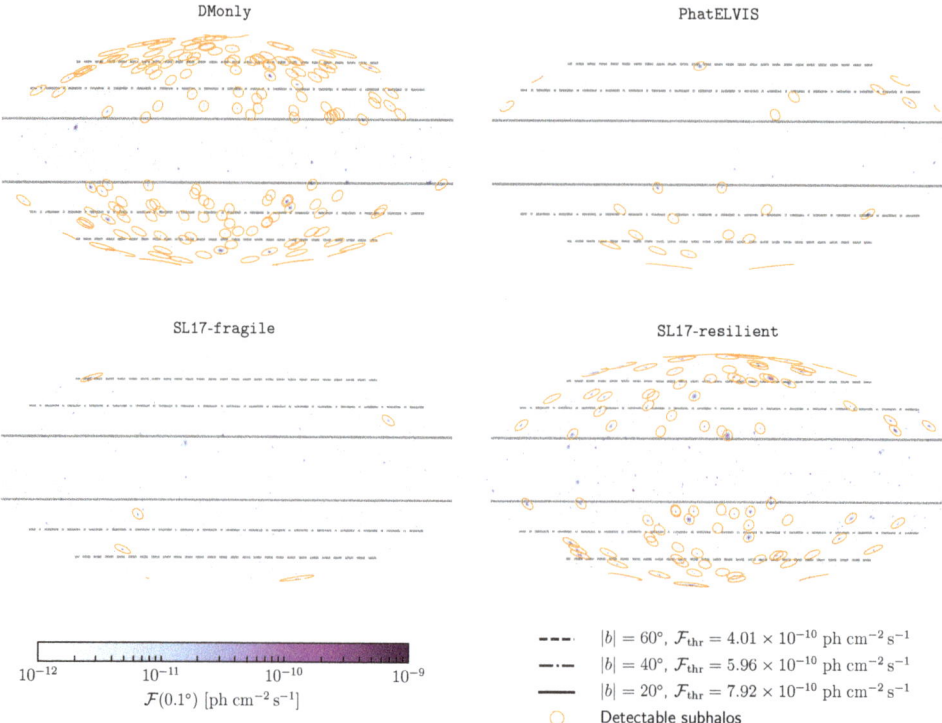

Figure 4. For the same realisations as in Figure 2, we display the corresponding all-sky gamma-ray maps of the selected haloes. Fluxes are computed assuming a DM mass of 100 GeV and an annihilation cross section into $b\bar{b}$ of 5×10^{-24} cm^3/s, for the 3FGL catalogue set-up. We overlay the LAT sensitivity threshold curves at fixed latitude values. The orange circles indicate the subhaloes that are above threshold, and that would therefore be detectable in the 3FGL catalogue. **Top left**: DMonly; **top right**: Phat-ELVIS; **bottom left**: SL17-fragile; **bottom right**: SL17-resilient. The orange circles indicating the detectable subhaloes have a diameter of 7°.

Dedicated searches for DM subhalo candidates among Fermi-LAT unassociated sources have been performed in the past through spectral and spatial analyses, often based on machine learning classification algorithms, see the latest analysis in [16]. The most recent analysis found 16 (4, 24) DM subhalo candidates in the 3FGL (2FHL, 3FHL) catalogue [16]. The flux sensitivity threshold inferred from Figure 9 of [16] was quite similar to the corresponding sensitivity curves of the 2FHL—so we will provide predictions for the 2FHL in the present work. Also, the limits from the 3FGL and 2FHL were completely complementary and the strongest over the full DM mass range considered in [16].

Knowing the number of DM subhalo candidates in Fermi-LAT catalogues, it is possible to infer an upper limit on the DM annihilation cross-section: For a given DM mass, this would be the value of $\langle \sigma v \rangle$ giving a number of detectable subhaloes equal to the number of DM subhalo candidates, N_c. The strongest bounds on DM would of course correspond to the case in which $N_c = 0$ (and therefore the subhalo to be used to set the bound is the one with the largest $\mathcal{J}$-factor). On the other hand, the most conservative limits come from the case where N_c is equal to the number of unassociated sources—which is anyhow unrealistic since most likely the largest fraction of these is indeed made up by standard astrophysical objects.

In Figure 5, we present upper limits on the DM annihilation cross-section as a function of the particle mass that comes from comparing the number of detectable subhaloes in the four models under consideration with the number of DM subhalo candidates from [16]. The upper limit on the cross section was defined as the maximum value of $\langle \sigma v \rangle$ for which the predicted subhalo gamma-ray fluxes were equal to the catalogue sensitivity flux threshold. The uncertainty bands corresponded to the uncertainty in the subhalo modelling, propagating the spread in the 1000 Monte Carlo realisations of the subhalo models (namely, the "galactic subhaloes variance"). We found that the DMonly configuration led to the strongest bounds on the annihilation cross-section. The bound from SL17-resilient was a factor of ∼2 weaker, while the bounds from Phat-ELVIS and SL17-fragile are similar and are ∼5–6 times weaker. Unsurprisingly, configurations where tidal disruption was not very efficient led to the strongest bounds. We could compare our bounds from the DMonly model with the limits obtained by [16] for the 3FGL and 2FHL catalogues (the authors also computed a limit for the updated 3FHL catalogue). Their limits were a factor of ∼3 stronger for both catalogues. For the origin of this difference there could be various reasons: for example, we recall that our DMonly model was based on the Aquarius cosmological simulation while the subhalo model used in [16] is based on Via Lactea II [10]. Subhaloes in these simulations have a different spatial distribution and the total number of resolved objects within the virial radius of the galactic halo also differs, hence there is no reason to expect the exact same gamma-ray prediction from both models. Also, [16] consider $\mathcal{J}$-factor integration angles equal to r_s, while we integrate only up to 0.1°. We note that in the case of the 2FHL our limits were cut at 100 GeV masses; below this mass the limits steeply increased because of a loss of sensitivity of the 2FHL catalogue.

In Figure 6, we put together the limits from the 3FGL and 2FHL catalogues and compared them to existing limits from gamma-ray observations of dwarf spheroidal galaxies [9,38]. We also display the "sensitivity reach" of DM searches towards unassociated gamma-ray sources, namely the limit on the annihilation cross section one gets imposing that no DM subhalo candidate remains among unassociated gamma-ray sources in the 3FGL and 2FHL catalogues. We stress that cutting the integration radius up to 0.1° leads to less strong bounds on the annihilation cross section (about a factor of 2 at all masses). Again, we believe our choice to be truly conservative, against what was done in the past. We can therefore see that the limits on the DM parameter space from the dark subhalo search are not as competitive as the search towards dwarf spheroidal galaxies—at least with present catalogues (and current sensitivity threshold). Indeed, the sensitivity reach for the 3FGL and 2FHL catalogues is always above the current limits from dwarf galaxies.

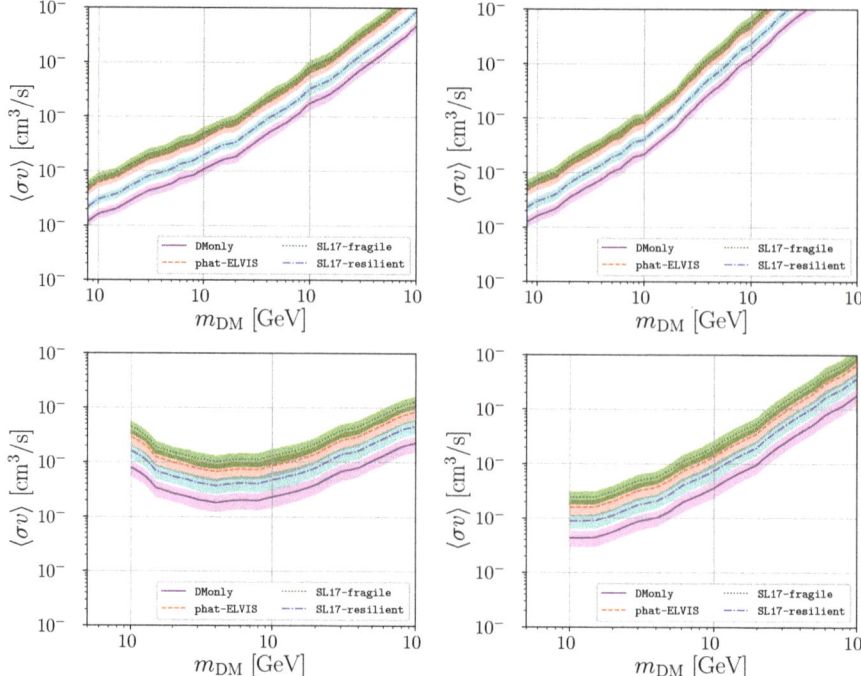

Figure 5. Upper limits on the dark matter (DM) annihilation cross-section, $\langle \sigma v \rangle$, from the observation of 16 (4) DM subhalo candidates, N_{cand}, in the 3FGL (2FHL) catalogue (the number of candidates is taken from [16]). We show the limit for DMonly (purple curve), Phat-ELVIS (red dashed curve), SL17-resilient (blue dotted-dashed curve) and SL17-fragile (green dotted curve). The same colour-code applies to uncertainty bands which represent the spread due to the 1000 Monte Carlo realisations for each subhalo model. *Top left (right) panel*: Annihilation into $b\bar{b}$ ($\tau^+\tau^-$) for the 3FGL catalogue. *Bottom left (right) panel*: Annihilation into $b\bar{b}$ ($\tau^+\tau^-$) for the 2FHL catalogue.

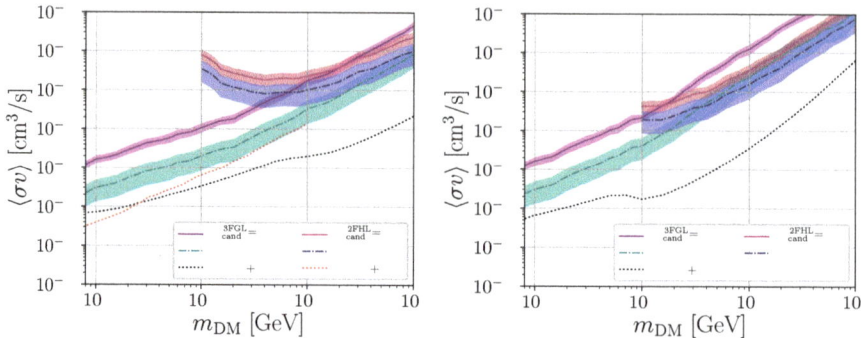

Figure 6. Upper limits on the DM annihilation cross section, $\langle \sigma v \rangle$, from the observation of 16 (4) DM subhalo candidates, N_{cand}, in the 3FGL—purple solid—(2FHL—red solid) catalogue, for the DMonly. The sensitivity reach ($N_{\text{cand}} = 0$) of the 3FGL (2FHL) is also shown by the turquoise dashed-dotted line (blue dashed-dotted line). The same colour-code applies to uncertainty bands which represent the spread due to the 1000 Monte Carlo realisations of the subhalo model. Left (Right) panel: Annihilation into b-quark (τ lepton) finale states. Overlaid, the limits from gamma-ray observations towards dwarf spheroidal galaxies from Albert et al. 2016 [38] (black dotted), and Calore et al. 2018 [9] (red dotted).

6. Discussion and Conclusions

In this work we have assessed the detectability of DM subhaloes in Fermi-LAT catalogues taking into account the uncertainties associated to the modelling of the galactic subhalo population. We have investigated four different subhalo models: one based on the Aquarius DMonly simulation [26], one on the Phat-ELVIS DM simulation which incorporates a disc potential [28], and two configurations based on an analytical model [29]. The incorporation of these models in CLUMPY [34–36] allowed us to perform 1000 Monte Carlo realisations for each configuration. We then identified among each realisations the detectable subhaloes according to the criterion derived by Calore et al. [22] for the 3FGL and 2FHL *Fermi* point-source catalogues. We obtained the DM annihilation cross-section required to detect at least one subhalo, see Table 1, to be a few $\times 10^{-25}$ ($\times 10^{-23}$) for the 3FGL (2FHL) catalogue set-up. We found that, irrespective of the subhalo model, the minimal cross section was already ruled out by gamma-ray observation of dwarf galaxies. Using the unassociated point-sources in the Fermi-LAT catalogues, we could derive upper limits on the annihilation cross section as a function of the DM mass. We have done so using the number of subhalo candidates found by Coronado-Blazquez et al. [16] to get a conservative limit and we have shown the result in Figure 5. A more stringent bound was obtained if we assumed all the unassociated sources were in fact explained by conventional astrophysical objects. We showed the corresponding bound along with existing limits from dwarf galaxies in Figure 6. We found that even for the DMonly configuration, which did not include tidal disruption from baryons, the subhalo bound was less stringent than the dwarf galaxy limit.

Comparing the results obtained with the different subhalo configurations, we found that baryonic effects on the subhalo population were significant and lead to DM constraints that were less stringent by a factor of ∼2 to ∼5. This uncertainty came from the unknown resilience of DM subhaloes to tidal disruption.

We note that, compared to previous works, we conservatively adopted a radius of $0.1°$ for the $\mathcal{J}$-factor integration. This choice was fully consistent with the computation of the Fermi-LAT threshold to subhalo signals as point-like sourcessubhaloes. Unavoidably, this led to limits on the annihilation cross section which were a factor of a few less stringent than what we found in the literature towards dark subhaloes. Nevertheless, we mention that stronger constraints can be set by looking at extended Fermi-LAT-unassociated sources. Spatial extension of a gamma-ray-unassociated source at high-latitude is generally considered a very promising hint for the DM nature of that emission. So far, however, no source has been flagged as extended [16], and, in general, only a few subhaloes are found to be extended in galaxy formation simulations [20,22]. The DM subhalo models studied in the present work instead, depending on the model, predicted from a few up to tens of subhaloes with significant angular extension ($>1^{\text{deg}}$). This means that, in order to properly assess their detectability, the sensitivity to the LAT to such a type of extended signals needs to be computed. We leave this work for a future publication, where we will also derive corresponding constraints on the DM parameter space.

In the future, CTA [6] is expected to boost the search for DM particles with high-energy gamma rays. Sensitivity studies show that we expect at least factor of 10 improvements in the limits from the galactic center analysis [43,44]. Also, searches towards dark subhaloes can be competitive, for example, by exploiting the data from a foreseen large-sky survey [43,45]. In particular, CTA deep follow-up observations of subhalo candidates or of hints of weak signals in gamma-ray surveys will provide an unprecedented discovery potential for indirect DM signals. Limits from known dwarf satellites with future telescopes will be very promising also because of the revolutionary results promised by the large synoptic survey telescope (LSST) [46]: tens to hundreds of new faint satellites of the Milky Way are expected to be discovered and their stellar kinematics to be measured with high accuracy, characterising their DM content. This will further accelerate the DM-constraining power of already existing data, such as the ones collected by Fermi-LAT and future CTA observations [47]. Finally, lower gamma-ray energies (i.e., <100 MeV) represent an almost unexplored territory. Advanced proposals for MeV telescopes exist [48,49], and future prospects look very promising, offering new opportunities to discover and/or constrain DM particle models [50].

Author Contributions: Conceptualization, F.C.; formal analysis, F.C.; writing—original draft preparation, F.C.; writing—review and editing, all authors.

Funding: The work of M.H. is supported by the Max Planck society (MPG). M.S. acknowledges support from the ANR project GaDaMa (ANR-18-CE31-0006), the CNRS IN2P3-Theory/INSU-PNHE-PNCG project "Galactic Dark Matter", and European Union's Horizon 2020 research and innovation program under the Marie Skłodowska-Curie grant agreements N° 690575 and N° 674896.

Acknowledgments: We thank Maurin for the reading of the manuscript and constructive feedback. We also thank M. Doro and M. A. Sánchez-Conde for inviting us to contribute to the Special Issue "The Role of Halo Substructure in Gamma-Ray Dark Matter Searches".

Conflicts of Interest: The authors declare no conflict of interest. The funders had no role in the design of the study; in the collection, analyses, or interpretation of data; in the writing of the manuscript, or in the decision to publish the results.

References

1. Aghanim, N.; Akrami, Y.; Ashdown, M.; Aumont, J.; Baccigalupi, C.; Ballardini, M.; Banday, A.J.; Barreiro, R.B.; Bartolo, N.; Basak, S.; et al. Planck 2018 results. VI. Cosmological parameters. *arXiv* **2018**, arXiv:1807.06209.
2. Bertone, G.; Hooper, D.; Silk, J. Particle dark matter: Evidence, candidates and constraints. *Phys. Rep.* **2005**, *405*, 279–390, doi:10.1016/j.physrep.2004.08.031. [CrossRef]
3. Bullock, J.S.; Boylan-Kolchin, M. Small-Scale Challenges to the ΛCDM Paradigm. *Ann. Rev. Astron. Astrophys.* **2017**, *55*, 343–387, doi:10.1146/annurev-astro-091916-055313. [CrossRef]
4. Bringmann, T.; Weniger, C. Gamma Ray Signals from Dark Matter: Concepts, Status and Prospects. *Phys. Dark Univ.* **2012**, *1*, 194–217, doi:10.1016/j.dark.2012.10.005. [CrossRef]
5. Gaskins, J.M. A review of indirect searches for particle dark matter. *Contemp. Phys.* **2016**, *57*, 496–525, doi:10.1080/00107514.2016.1175160. [CrossRef]
6. Acharya, B.S.; Actis, M.; Aghajani, T.; Agnetta, G.; Aguilar, J.; Aharonian, F.; Ajello, M.; Akhperjanian, A.; Alcubierre, M.; Aleksić, J.; et al. Introducing the CTA concept. *Astropart. Phys.* **2013**, *43*, 3–18. doi:10.1016/j.astropartphys.2013.01.007. [CrossRef]
7. Strigari, L.E. Dark matter in dwarf spheroidal galaxies and indirect detection: A review. *Rep. Prog. Phys.* **2018**, *81*, 056901, doi:10.1088/1361-6633/aaae16. [CrossRef]
8. Ackermann, M.; Albert, A.M.; Anderson, B.; Atwood, W.B.; Baldini, L.; Barbiellini, G.; Bastieri, D.; Bechtol, K.; Bellazzini, R.; Bissaldi, E.; et al. Searching for Dark Matter Annihilation from Milky Way Dwarf Spheroidal Galaxies with Six Years of Fermi Large Area Telescope Data. *Phys. Rev. Lett.* **2015**, *115*, 231301, doi:10.1103/PhysRevLett.115.231301. [CrossRef]
9. Calore, F.; Serpico, P.D.; Zaldivar, B. Dark matter constraints from dwarf galaxies: A data-driven analysis. *J. Cosmol. Astropart. Phys.* **2018**, *1810*, 029, doi:10.1088/1475-7516/2018/10/029. [CrossRef]
10. Diemand, J.; Kuhlen, M.; Madau, P.; Zemp, M.; Moore, B.; Potter, D.; Stadel, J. Clumps and streams in the local dark matter distribution. *Nature* **2008**, *454*, 735–738, doi:10.1038/nature07153. [CrossRef]
11. Zavala, J.; Frenk, C.S. Dark Matter Haloes and Subhaloes. *Galaxies* **2019**, *7*, 81, doi:10.3390/galaxies7040081. [CrossRef]
12. Acero, F.; Ackermann, M.; Ajello, M.; Albert, A.; Atwood, W.B.; Axelsson, M.; Baldini, L.; Ballet, J.; Barbiellini, G.; Bastieri, D.; et al. Fermi Large Area Telescope Third Source Catalog. *Astrophys. J. Suppl.* **2015**, *218*, 23, doi:10.1088/0067-0049/218/2/23. [CrossRef]
13. Ackermann, M.; Ajello, M.; Atwood, W.B.; Baldini, L.; Ballet, J.; Barbiellini, G.; Bastieri, D.; Gonzalez, J.B.; Bellazzini, R.; Bissaldi, E.; et al. 2FHL: The Second Catalog of Hard Fermi-LAT Sources. *Astrophys. J. Suppl.* **2016**, *222*, 5, doi:10.3847/0067-0049/222/1/5. [CrossRef]
14. Mirabal, N.; Charles, E.; Ferrara, E.C.; Gonthier, P.L.; Harding, A.K.; Sánchez-Conde, M.A.; Thompson, D.J. 3FGL Demographics Outside the Galactic Plane using Supervised Machine Learning: Pulsar and Dark Matter Subhalo Interpretations. *Astrophys. J.* **2016**, *825*, 69, doi:10.3847/0004-637X/825/1/69. [CrossRef]

15. Salvetti, D.; Chiaro, G.; La Mura, G.; Thompson, D.J. 3FGLzoo: Classifying 3FGL unassociated Fermi-LAT gamma-ray sources by artificial neural networks. *Mon. Not. R. Astron. Soc.* **2017**, *470*, 1291–1297, doi:10.1093/mnras/stx1328. [CrossRef]
16. Coronado-Blazquez, J.; Sanchez-Conde, M.A.; Dominguez, A.; Aguirre-Santaella, A.; Di Mauro, M.; Mirabal, N.; Nieto, D.; Charles, E. Unidentified Gamma-ray Sources as Targets for Indirect Dark Matter Detection with the Fermi-Large Area Telescope. *J. Cosmol. Astropart. Phys.* **2019**, *1907*, 020, doi:10.1088/1475-7516/2019/07/020. [CrossRef]
17. Belikov, A.V.; Hooper, D.; Buckley, M.R. Searching For Dark Matter Subhalos In the Fermi-LAT Second Source Catalog. *Phys. Rev. D* **2012**, *86*, 043504, doi:10.1103/PhysRevD.86.043504. [CrossRef]
18. Berlin, A.; Hooper, D. Stringent Constraints on the Dark Matter Annihilation Cross Section from Subhalo Searches with the Fermi Gamma-Ray Space Telescope. *Phys. Rev. D* **2014**, *89*, 016014, doi:10.1103/PhysRevD.89.016014. [CrossRef]
19. Bertoni, B.; Hooper, D.; Linden, T. Examining The Fermi-LAT Third Source Catalog in Search of Dark Matter Subhalos. *J. Cosmol. Astropart. Phys.* **2015**, *1512*, 035, doi:10.1088/1475-7516/2015/12/035. [CrossRef]
20. Schoonenberg, D.; Gaskins, J.; Bertone, G.; Diemand, J. Dark matter subhalos and unidentified sources in the Fermi 3FGL source catalog. *J. Cosmol. Astropart. Phys.* **2016**, *1605*, 028, doi:10.1088/1475-7516/2016/05/028. [CrossRef]
21. Hooper, D.; Witte, S.J. Gamma Rays From Dark Matter Subhalos Revisited: Refining the Predictions and Constraints. *J. Cosmol. Astropart. Phys.* **2017**, *1704*, 018, doi:10.1088/1475-7516/2017/04/018. [CrossRef]
22. Calore, F.; De Romeri, V.; Di Mauro, M.; Donato, F.; Marinacci, F. Realistic estimation for the detectability of dark matter sub-halos with Fermi-LAT. *Phys. Rev.* **2017**, *D96*, 063009, doi:10.1103/PhysRevD.96.063009. [CrossRef]
23. Hütten, M.; Stref, M.; Combet, C.; Lavalle, J.; Maurin, D. γ-ray and ν Searches for Dark-Matter Subhalos in the Milky Way with a Baryonic Potential. *Galaxies* **2019**, *7*, 60, doi:10.3390/galaxies7020060. [CrossRef]
24. Green, A.M.; Hofmann, S.; Schwarz, D.J. The First wimpy halos. *J. Cosmol. Astropart. Phys.* **2005**, *508*, 3, doi:10.1088/1475-7516/2005/08/003. [CrossRef]
25. Bringmann, T. Particle Models and the Small-Scale Structure of Dark Matter. *New J. Phys.* **2009**, *11*, 105027, doi:10.1088/1367-2630/11/10/105027. [CrossRef]
26. Springel, V.; Wang, J.; Vogelsberger, M.; Ludlow, A.; Jenkins, A.; Helmi, A.; Navarro, J.F.; Frenk, C.S.; White, S.D.M. The Aquarius Project: the subhaloes of galactic haloes. *Mon. Not. R. Astron. Soc.* **2008**, *391*, 1685–1711. [CrossRef]
27. Moliné, A.; Sánchez-Conde, M.A.; Palomares-Ruiz, S.; Prada, F. Characterization of subhalo structural properties and implications for dark matter annihilation signals. *Mon. Not. R. Astron. Soc.* **2017**, *466*, 4974–4990, doi:10.1093/mnras/stx026. [CrossRef]
28. Kelley, T.; Bullock, J.S.; Garrison-Kimmel, S.; Boylan-Kolchin, M.; Pawlowski, M.S.; Graus, A.S. Phat ELVIS: The inevitable effect of the Milky Way's disk on its dark matter subhaloes. *arXiv* **2018**, arXiv:1811.12413.
29. Stref, M.; Lavalle, J. Modeling dark matter subhalos in a constrained galaxy: Global mass and boosted annihilation profiles. *Phys. Rev. D* **2017**, *95*, 063003, doi:10.1103/PhysRevD.95.063003. [CrossRef]
30. van den Bosch, F.C.; Ogiya, G.; Hahn, O.; Burkert, A. Disruption of Dark Matter Substructure: Fact or Fiction? *Mon. Not. R. Astron. Soc.* **2018**, *474*, 3043–3066, doi:10.1093/mnras/stx2956. [CrossRef]
31. van den Bosch, F.C.; Ogiya, G. Dark Matter Substructure in Numerical Simulations: A Tale of Discreteness Noise, Runaway Instabilities, and Artificial Disruption. *Mon. Not. R. Astron. Soc.* **2018**, *475*, 4066–4087, doi:10.1093/mnras/sty084. [CrossRef]
32. Errani, R.; Peñarrubia, J. Can tides disrupt cold dark matter subhaloes? *arXiv* **2019**, arXiv:1906.01642,
33. Stref, M.; Lacroix, T.; Lavalle, J. Remnants of Galactic Subhalos and Their Impact on Indirect Dark-Matter Searches. *Galaxies* **2019**, *7*, 65, doi:10.3390/galaxies7020065. [CrossRef]
34. Charbonnier, A.; Combet, C.; Maurin, D. CLUMPY: A code for gamma-ray signals from dark matter structures. *Comput. Phys. Commun.* **2012**, *183*, 656–668, doi:10.1016/j.cpc.2011.10.017. [CrossRef]
35. Bonnivard, V.; Hütten, M.; Nezri, E.; Charbonnier, A.; Combet, C.; Maurin, D. CLUMPY: Jeans analysis, gamma-ray and neutrino fluxes from dark matter (sub-)structures. *Comput. Phys. Commun.* **2016**, *200*, 336–349, doi:10.1016/j.cpc.2015.11.012. [CrossRef]
36. Hütten, M.; Combet, C.; Maurin, D. CLUMPY v3: Gamma-ray and neutrino signals from dark matter at all scales. *Comput. Phys. Commun.* **2019**, *235*, 336–345, doi:10.1016/j.cpc.2018.10.001. [CrossRef]

37. Cirelli, M.; Corcella, G.; Hektor, A.; Hutsi, G.; Kadastik, M.; Panci, P.; Raidal, M.; Sala, F.; Strumia, A. PPPC 4 DM ID: A Poor Particle Physicist Cookbook for Dark Matter Indirect Detection. *J. Cosmol. Astropart. Phys.* **2011**, *1103*, 051, doi:10.1088/1475-7516/2012/10/E01. [CrossRef]
38. Albert, A.; Anderson, B.; Bechtol, K.; Drlica-Wagner, A.; Meyer, M.; Sánchez-Conde, M.; Strigari, L.; Wood, M.; Abbott, T.M.C.; Abdalla, F.B.; et al. Searching for Dark Matter Annihilation in Recently Discovered Milky Way Satellites with Fermi-LAT. *Astrophys. J.* **2017**, *834*, 110, doi:10.3847/1538-4357/834/2/110. [CrossRef]
39. Bonnivard, V.; Combet, C.; Maurin, D.; Walker, M.G. Spherical Jeans analysis for dark matter indirect detection in dwarf spheroidal galaxies—Impact of physical parameters and triaxiality. *Mon. Not. R. Astron. Soc.* **2015**, *446*, 3002–3021, doi:10.1093/mnras/stu2296. [CrossRef]
40. Klop, N.; Zandanel, F.; Hayashi, K.; Ando, S. Impact of axisymmetric mass models for dwarf spheroidal galaxies on indirect dark matter searches. *Phys. Rev. D* **2017**, *95*, 123012, doi:10.1103/PhysRevD.95.123012. [CrossRef]
41. Ullio, P.; Valli, M. A critical reassessment of particle Dark Matter limits from dwarf satellites. *J. Cosmol. Astropart. Phys.* **2016**, *1607*, 25, doi:10.1088/1475-7516/2016/07/025. [CrossRef]
42. Zhu, Q.; Marinacci, F.; Maji, M.; Li, Y.; Springel, V.; Hernquist, L. Baryonic impact on the dark matter distribution in Milky Way-sized galaxies and their satellites. *Mon. Not. R. Astron. Soc.* **2016**, *458*, 1559–1580, doi:10.1093/mnras/stw374. [CrossRef]
43. Acharya, B.S.; Agudo, I.; Al Samarai, I.; Alfaro, R.; Alfaro, J.; Alispach, C.; Batista, R.A.; Amans, J.-P.; Amato, E.; Ambrosi, G.; et al. *Science with the Cherenkov Telescope Array*; WSP: Singapore, 2018, doi:10.1142/10986. [CrossRef]
44. Silverwood, H.; Weniger, C.; Scott, P.; Bertone, G. A realistic assessment of the CTA sensitivity to dark matter annihilation. *J. Cosmol. Astropart. Phys.* **2015**, *1503*, 055, doi:10.1088/1475-7516/2015/03/055. [CrossRef]
45. Hütten, M.; Combet, C.; Maier, G.; Maurin, D. Dark matter substructure modelling and sensitivity of the Cherenkov Telescope Array to Galactic dark halos. *J. Cosmol. Astropart. Phys.* **2016**, *1609*, 047, doi:10.1088/1475-7516/2016/09/047. [CrossRef]
46. LSST Science Collaboration; Abell, P.A.; Allison, J.; Anderson, S.F.; Andrew, J.R.; Angel, J.R.P.; Armus, L.; Arnett, D.; Asztalos, S.J.; Axelrod, T.S.; et al. LSST Science Book, Version 2.0. *arXiv* **2009**, arXiv:0912.0201.
47. Ando, S.; Kavanagh, B.J.; Macias, O.; Alves, T.; Broersen, S.; Delnoij, S.; Goldman, T.; Groefsema, J.; Kleverlaan, J.; Lenssen, J.; et al. Discovery prospects of dwarf spheroidal galaxies for indirect dark matter searches. *arXiv* **2019**, arXiv:1905.07128.
48. Moiseev, A.A.; Ajello, M.; Buckley, J.H.; Caputo, R.; Ferrara, E.C.; Hartmann, D.H.; Hays, E.; McEnery, J.E.; Mitchell, J.W.; Ojha, R.; et al. Compton-Pair Production Space Telescope (ComPair) for MeV Gamma-ray Astronomy. *arXiv* **2015**, arXiv:1508.07349.
49. De Angelis, A.; Tatischeff, V.; Tavani, M.; Oberlack, U.; Grenier, I.A.; Hanlon, L.; Walter, R.; Argan, A.; von Ballmoos, P.; Bulgarelli, A.; et al. The e-ASTROGAM mission (exploring the extreme Universe in the MeV-GeV range). *arXiv* **2016**, arXiv:1611.02232.
50. De Angelis, A.; Tatischeff, V.; Grenier, I.A.; McEnery, J.; Mallamaci, M.; Tavani, M.; Oberlack, U.; Hanlon, L.; Walter, R.; Argan, A.; et al. Science with e-ASTROGAM: A space mission for MeV–GeV gamma-ray astrophysics. *J. High Energy Astrophys.* **2018**, *19*, 1–106, doi:10.1016/j.jheap.2018.07.001. [CrossRef]

 © 2019 by the authors. Licensee MDPI, Basel, Switzerland. This article is an open access article distributed under the terms and conditions of the Creative Commons Attribution (CC BY) license (http://creativecommons.org/licenses/by/4.0/).

Article

Constraints to Dark Matter Annihilation from High-Latitude HAWC Unidentified Sources

Javier Coronado-Blázquez [1,2,*] and Miguel A. Sánchez-Conde [1,2,*]

1. Instituto de Física Teórica UAM-CSIC, Universidad Autónoma de Madrid, C/ Nicolás Cabrera, 13-15, 28049 Madrid, Spain
2. Departamento de Física Teórica, M-15, Universidad Autónoma de Madrid, E-28049 Madrid, Spain
* Correspondence: javier.coronado@uam.es (J.C.-B.); miguel.sanchezconde@uam.es (M.A.S.-C.)

Received: 12 November 2019; Accepted: 25 December 2019; Published: 30 December 2019

Abstract: The ΛCDM cosmological framework predicts the existence of thousands of subhalos in our own Galaxy not massive enough to retain baryons and become visible. Yet, some of them may outshine in gamma rays provided that the dark matter is made of weakly interacting massive particles (WIMPs), which would self-annihilate and would appear as unidentified gamma-ray sources (unIDs) in gamma-ray catalogs. Indeed, unIDs have proven to be competitive targets for dark matter searches with gamma rays. In this work, we focus on the three high-latitude ($|b| \geq 10$) sources present in the 2HWC catalog of the High Altitude Water Cherenkov (HAWC) observatory with no clear associations at other wavelengths. Indeed, only one of these sources, 2HWC J1040+308, is found to be above the HAWC detection threshold when considering 760 days of data, i.e., a factor 1.5 more exposure time than in the original 2HWC catalog. Other gamma-ray instruments, such as Fermi-LAT or VERITAS at lower energies, do not detect the source. Also, this unID is reported as spatially extended, making it even more interesting in a dark matter search context. While waiting for more data that may shed further light on the nature of this source, we set competitive upper limits on the annihilation cross section by comparing this HAWC unID to expectations based on state-of-the-art N-body cosmological simulations of the Galactic subhalo population. We find these constraints to be particularly competitive for heavy WIMPs, i.e., masses above $\sim$25 (40) TeV in the case of the $b\bar{b}$ ($\tau^+\tau^-$) annihilation channel, reaching velocity-averaged cross section values of 2×10^{-25} (5×10^{-25}) cm$^3 \cdot$s^{-1}. Although far from testing the thermal relic cross section value, the obtained limits are independent and nicely complementary to those from radically different DM analyses and targets, demonstrating once again the high potential of this DM search approach.

Keywords: dark matter; subhalos; dark matter searches; gamma-rays

1. Introduction

As of today, we believe that about 85% of all the matter in the Universe is of a non-baryonic nature [1–3]. Despite its large abundance, the ultimate nature of this so-called dark matter (DM) remains unknown, being at present one of the most puzzling questions in modern physics.

N-body cosmological simulations reveal that DM structures form hierarchically in a bottom-up scenario: the DM particles first collapse into small gravitationally bound systems (known as halos), and then form more massive structures through a complex history of merging and accretion. As a result, DM halos contain a very large number of smaller *subhalos* [4,5].

For DM candidates with weak-scale masses and interactions, as those preferred by supersymmetric particle physics theories [6], subhalos with masses between approximately 10^{-11}–10^{-3} M$_\odot$ [7–9] (depending on the specific DM particle model) up to roughly 10^{10} M$_\odot$ are expected to exist in a Milky Way-like galaxy. The vast majority of the Galactic DM subhalos are not expected to be massive enough to host baryons and therefore remain completely dark, with no visible counterparts [10–14]. Yet, many of these light subhalos or *dark satellites* will lie much closer to the Earth compared to the most massive ones, given both their much larger number density and higher survival probability at small Galactocentric radii against tidal disruption [15,16].

If the DM is made of the so-called *Weakly Interacting Massive Particles* (WIMPs) DM [17,18], these small subhalos may outshine in gamma rays. WIMPs can achieve the correct relic DM abundance (the so-called "WIMP miracle"), through self-annihilation in the early Universe. This process gives rise to a Standard Model (SM) particle-antiparticle pair which, among other possible subsequent by-products, may yield gamma-ray photons [19,20]. Small subhalos may be potentially interesting for this kind of *indirect* dark matter search since, as mentioned, many of them may be located close enough to Earth to yield high enough gamma-ray fluxes [21–24]. Interestingly, an important fraction of objects (typically between 30–40%, depending on the catalog) in the very high energy (VHE, $E > 100$ GeV) sky are unidentified sources (unIDs), i.e., objects with no clear single association or counterpart. Some of these unIDs may actually be DM subhalos [22,23,25–33]. In our work, we will explore this possibility focusing on the unIDs recently detected in the VHE regime by the High Altitude Water Cherenkov Observatory (HAWC) [34,35].

One of the greatest challenges when using dark satellites to search for DM is to come up with a reliable characterization of the low-mass subhalo population that would allow for robust predictions of their DM annihilation fluxes. Indeed, there is no N-body cosmological simulation at present able to resolve the entire Galactic subhalo population all the way down to the expected minimum subhalo mass. In this paper, we will adopt the results in [24], where the authors devised an algorithm to repopulate the original Via Lactea II (VL-II) [15] N-body cosmological simulation with low-mass subhalos below its resolution limit, of about $\sim 10^5 M_\odot$. To do so, they first studied what found for the abundance and distribution of resolved subhalos in the simulation, and extrapolated the relevant quantities to smaller subhalo masses. In addition, they adopted state-of-the-art models to describe the subhalo structural properties [36]. With the predictions of [24] at hand, and in the absence of an obvious DM subhalo candidate among the pool of HAWC unIDs, which is actually reduced to just one source in the end, 2HWC J1040+308, we will show in this work that it is possible to place competitive constraints on the WIMP DM parameter space following the procedure described in [22,23].

The structure of this paper is as follows. In Section 2 we briefly describe the observational status of the VHE sky, paying special attention to the HAWC observatory. We also address our specific target selection in this section. Section 3 is devoted to the computation of the DM constraints taking into account the different instrumental parameters and the theoretical predictions from N-body simulations. We conclude in Section 4, where we also make explicit the caveats in our analysis, compare our results to previous work and make qualitative assessments for future unID-based DM search strategies and experiments.

2. HAWC Unidentified Sources

In this section, we briefly review the HAWC observatory and discuss the target selection we made for this work.

2.1. HAWC and the TeV Sky

The High Altitude Water Cherenkov Observatory (HAWC) is a VHE gamma-ray detector with a one-year sensitivity of 5–10% of the flux of the Crab Nebula. Unlike imaging atmospheric Cherenkov telescopes (IACTs), such as H.E.S.S. [37], MAGIC [38] or VERITAS [39], which detect the Cherenkov light emitted by the extensive air showers produced by the gamma rays in the atmosphere, HAWC detects particles of these air showers that reach ground level with water-based Cherenkov detectors. Therefore, HAWC is able to operate continuously and to achieve a field of view (FoV) larger than 1.5 sr [40]. HAWC is located on the flanks of the Sierra Negra volcano near Puebla, Mexico, and consists on a large pond of water tanks, each of 7.3 m of diameter, located at 4100 m elevation. The pond contains 900 photomultiplier tubes (PMTs), while each tank contains 200,000 liters of purified water [41]. Starting operations in 2012 with 30 tanks, they were progressively increased to 300 in 2015, detecting TeV gamma and cosmic rays with a high duty cycle (>95%) [42,43].

Currently, there are about 200 known VHE gamma-ray sources detected by a handful of observatories, compiled within the TeVCat[1] catalog [44]. A Galactic map in Hammer-Aitoff projection with all TeVCat sources and their class is plotted in Figure 1. From these, roughly 60 sources remain with no clear association. Most of them are probably astrophysical sources of Galactic origin, since many were discovered after the Galactic plane survey by H.E.S.S [45]. Yet, there are several unIDs located at high Galactic latitudes, in particular three of them reported in the 2HWC catalog—the second HAWC catalog [46] comprising 507 days of instrument operation. In this work, we will focus on these sources.

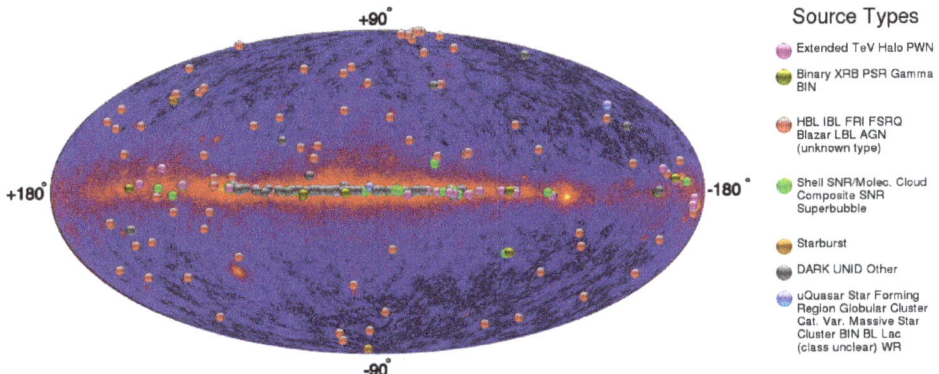

Figure 1. VHE sources listed in the TeVCat online catalog as of November 2019, here shown in Hammer-Aitoff projection and Galactic coordinates. Background is the gamma-ray sky as seen by Fermi-LAT in 4 years of operation. Note the higher density of sources, in particular of unIDs, along the Galactic plane. The figure was generated with the TeVCat online tool [44].

2.2. Target Selection

We chose to work only with the high-latitude unIDs listed in the 2HWC catalog mainly for two reasons:

- **Why HAWC?**—HAWC is currently the only VHE observatory able to survey a significant fraction of the whole sky. Indeed, the combination of its 1.5 sr FoV, Earth rotation and location in Mexican soil, 19° above the Equator, makes it possible to observe 2/3 of the sky every day.[2] In contrast, IACTs are pointing telescopes, i.e., they need to know the location of the sources in advance and, thus, just point towards specific targets. For the purposes of this work, it becomes critical to have observations of a large fraction of the sky. This is so because we adopt the methodology introduced in [23] to set constraints on the DM annihilation parameter space using unIDs. The method is based on a comparison between the number of catalogued unID sources and subhalo predictions derived from N-body cosmological simulations. Large fractions of the sky, observed with a nearly uniform exposure, will allow for a more accurate comparison to N-body simulations, as we can

---

[1] tevcat2.uchicago.edu.
[2] In fact, HAWC applies a zenith cut of 45°, which for a daily exposure translates in 8.4sr, corresponding exactly to 66.85% of the sky.

derive a more robust statistical determination of the subhalo annihilation fluxes for significantly large sky areas. We remind that these fluxes are proportional to the so-called J-factor:

$$J = \int_{\Delta\Omega} d\Omega \int_{l.o.s} \rho_{DM}^2 \left[r(\lambda)\right] d\lambda, \tag{1}$$

where the first integral is performed over the solid angle of observation ($\Delta\Omega$), the second one along the line of sight (l.o.s, λ), and ρ_{DM} is the dark matter density profile of a given subhalo.

- **Why high latitude?** ΛCDM predicts the existence of DM subhalos at all Galactic latitudes distributed nearly isotropically from the Sun's position in the Galaxy. On the other hand, the majority of Galactic VHE astrophysical sources, such as pulsars, binaries, supernova remnants and pulsar wind nebulae, are expected to cluster heavily along the Galactic plane, where most of the stars and gas reside. Since many of these objects are expected to also be hidden among the pool of unIDs awaiting for a proper classification, we expect the distribution of unIDs to peak around zero Galactic latitude as well (as already discussed in Figure 1). For our purposes, these low-latitude sources only add contamination to our sample of potential DM subhalo candidates. Therefore, we apply a cut in our analysis at Galactic latitudes $|b| \leq 10°$. For consistency, this cut will also need to be done on our predicted subhalo distribution. On average, our Galactic cut removes only $\sim$11% of the simulated subhalos, while 87% of 2HWC unIDs are left out. Actually, in practice we cut a larger region $|b| \leq 30°$ in the N-body simulation in order to match the 2/3 sky coverage of HAWC, and in this way we have a totally fair one-to-one comparison among observed and simulated sky.

After this selection, we are left with only three candidates in the 2HWC catalog, namely 2HWC J1309−054, 2HWC J1040+308 and 2HWC J0819+157. All three candidates are reported as very faint, i.e., near the source detection threshold of the catalog. In fact, [35] recently analyzed 760 days of HAWC data, i.e., approximately 50% more integration time than the original 2HWC catalog, and found both 2HWC J0819+157 and 2HWC J1309−054 to be under the detection threshold. This behaviour of source flux versus integration time is not expected for actual sources but from spurious ones instead. Thus, in the following we only consider the remaining candidate for our analysis, 2HWC J1040+308. We note that a spectral analysis was already performed for this source in [35], no preference for DM being found. Yet, more data are probably needed before being able to definitely reject this unID as a DM subhalo. Thus, we conservatively keep this source as the only surviving target in our list of potential subhalo candidates and set DM limits by assuming the existence of one single DM subhalo candidate in HAWC data.

2.3. 2HWC J1040+308 as a DM Subhalo Candidate

Before proceeding to set DM limits in the next section, we would like to highlight here some properties of 2HWC J1040+308 that make this unID particularly appealing in a DM context, as such properties may be compatible with a DM origin for the observed emission:

- **Spatial extension**—Spatial extension has been hailed as a "smoking gun" for DM annihilation [24–26,47]. Indeed, low-mass yet sufficiently close DM subhalos may be expected to appear as extended unID sources. 2HWC J1040+308 is found to be spatially extended in HAWC data with a radius of 0.5° [46]. This fact is even more notable for the case of very high latitude sources like this one ($b = 61°$), as sources at high latitudes are typically extragalactic and thus the majority of them are expected to be point-like. Indeed, it would be hard to explain all these features with a single astrophysical source – AGNs would appear as point-like sources, while Galactic sources could appear as extended, yet this source's high latitude would imply a small distance and thus it would be surprising not to have a detection in any other wavelengths. On the other hand, as these unIDs correspond to very faint sources near the detection threshold, it is

currently unclear whether they would appear as extended, even if being actual DM subhalos, as the DM annihilation flux profile decreases very rapidly with distance to the subhalo center.
- **Multi-wavelength search**—As already mentioned, dark satellites are not massive enough to retain baryons and, as a result, they are not expected to shine at any wavelengths due to astrophysical processes. However, gamma-ray emission is expected to happen should these objects be composed of WIMP DM.[3] A dedicated search at different wavelengths was performed for 2HWC J1040+308, with null results. In particular, a combined search in less energetic gamma rays with *Fermi*-LAT and VERITAS was performed in [52], with no detection. An additional search can be performed with the SSDC online tool,[4] where no significant emission at lower energies is found.
- **Heavy WIMP mass**—It is interesting to note that a joint analysis of Fermi LAT [53], VERITAS [54] and HAWC data was recently done in [52] and no gamma-ray emission was reported for 2HWC J1040+308 in the comparatively lower energy range covered by the LAT and VERITAS. This is so despite the fact that 2HWC J1040+308 exhibits a *hard* spectrum with a photon spectral index of $\Gamma = 2.08 \pm 0.25$ in the HAWC energy range [46], which would in principle make a detection at low energies easier to realize. If interpreted in a DM scenario, these results suggest heavy, TeV WIMP masses as we would not expect a detection by Fermi LAT or VERITAS only in case of significantly high values of the WIMP mass, for which a DM spectrum beyond the range of sensitivity of these two instruments would be generated. Also, interestingly, this unID slightly prefers a fit to a "Cutoff Power Law" instead of a "Simple Power Law" [35], which is a parametric form that better reproduces a typical DM annihilation spectrum [22,23]. Heavy WIMPs are well motivated [55–57] and, probably, favoured in the light of current DM constraints in the usual $\langle \sigma v \rangle$ (velocity-averaged annihilation cross section) vs. m_χ (WIMP mass) parameter space: IACTs are still far from being able to probe the *thermal relic* cross section values [58] for large, TeV WIMP masses, while for much lighter, $\mathcal{O}(\text{GeV})$ masses there are already robust and strong constraints, e.g., [23,59].

All the above considerations make 2HWC J1040+308 an excellent candidate for being a DM subhalo. Yet, lacking a definitive answer for the moment we will proceed and will set constraints to DM annihilation using HAWC unIDs, just assuming that we cannot discard 2HWC J1040+308 as a DM subhalo.

3. DM Constraints

The methodology we follow to set our DM constraints is explained in full detail in [23]. In short, these are computed according to (i) the number of remaining HAWC unIDs as potential DM subhalo candidates, i.e., just one source (2HWC J1040+308), as discussed in Section 2; (ii) the flux sensitivity of the instrument to DM annihilation; and iii) the J-factor predictions for the Galactic DM subhalo population as derived from N-body cosmological simulations. The annihilation cross section and WIMP mass $\langle \sigma v \rangle$ and m_χ are related by [23]:

$$\langle \sigma v \rangle = \frac{8\pi \cdot m_\chi^2 \cdot F_{min}}{J \cdot \int_{E_{th}}^{\infty} \left(\frac{dN}{dE}\right) dE} \qquad (2)$$

where F_{min} is the minimum flux needed for a point-source detection, J is the astrophysical J-factor (see Equation (1)), and $\int_{E_{th}}^{\infty} \left(\frac{dN}{dE}\right) dE = N_\gamma$ is the integrated DM spectrum. This DM spectrum is obtained from the PPPC4 DM ID tables [60], including electroweak corrections, and integrated from

---

[3] DM-induced emission may also be expected in radio due to synchrotron radiation from secondary particles [48–51]; yet the high density of radio sources would make any potential association probably very challenging.

[4] https://tools.ssdc.asi.it/

$E_{th} = 300$ GeV for a variety of annihilation channels. Hereafter, though, and for the sake of clarity, we will focus on $b\bar{b}$ and $\tau^+\tau^-$ channels, considering only pure annihilations (unity branching ratio).

3.1. Minimum Detection Flux

As mentioned previously, the minimum detection flux (F_{min}) is defined as the one required by the instrument to have a detection. More specifically, it is the source flux that is needed to reach $TS = 25$ over the background, where TS is the Test Statistic, defined as,

$$TS = -2\log\left[\frac{\mathcal{L}(H_1)}{\mathcal{L}(H_0)}\right] \quad (3)$$

$\mathcal{L}(H_1)$ and $\mathcal{L}(H_0)$ are, respectively, the likelihoods under the source and no source (null hypothesis) assumptions. Ideally, as done in [23], one would need to account for the variations of this quantity across different instrumental setups (integration time, energy threshold, instrument response functions, etc.) and adopting different DM spectra.[5] Finally, the source latitude should also be taken into account; however this is a second-order effect once the Galactic plane is removed from the analysis, as it is the case.

Unfortunately, the VHE sky still lacks a proper model for the all-sky Galactic diffuse gamma-ray emission, which exists at lower energies [61]. This, and the fact that we do not possess the HAWC instrumental response functions (IRFs), private for this Collaboration's internal use at the moment, precludes a precise characterization of the HAWC F_{min}. Instead, we will have to rely only on the differential sensitivity from Figure 16 of [62].[6] At each energy, the F_{min} can be obtained by dividing the differential sensitivity in TeV · cm^{-2} · s^{-1}) value by the log-central energy value in each considered bin to obtain ph · cm^{-2} · s^{-1}). In our case, we need the F_{min} corresponding to a declination of 30.87°, which turns out to be very near the absolute minimum of the curve shown in Figure 16 of [62].[7] Note, however, that the reported flux in [62] assumes a power law spectrum with a photon spectral index $\Gamma = 2.63$ (Crab-like), while our DM subhalo candidate exhibits a considerably harder spectrum, $\Gamma = 2.08$, and therefore it should be comparatively easier to detect. Also, the sensitivity is expected to scale as $\sim\sqrt{t}$ (assuming Poisson photon statistics), where t is the exposure time. Thus, as the reported flux in [46] was computed for 507 days of operation, the performance of HAWC would be underestimated for the 760 days of the current analysis.

With these limitations in mind, it is actually possible to compute a better estimate for F_{min} than the one described above. We will define the improvement factor as the ratio between our estimation of F_{min} and the official one. First, we take into account the improvement due to the photon spectral index—a source with smaller index (harder spectrum), will be easier to detect than a source with larger index (softer spectrum). We can use the scaling relation reported in [46], i.e., the variation of F_{min} with the source spectral index, for an energy value of 7 TeV, used as the pivot energy. By calibrating the improvement at this energy, we can extend it to all the considered energy range (300 GeV–100 TeV), taking into account the spectral shapes of both power laws. Then, we rescale the obtained F_{min} by differences in exposure times, i.e., a factor $\sqrt{507/760} = 0.82$.

Also, as our unID is an extended source, with a reported extension of 0.5°, the HAWC sensitivity is expected to worsen with respect to that for a point-like source. Indeed, according to [34], a 0.5° source would require a flux 1.93 times larger to be detected at the same significance, independently

---

[5] This is so because, by default, this quantity is computed for a source spectrum with a power law index $\Gamma \sim 2$, while the DM is better parametrized by a "Power Law with SuperExponential Cutoff" [23], where the index and cutoff energy vary over the annihilation channel and WIMP mass.
[6] This sensitivity is computed for the Crab, which is located at DEC$\sim 22°$, while our source is at $\sim 30°$ instead. Fortunately, the HAWC sensitivity for both locations is expected to be very similar [34].
[7] We note that, should we have had more than one unID, we would have computed a mean value after averaging the corresponding differential sensitivities for each source's declination.

of the assumed spectral index. As we are tailoring the sensitivity to our candidate, we will take this correction factor into account. As a result, the disagreement between our results and those in [35] is reduced to a factor ~ 25.

A comparison between both the originally reported F_{min} in [62] ($\Gamma = 2.63$, $t = 507$ days, point-like) and the "improved" F_{min} here described for our candidate ($\Gamma = 2.08$, $t = 760$ days, $0.5°$) is shown in Figure 2 as the blue and red curves, respectively. The latter is the one that will be used to set our DM constraints in Section 3.3.[8]

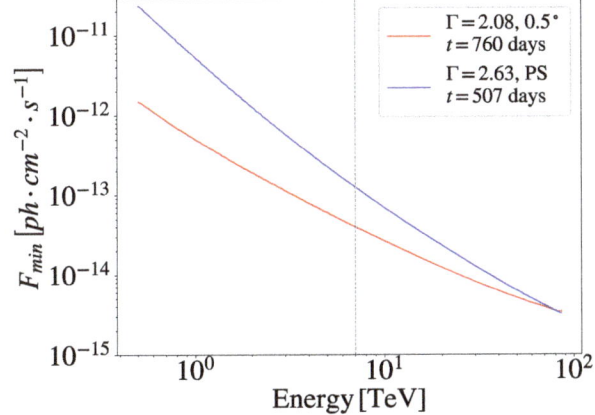

Figure 2. HAWC minimum flux, F_{min}, needed to reach source detection ($TS = 25$) at a sky declination equal to $30.87°$. In blue, the F_{min} as originally reported in [62] for the case of a point source described by a power law spectrum with index $\Gamma = 2.63$ and 507 days of exposure. The red curve shows the F_{min} for our DM subhalo candidate instead (with a spatial extension of $0.5°$, $\Gamma = 2.08$ and 760 days of exposure). This one is computed taking into account (i) the improvement factor of ~ 5 derived for this unID at a pivot energy of 7 TeV [46], marked in the figure as a dashed grey vertical line; (ii) the difference in exposure times, which makes the F_{min} improve by a factor $\sqrt{507/760} = 0.82$; and (iii) a factor ~ 2 worsening due to the extension of the source, according to [34]. See text for further details on these computations. The overall F_{min} improvement is roughly an order of magnitude at low energies, while at the highest energies there is no improvement at all.

3.2. J-Factor

The last ingredient needed to set the DM constraints is the J-factor. In the case of unID-based DM searches, we conservatively assume that all N DM subhalo candidates in our unID list ($N = 1$ in this work) are in fact DM subhalos [22,23]. Then, we require consistency of this supposedly observed number of DM subhalos with that obtained from N-body simulations, by assuming that the J-factor of the Nth (1st, in this work) most brilliant DM subhalo in our simulation sets the border between the population of detected and non-detectable subhalos.

As input, we use the N-body simulation work by [24], specifically designed to assess the relevance of low-mass subhalos for this kind of studies. The authors of [24] *repopulate* the original Via Lactea II DM-only N-body cosmological simulation of a Milky-Way-size halo [15] with subhalos well below its resolution limit by applying bootstrapping techniques and semi-analytical extrapolations of the relevant quantities. In particular, the subhalo mass function is extended down to $10^3 M_\odot$, the subhalo distribution within the host halo is assumed to be similar to that of the resolved subhalos in the parent

---

[8] The spectral index is reported to have an uncertainty $\Gamma = 2.08 \pm 0.25$. We checked that this uncertainty translates into a factor ~ 4 uncertainty in F_{min} at most (factor 2 when compared to the benchmark $\Gamma = 2.08$).

simulation, and a state-of-the-art concentration model [36] is used to model the subhalo structural properties and, ultimately, to compute the J-factors. Actually, hundreds of realizations of the parent simulation are performed that allow deriving statistical meaningful J-factor results. More precisely, as done in [23], in order to derive 95% C.L. upper limits on the DM annihilation cross section, we obtain the distribution of J-factor values for the Nth subhalo under consideration and then as reference J-factor the one above which 95% of the J-factor distribution is contained; see Figure 3.

In this particular case, we cut out from the simulation results the Galactic region $|b| \leq 10°$ in order to match the cut that we applied on the data to avoid the Galactic plane contamination (see Section 2.2). We also take into account that HAWC, unlike Fermi-LAT, does not achieve a uniform exposure for the whole sky but only for ∼2/3 of it. Specifically, to mimic this effect in the simulation side, we cut out the $|b| \leq 30°$ region (i.e., we only keep 2/3 of the whole sky) in every realization. Finally, as we are interested in dark satellites alone, only M$\leq 10^7 M_\odot$ subhalos (i.e., not massive enough to retain baryons) are considered. We checked that the resulting distribution looks isotropic as seen from the Earth. The result of applying and of not applying cuts on the J-factor distributions for the most brilliant subhalo in the simulations is shown in Figure 3. The value used to set the constraints, above which 95% of the distribution is contained, is $\log_{10}(J) = 19.47$, a factor 4.26 smaller than the value before cuts.[9]

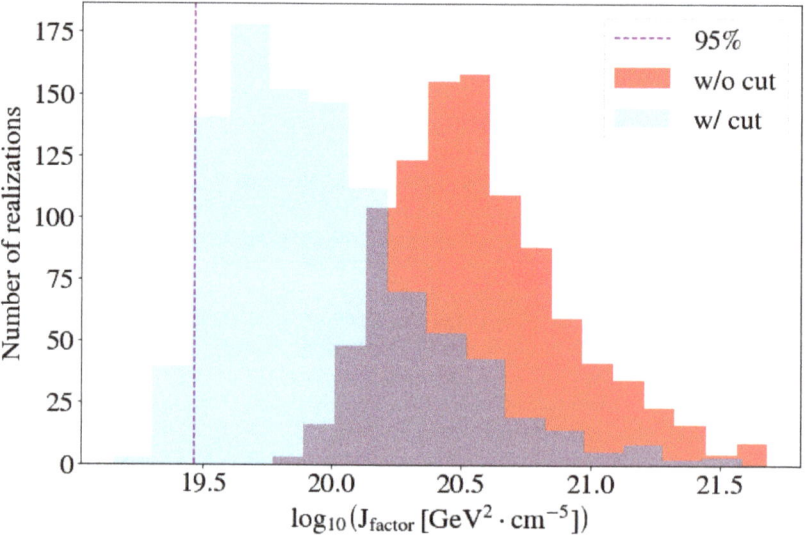

Figure 3. J-factor distributions for the brightest subhalo across 1000 realizations of the DM subhalo population as derived in [23] from the Via Lactea II N-body simulation. Red and blue histograms show, respectively, the values before and after applying our selection cuts on the simulated DM subhalos, namely a mass cut ($M \leq 10^7 M_\odot$) and a coordinate cut ($|b| > 30°$). The dashed, purple vertical line marks the value of the J-factor that will be used to set the 95% C.L. DM limits in Section 3, and it is defined as the one above which 95% of the J-factor distribution is contained.

---

[9] The J-factor of our brightest subhalo is comparable to the one typically quoted for dwarf spheroidal satellite galaxies (see e.g., [63]), which are expected to be well above the extragalactic isotropic DM-induced gamma ray background. On the other hand, the value of the DM-induced Galactic diffuse emission is comparable to the isotropic contribution at high latitudes [64], and therefore also negligible compared to the brightest subhalo J-factor. Finally, the boost due to Galactic unresolved substructure contributing to the diffuse emission in the line of sight of this unID is expected to be at the level of a few percent and, thus, not relevant here. The boost due to substructures is only particularly important when integrating the subhalo signal for the whole host halo; see e.g., the discussions and results in [36].

3.3. Results

Once all involved quantities in Equation (2) have been properly characterized, we can set constraints on the DM parameter space at 95% confidence level. These are plotted in Figure 4 for the $b\bar{b}$ and $\tau^+\tau^-$ annihilation channels. We remind, once again, that these limits adopt a single HAWC unID as being a potential DM subhalo (2HWC J1040+308).

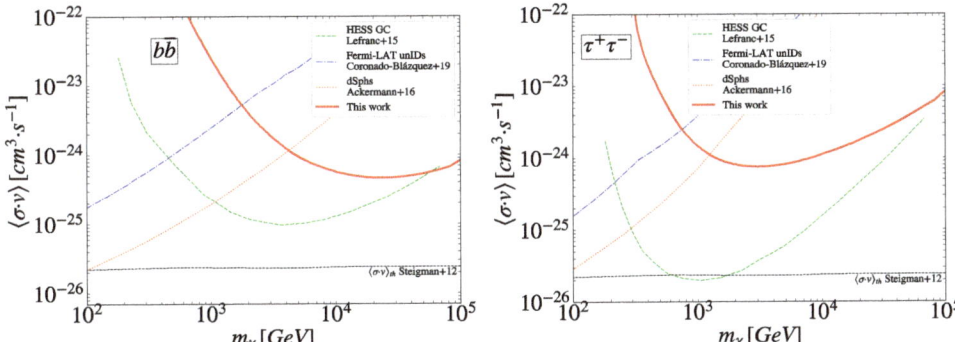

Figure 4. 95% upper limits on the DM annihilation cross section as derived from HAWC unIDs and predictions from N-body cosmological simulations. Indeed, these constraints use 2HWC J1040+308 alone, i.e., the only HAWC unID located at $|b| \geq 10°$ and surviving our selection criteria in Section 2.2. Left (right) panel shows the 95% C.L. upper limits for the $b\bar{b}$ ($\tau^+\tau^-$) annihilation channel. For comparison, we also show as a blue, dot-dashed line the DM constraints obtained from Fermi-LAT unIDs using the same methodology [47]. Limits from the observation of the Galactic center region by H.E.S.S., i.e., the best DM constraints achieved from IACT observations at present, are included in both panels as well as a green, dashed line [65]. Finally, a dotted, orange line shows the constraints from Fermi-LAT observations of dSphs [59]. Please note that we did not include in this Figure the results in [35], as a one-to-one comparison would be misleading given the fact that the methodology and underlying assumptions in both works are significantly different; see discussion in Section 3 for details.

The obtained 95% upper limits to the DM annihilation cross section are most stringent for masses of ∼20 (2) TeV for the $b\bar{b}$ ($\tau^+\tau^-$) annihilation channel, reaching cross section values around 5×10^{-25} (8×10^{-25}) cm$^3 \cdot$s^{-1}. This is roughly one order of magnitude far from the thermal relic cross section value, yet they are competitive with today's IACTs best constraints (derived from H.E.S.S. observations of the Galactic center [65]) for very heavy WIMPs annihilating to $b\bar{b}$ above ∼50 TeV. It is worth emphasizing that our DM constraints are independent and complementary to the ones obtained for other astrophysical targets and by means of different analysis techniques.

We must note here that we observe a mismatch between the DM constraints obtained in this work and the ones shown in [35]. This disagreement, that reaches a factor ∼25 for $\tau^+\tau^-$, is hard to understand in detail but it is probably due to several factors. In particular, our F_{min} is expected to be a fair approximation while [35] adopts the actual sensitivity of HAWC to a DM spectrum (only computable with the IRFs). In that work, the structural properties of subhalos were described according to [66], which is well-suited for field halos, not subhalos. Indeed, we use [36], where subhalos exhibit concentration values a factor 2-3 larger than field halos of the same mass. As the J-factor is roughly proportional to the third power of the concentration, this correction would thus yield J-factors about a factor 8-9 times larger, leading to better constraints. Other possible systematics, such as those coming from the Galactic diffuse emission model and the observation strategy adopted in [35], may turn out to be particularly relevant. Also, we note that they show cross-section values needed for detection, while we here show 95‰ upper limits. All in all, a precise one-to-one comparison with the results in [35] is not possible, as the methodology and underlying assumptions are different.

For the sake of comparison, we also show in Figure 4 the constraints found by [23] following a similar methodology, but using unIDs in *Fermi*-LAT catalogs instead. Both these low and high energy unID-based DM limits meet at roughly 2 (0.7) TeV for $b\bar{b}$ ($\tau^+\tau^-$) annihilation channel.

Finally, it is important to note that our DM limits are subject to potentially large uncertainties. In addition to the already mentioned caveats in the computation of the minimum detection flux (see Section 3.1), [10] there exist important theoretical uncertainties in the N-body simulation predictions, such as the survival probability of the lightest subhalos near the Galactic center [67–69],[11] the precise subhalo structural properties, the impact of baryons on the subhalo population [22,70] or the value of the minimum subhalo mass that is adopted to separate between visible and dark satellites [71]. These and other issues that affect the Galactic subhalo population will be addressed elsewhere. Also, for WIMP masses above ∼10 TeV, the theoretical spectra of [60] should include higher-order EW corrections, although these are expected to be potentially relevant only for leptonic channels such as e^+e^- and $\mu^+\mu^-$.

4. Summary and Conclusions

In this work, we studied the HAWC unIDs reported in the 2HWC catalog to search for potential DM subhalos. HAWC proves to be the best VHE gamma-ray observatory to perform this kind of search at present, as it is the only one that surveys a significant portion of the sky in this energy window.

Starting from a total of 23 unIDs in the 2HWC, we apply several selection criteria in Section 2.2 based on expected DM signal and subhalo properties. More precisely, at first we only select those HAWC unIDs located at Galactic latitudes $|b| \geq 10°$ to avoid contamination from astrophysical sources along the Galactic plane. 87% of the 2HWC unIDs do not pass this cut, i.e., we are left with only three unIDs. We then consider these three unIDs as DM subhalo candidates, yet two of them do not even reach the flux detection threshold when adding ∼250 more days of data [35]. Therefore, only one unID, labeled as 2HWC J1040+308, remains as a potential DM subhalo after our filtering procedure. Indeed, the measured flux of this unID grows as expected when considering more integration time, which reinforces its actual existence.

When scrutinized in further detail, 2HWC J1040+308 turns out to be especially interesting in a DM context, as it is reported in the 2HWC catalog as spatially extended, it exhibits a hard spectrum and does not possess visible counterparts neither at other wavelengths nor for other gamma-ray observatories operating at lower energies such as *Fermi*-LAT and VERITAS. The combination of the latter with the hard source spectrum might point towards particularly heavy WIMP masses, if the gamma-ray emission was actually produced by DM annihilation.

While waiting for more data that can help to shed further light on the true nature of 2HWC J1040+308, we proceed in Section 3 and set limits on the DM annihilation cross section by conservatively assuming that this source is actually a subhalo. To do so, we follow the methodology in [23] and compare the HAWC unID observations with predictions for the Galactic subhalo population as derived from N-body cosmological simulations of a galaxy such as our own. In particular, we follow the simulation work in [23,24], where the VL-II DM-only simulation was repopulated to include subhalos with masses down to $10^3 M_\odot$. This work provides the expected subhalo J-factors, which we combine with our own estimate of the HAWC instrumental sensitivity to DM subhalos in order to set 95% C.L. upper limits in the $\langle \sigma v \rangle$-m_χ parameter space.

---

[10] Namely the lack of both the HAWC IRFs and a Galactic diffuse TeV emission model.
[11] We note that tidal stripping is implicitly included in our simulations. Yet, we note that the mentioned works claim that the disruption of a large fraction of low-mass subhalos in simulations may be artificial and numerical in origin—if this was indeed the case, the distribution of J-factors of the entire subhalo population would surely reach larger values, as we would expect more and closer-to-Earth subhalos. Therefore, our results are conservative, as the DM limits would become even more stringent for resilient subhalos.

Our DM limits are most stringent for masses of ∼20 (2) TeV for the $b\bar{b}$ ($\tau^+\tau^-$) annihilation channel, reaching cross section values around 5×10^{-25} (8×10^{-25}) cm$^3 \cdot$s^{-1}. Although roughly one order of magnitude far from the thermal relic cross section value, our limits are competitive with today's IACTs best constraints [65]) above ∼50 TeV for $b\bar{b}$ annihilation channel. Also, these constraints are independent and complementary, yet competitive, to those obtained for other astrophysical targets adopting diverse analysis techniques.

The analysis presented in this work is subject to some potentially important uncertainties. In particular, a proper determination of the minimum HAWC detection flux taking into account the specific instrumental setup used in the observations, as well as the DM spectrum, annihilation channel and WIMP mass, would be required. The HAWC IRFs are nevertheless not public, thus this study is beyond the scope of the current paper. Instead, we used public information in [46,62] to come up with a fair estimate of the HAWC sensitivity to DM subhalos (Section 3.1). Another source of uncertainty is the lack of a proper Galactic diffuse gamma-ray background model, which prevents us from including it in our estimate of the HAWC sensitivity to DM. Yet, as in the end only one unID survives our cuts, this single unID is located at high Galactic latitudes (where the diffuse emission is expected to be subdominant) and exhibits a hard spectrum well fitted by a power law, we expect this uncertainty from the Galactic diffuse model to be a second order effect. Finally, there are also theoretical uncertainties hidden behind the adopted N-body simulation work. In particular, the survival probability of the smallest subhalos near the Galactic center—thus subject to strong tidal forces—is a matter of fierce debate in the community at present, with different works [67–69] providing diverse answers. In addition, the impact of baryons on the subhalo population was not considered here. Baryons may induce a reduction of the number of subhalos near the Sun's position, which would worsen our DM constraints probably by a factor of a few [22,70].

Looking into the future, more data, preferably not only in the VHE regime, are needed in order to elucidate the true nature of 2HWC J1040+308, i.e., the only HAWC unID that survives our selection cuts and, indeed, a promising DM subhalo candidate. Related to this, we note that the simplest and most robust way to improve our DM constraints is to also associate 2HWC J1040+308 to a known astrophysical source. In fact, a simple detection of variability or multi-wavelength emission would be enough to discard it as a DM subhalo (see several examples of unID rejections in [23]). As HAWC accumulates exposure time, also its minimum detectable flux will be lowered, thus allowing for better DM limits for a fixed number of unIDs. New HAWC observations and additional exposure time will surely bring up new unIDs though, thus in the end the resulting DM limits may improve or worsen depending on the number of unIDs that will be potential DM subhalos [22,23].

Finally, the upcoming Cherenkov Telescope Array (CTA) [72], with its superb instrumental capabilities, enhanced sensitivity to DM, improved angular resolution to pinpoint source extension, and already planned surveys of large portions of the VHE sky with unprecedented sensitivity, should be able to set the most competitive DM limits in the TeV energy range by means of the unID-based method used in this work. Alternatively, it is also possible that CTA may bring the first robust DM subhalo discovery.

Author Contributions: Author Contributions: Conceptualization, J.C.-B and M.A.S.-C.; formal analysis, J.C.-B.; writing—original draft, J.C.-B.; writing—review and editing, J.C.-B and M.A.S.-C. All authors have read and agreed to the published version of the manuscript.

Funding: J.C.-B. and M.A.S.-C. are supported by the *Atracción de Talento* contract no. 2016-T1/TIC-1542 granted by the Comunidad de Madrid in Spain, by the Spanish Agencia Estatal de Investigación through the grants PGC2018-095161-B-I00, IFT Centro de Excelencia Severo Ochoa SEV-2016-0597, and Red Consolider MultiDark FPA2017-90566-REDC.

Acknowledgments: The authors would like to thank Hugo Alberto Ayala Solares, Viviana Gammaldi and Colas Rivière for valuable help and feedback.

Conflicts of Interest: The authors declare no conflict of interest. The funders had no role in the design of the study; in the collection, analyses, or interpretation of data; in the writing of the manuscript, or in the decision to publish the results.

References

1. The Planck Collaboration. Planck 2015 results. XIII. Cosmological parameters. *Astron. Astrophys.* **2016**. [CrossRef]
2. Garrett, K.; Duda, G. Dark Matter: A Primer. *Adv. Astron.* **2010**. [CrossRef]
3. Freese, K. *Review of Observational Evidence for Dark Matter in the Universe and in Upcoming Searches for Dark Stars*; EAS Publications Series; EAS: Frascati, Italy, 2009; Volume 36, pp. 113–126. [CrossRef]
4. Madau, P.; Diemand, J.; Kuhlen, M. Dark Matter Subhalos and the Dwarf Satellites of the Milky Way. *Astrophys. J.* **2008**, *679*, 1260–1271. [CrossRef]
5. Zavala, J.; Frenk, C.S. Dark Matter Haloes and Subhaloes. *Galaxies* **2019**, *7*, 81. [CrossRef]
6. Jungman, G.; Kamionkowski, M.; Griest, K. Supersymmetric dark matter. *Phys. Rep.* **1996**, *267*, 195–373. [CrossRef]
7. Bertschinger, E. Effects of cold dark matter decoupling and pair annihilation on cosmological perturbations. *Phys. Rev. D* **2006**, *74*. [CrossRef]
8. Profumo, S.; Sigurdson, K.; Kamionkowski, M. What Mass Are the Smallest Protohalos? *Phys. Rev. Lett.* **2006**, *97*. [CrossRef]
9. Bringmann, T. Particle Models and the Small-Scale Structure of Dark Matter. *New J. Phys.* **2009**. [CrossRef]
10. Gao, L.; White, S.D.M.; Jenkins, A.; Stoehr, F.; Springel, V. The subhalo populations of ΛCDM dark haloes. *Mon. Not. R. Astron. Soc.* **2004**, *355*, 819–834. [CrossRef]
11. Ocvirk, P.; Gillet, N.; Shapiro, P.R.; Aubert, D.; Iliev, I.T.; Teyssier, R.; Yepes, G.; Choi, J.H.; Sullivan, D.; Knebe, A.; et al. Cosmic Dawn (CoDa): The first radiation-hydrodynamics simulation of reionization and galaxy formation in the Local Universe. *Mon. Not. R. Astron. Soc.* **2016**, *463*, 1462–1485. [CrossRef]
12. Sawala, T.; Frenk, C.S.; Fattahi, A.; Navarro, J.F.; Theuns, T.; Bower, R.G.; Crain, R.A.; Furlong, M.; Jenkins, A.; Schaller, M.; et al. The chosen few: The low-mass haloes that host faint galaxies. *Mon. Not. R. Astron. Soc.* **2015**, *456*, 85–97. [CrossRef]
13. Sawala, T.; Frenk, C.S.; Fattahi, A.; Navarro, J.F.; Bower, R.G.; Crain, R.A.; Vecchia, C.D.; Furlong, M.; Jenkins, A.; McCarthy, I.G.; et al. Bent by baryons: The low-mass galaxy-halo relation. *Mon. Not. R. Astron. Soc.* **2015**, *448*, 2941–2947. [CrossRef]
14. Fitts, A.; Boylan-Kolchin, M.; Bullock, J.S.; Weisz, D.R.; El-Badry, K.; Wheeler, C.; Faucher-Giguère, C.A.; Quataert, E.; Hopkins, P.F.; Kereš, D.; et al. No assembly required: Mergers are mostly irrelevant for the growth of low-mass dwarf galaxies. *Mon. Not. R. Astron. Soc.* **2018**, *479*, 319–331. [CrossRef]
15. Diemand, J.; Kuhlen, M.; Madau, P.; Zemp, M.; Moore, B.; Potter, D.; Stadel, J. Clumps and streams in the local dark matter distribution. *Nature* **2008**. [CrossRef] [PubMed]
16. Springel, V.; Wang, J.; Vogelsberger, M.; Ludlow, A.; Jenkins, A.; Helmi, A.; Navarro, J.F.; Frenk, C.S.; White, S.D.M. The Aquarius Project: The subhaloes of galactic haloes. *Mon. Not. R. Astron. Soc.* **2008**, *391*, 1685–1711. [CrossRef]
17. Roszkowski, L.; Sessolo, E.M.; Trojanowski, S. WIMP dark matter candidates and searches - current issues and future prospects. *arXiv* **2017**, arXiv:1707.06277v1.
18. Bertone, G. The moment of truth for WIMP Dark Matter. *Nature* **2010**, *468*, 389–393. [CrossRef]
19. Bertone, G.; Hooper, D.; Silk, J. Particle dark matter: Evidence, candidates and constraints. *Phys. Rep.* **2005**, *405*, 279–390. [CrossRef]
20. Bertone, G. (Ed.) *Particle Dark Matter*; Cambridge University Press: Cambridge, UK, 2009. [CrossRef]
21. Anderson, B.; Kuhlen, M.; Diemand, J.; Johnson, R.P.; Madau, P. Fermi-LAT Sensitivity to Dark Matter Annihilation in VIA Lactea II Substructure. *Astrophys. J.* **2010**, *718*, 899–904. [CrossRef]
22. Calore, F.; Romeri, V.D.; Mauro, M.D.; Donato, F.; Marinacci, F. Realistic estimation for the detectability of dark matter sub-halos with Fermi-LAT. *Phys. Rev. D* **2016**. [CrossRef]
23. Coronado-Blázquez, J.; Sánchez-Conde, M.A.; Domínguez, A.; Aguirre-Santaella, A.; Mauro, M.D.; Mirabal, N.; Nieto, D.; Charles, E. Unidentified gamma-ray sources as targets for indirect dark matter detection with the Fermi-Large Area Telescope. *J. Cosmol. Astropart. Phys.* **2019**, *2019*, 20. [CrossRef]
24. Aguirre-Santaella, A.; Sánchez-Conde, M.; Moliné, A.; Coronado-Blázquez, J. *TBD* **2020**, in preparation.
25. Bertoni, B.; Hooper, D.; Linden, T. Examining The Fermi-LAT Third Source Catalog In Search Of Dark Matter Subhalos. *J. Cosmol. Astropart. Phys.* **2015**, *12*, 35. [CrossRef]

26. Bertoni, B.; Hooper, D.; Linden, T. Is The Gamma-Ray Source 3FGL J2212.5+0703 A Dark Matter Subhalo? *J. Cosmol. Astropart. Phys.* **2016**, *5*, 49. [CrossRef]
27. Buckley, M.R.; Hooper, D. Dark Matter Subhalos In the Fermi First Source Catalog. *Phys. Rev. D* **2010**. [CrossRef]
28. The Fermi LAT Collaboration. Search for Dark Matter Satellites using the FERMI-LAT. *Astrophys. J.* **2012**. [CrossRef]
29. Hooper, D.; Witte, S.J. Gamma Rays From Dark Matter Subhalos Revisited: Refining the Predictions and Constraints. *J. Cosmol. Astropart. Phys.* **2017**, *4*, 18. [CrossRef]
30. Egorov, A.E.; Galper, A.M.; Topchiev, N.P.; Leonov, A.A.; Suchkov, S.I.; Kheymits, M.D.; Yurkin, Y.T. Detactability of Dark Matter Subhalos by Means of the GAMMA-400 Telescope. *Phys. At. Nucl.* **2018**, *81*, 373–378. [CrossRef]
31. De Angelis, A.; Tatischeff, V.; Grenier, I.A.; McEnery, J. Mallamaci, M.; Tavani, M.; Oberlack, U.; Hanlon, L.; Walter, R.; Argan, A.; et al. Science with e-ASTROGAM. *J. High Energy Astrophys.* **2018**, *19*, 1–106. [CrossRef]
32. Zechlin, H.S.; Fernandes, M.V.; Elsaesser, D.; Horns, D. Dark matter subhaloes as gamma-ray sources and candidates in the first Fermi-LAT catalogue. *Astron. Astrophys.* **2011**. [CrossRef]
33. Zechlin, H.S.; Horns, D. Unidentified sources in the Fermi-LAT second source catalog: The case for DM subhalos. *J. Cosmol. Astropart. Phys.* **2012**. [CrossRef]
34. The HAWC Collaboration. Sensitivity of the high altitude water Cherenkov detector to sources of multi-TeV gamma rays. *Astropart. Phys.* **2013**, *50–52*, 26–32. [CrossRef]
35. The HAWC Collaboration. Searching for dark matter sub-structure with HAWC. *J. Cosmol. Astropart. Phys.* **2019**, *2019*, 22. [CrossRef]
36. Ángeles Moliné.; Sánchez-Conde, M.A.; Palomares-Ruiz, S.; Prada, F. Characterization of subhalo structural properties and implications for dark matter annihilation signals. *Mon. Not. R. Astron. Soc.* **2017**, *466*, 4974–4990. [CrossRef]
37. The H.E.S.S. Collaboration. Calibration of cameras of the H.E.S.S. detector. *Astropart. Phys.* **2004**, *22*, 109–125. [CrossRef]
38. The MAGIC Collaboration. The major upgrade of the MAGIC telescopes, Part II: A performance study using observations of the Crab Nebula. *Astropart. Phys.* **2016**, *72*, 76–94. [CrossRef]
39. The VERITAS Collaboration. The first VERITAS telescope. *Astropart. Phys.* **2006**, *25*, 391–401. [CrossRef]
40. Andrew J. Smith for the HAWC Collaboration. HAWC: A next generation all-sky gamma-ray telescope. *J. Phys. Conf. Ser.* **2007**, *60*, 131–134. [CrossRef]
41. Andrew J. Smith for the HAWC Collaboration. HAWC: Design, Operation, Reconstruction and Analysis. *arXiv* **2015**, arXiv:1508.05826.
42. Dingus, B.L. HAWC (High Altitude Water Cherenkov) Observatory for Surveying the TeV Sky. In *AIP Conference Proceedings*; AIP: College Park, MD, USA, 2007. [CrossRef]
43. The HAWC Collaboration. The HAWC Gamma-Ray Observatory: Design, Calibration, and Operation. *arXiv* **2013**, arXiv:1310.0074v1.
44. Wakely, S.P.; Horan, D. TeVCat: An online catalog for Very High Energy Gamma-Ray Astronomy. *Proc. ICRC 2008* **2008**, *3*, 1341–1344.
45. The H.E.S.S. Collaboration. The H.E.S.S. Galactic plane survey. *Astron. Astrophys.* **2018**, *612*, A1. [CrossRef]
46. The HAWC Collaboration. The 2HWC HAWC Observatory Gamma-Ray Catalog. *Astrophys. J.* **2017**, *843*, 40. [CrossRef]
47. Coronado-Blázquez, J.; Sánchez-Conde, M.A.; Mauro, M.D.; Aguirre-Santaella, A.; Ciucă, I.; Domínguez, A.; Kawata, D.; Mirabal, N. Spectral and spatial analysis of the dark matter subhalo candidates among Fermi Large Area Telescope unidentified sources. *J. Cosmol. Astropart. Phys.* **2019**, *2019*, 45. [CrossRef]
48. Colafrancesco, S.; Profumo, S.; Ullio, P. Multi-frequency analysis of neutralino dark matter annihilations in the Coma cluster. *Astron. Astrophys.* **2006**, *455*, 21–43. [CrossRef]
49. Colafrancesco, S.; Profumo, S.; Ullio, P. Detecting dark matter WIMPs in the Draco dwarf: A multiwavelength perspective. *Phys. Rev. D* **2007**, *75*. [CrossRef]
50. Marchegiani, P.; Colafrancesco, S.; Khanye, N.F. The role of dark matter annihilation in the radio emission of the galaxy cluster A520. *Mon. Not. R. Astron. Soc.* **2018**, *483*, 2795–2800. [CrossRef]
51. Cembranos, J.A.R.; de la Cruz-Dombriz, A.; Gammaldi, V.; Mendez-Isla, M. SKA-Phase 1 sensitivity for synchrotron radio emission from multi-TeV Dark Matter candidates. *arXiv* **2019**, arXiv:1905.11154v1.

52. The Fermi-LAT, HAWC and VERITAS Collaborations. VERITAS and Fermi-LAT Observations of TeV Gamma-Ray Sources Discovered by HAWC in the 2HWC Catalog. *Astrophys. J.* **2018**, *866*, 24. [CrossRef]
53. The Fermi-LAT Collaboration; Atwood, W.B. The Large Area Telescope on the Fermi Gamma-ray Space Telescope Mission. *Astrophys. J.* **2009**. [CrossRef]
54. Nahee Park for the VERITAS Collaboration. Performance of the VERITAS experiment. *arXiv* **2015**, arXiv:1508.07070v2.
55. Gammaldi, V. Multimessenger Multi-TeV Dark Matter. *Front. Astron. Space Sci.* **2019**, *6*. [CrossRef]
56. Cembranos, J.A.R.; Dobado, A.; Maroto, A.L. Brane-World Dark Matter. *Phys. Rev. Lett* **2003**, *90*. [CrossRef] [PubMed]
57. Cembranos, J.A.R.; de la Cruz-Dombriz, A.; Gammaldi, V.; Maroto, A.L. Detection of branon dark matter with gamma ray telescopes. *Phys. Rev. D* **2012**, *85*. [CrossRef]
58. Steigman, G.; Dasgupta, B.; Beacom, J.F. Precise Relic WIMP Abundance and its Impact on Searches for Dark Matter Annihilation. *Phys. Rev. D* **2012**. [CrossRef]
59. The Fermi-LAT Collaboration. Searching for Dark Matter Annihilation from Milky Way Dwarf Spheroidal Galaxies with Six Years of Fermi-LAT Data. *Phys. Rev. Lett* **2015**. [CrossRef]
60. Cirelli, M.; Corcella, G.; Hektor, A.; Hütsi, G.; Kadastik, M.; Panci, P.; Raidal, M.; Sala, F.; Strumia, A. PPPC 4 DM ID: A Poor Particle Physicist Cookbook for Dark Matter Indirect Detection. *J. Cosmol. Astropart. Phys.* **2010**. [CrossRef]
61. The Fermi-LAT Collaboration. Fermi-LAT Observations of the Diffuse Gamma-Ray Emission: Implications for Cosmic Rays and the Interstellar Medium. *Astrophys. J.* **2012**. [CrossRef]
62. The HAWC Collaboration. Observation of the Crab Nebula with the HAWC Gamma-Ray Observatory. *Astrophys. J.* **2017**, *843*, 39. [CrossRef]
63. The Fermi-LAT Collaboration. Searching for Dark Matter Annihilation in Recently Discovered Milky Way Satellites with Fermi-Lat. *Astrophys. J.* **2017**, *834*, 110. [CrossRef]
64. The Fermi-LAT Collaboration. Limits on dark matter annihilation signals from the Fermi LAT 4-year measurement of the isotropic gamma-ray background. *J. Cosmol. Astropart. Phys.* **2015**, *2015*, 8. [CrossRef]
65. Lefranc, V.; Moulin, E.; for the H. E. S. S. Collaboration. Dark matter search in the inner Galactic halo with H.E.S.S. I and H.E.S.S. II. *arXiv* **2015**, arXiv:1509.04123v2.
66. Sánchez-Conde, M.A.; Prada, F. The flattening of the concentration-mass relation towards low halo masses and its implications for the annihilation signal boost. *Mon. Not. R. Astron. Soc.* **2014**, *442*, 2271–2277. [CrossRef]
67. van den Bosch, F.C.; Ogiya, G.; Hahn, O.; Burkert, A. Disruption of dark matter substructure: Fact or fiction? *Mon. Not. R. Astron. Soc.* **2017**, *474*, 3043–3066. [CrossRef]
68. Garrison-Kimmel, S.; Wetzel, A.; Bullock, J.S.; Hopkins, P.F.; Boylan-Kolchin, M.; Faucher-Giguère, C.A.; Kereš, D.; Quataert, E.; Sanderson, R.E.; Graus, A.S.; et al. Not so lumpy after all: Modelling the depletion of dark matter subhaloes by Milky Way-like galaxies. *Mon. Not. R. Astron. Soc.* **2017**, *471*, 1709–1727. [CrossRef]
69. Errani, R.; Peñarrubia, J. Can tides disrupt cold dark matter subhaloes? *arXiv* **2019**, arXiv:1906.01642v1.
70. Kelley, T.; Bullock, J.S.; Garrison-Kimmel, S.; Boylan-Kolchin, M.; Pawlowski, M.S.; Graus, A.S. Phat ELVIS: The inevitable effect of the Milky Way's disc on its dark matter subhaloes. *Mon. Not. R. Astron. Soc.* **2019**, *487*, 4409–4423. [CrossRef]
71. Sawala, T.; Pihajoki, P.; Johansson, P.H.; Frenk, C.S.; Navarro, J.F.; Oman, K.A.; White, S.D.M. Shaken and stirred: The Milky Way's dark substructures. *Mon. Not. R. Astron. Soc.* **2017**, *467*, 4383–4400. [CrossRef]
72. The Cherenkov Telescope Array Consortium. Science with the Cherenkov Telescope Array. *arXiv* **2017**, arXiv:1709.07997.

© 2019 by the authors. Licensee MDPI, Basel, Switzerland. This article is an open access article distributed under the terms and conditions of the Creative Commons Attribution (CC BY) license (http://creativecommons.org/licenses/by/4.0/).

Article

Properties of Subhalos in the Interacting Dark Matter Scenario

Ángeles Moliné [1,2,*], Jascha A. Schewtschenko [3], Miguel A. Sánchez-Conde [1,2], Alejandra Aguirre-Santaella [1,2], Sofía A. Cora [4,5,6] and Mario G. Abadi [7,8]

1. Instituto de Física Teórica UAM-CSIC, Universidad Autónoma de Madrid, C/ Nicolás Cabrera, 13-15, 28049 Madrid, Spain; miguel.sanchezconde@uam.es (M.A.S.-C.); alejandra.aguirre@uam.es (A.A.-S.)
2. Departamento de Física Teórica, M-15, Universidad Autónoma de Madrid, E-28049 Madrid, Spain
3. Institute of Cosmology and Gravitation, University of Portsmouth, Dennis Sciama Building, Portsmouth PO1 3FX, UK; jascha.schewtschenko@port.ac.uk
4. Instituto de Astrofísica de La Plata (CCT La Plata, CONICET, UNLP), Observatorio Astronómico, Paseo del Bosque, B1900FWA La Plata, Argentina; sacora@fcaglp.unlp.edu.ar
5. Facultad de Ciencias Astronómicas y Geofísicas, Universidad Nacional de La Plata, Observatorio Astronómico, Paseo del Bosque, B1900FWA La Plata, Argentina
6. Consejo Nacional de Investigaciones Científicas y Técnicas (CONICET), Godoy Cruz 2290, C1425FQB CABA, Argentina
7. Observatorio Astronómico, Universidad Nacional de Córdoba, Laprida 854, X5000BGR Córdoba, Argentina; mario.abadi@unc.edu.ar
8. Instituto de Astronomía Teórica y Experimental (IATE), CONICET-Universidad Nacional de Córdoba, Córdoba, Argentina
* Correspondence: angie.moline@uam.es; Tel.: +34-91-299-9873

Received: 26 July 2019; Accepted: 18 September 2019; Published: 21 September 2019

Abstract: One possible and natural derivation from the collisionless cold dark matter (CDM) standard cosmological framework is the assumption of the existence of interactions between dark matter (DM) and photons or neutrinos. Such a possible interacting dark matter (IDM) model would imply a suppression of small-scale structures due to a large collisional damping effect, even though the weakly-interacting massive particle (WIMP) can still be the DM candidate. Because of this, IDM models can help alleviate alleged tensions between standard CDM predictions and observations at small mass scales. In this work, we investigate the properties of the DM halo substructure or subhalos formed in a high-resolution cosmological N-body simulation specifically run within these alternative models. We also run its CDM counterpart, which allowed us to compare subhalo properties in both cosmologies. We show that, in the lower mass range covered by our simulation runs, both subhalo concentrations and abundances are systematically lower in IDM compared to the CDM scenario. Yet, as in CDM, we find that median IDM subhalo concentration values increase towards the innermost regions of their hosts for the same mass subhalos. Similarly to CDM, we find IDM subhalos to be more concentrated than field halos of the same mass. Our work has a direct application to studies aimed at the indirect detection of DM where subhalos are expected to boost the DM signal of their host halos significantly. From our results, we conclude that the role of the halo substructure in DM searches will be less important in interacting scenarios than in CDM, but is nevertheless far from being negligible.

Keywords: dark matter halos; subhalos; indirect dark matter searches; cosmological model

1. Introduction

The current standard model of cosmology, ΛCDM, is based on a cosmological constant to explain the late-time accelerated expansion of the Universe and a cold dark matter (CDM) component to account for the required additional gravitational attraction to form and support the galaxies and larger structures we observe today [1]. In this framework, the structure of the Universe is formed via a hierarchical, bottom-up scenario (see, e.g., [2]) with small primordial density perturbations growing to the point where they collapse into the filaments, walls, and eventually dark matter (DM) halos that form the underlying large-scale-structure filamentary web of the Universe. The galaxies are embedded in these massive, extended DM halos teeming with a self-bound substructure. Any viable cosmological model has to predict both the abundance and internal properties of these structures and their substructures successfully, and match the observational data on a wide range of scales. ΛCDM achieves this challenging feat well on the largest scales [3–7]. Yet, on small scales, tensions have been reported between its predictions and observations in our local cosmological neighbourhood. The abundance of DM substructures predicted by numerical simulations of structure formation exceeds significantly the number of satellite galaxies observed around the Milky Way and neighbouring Andromeda galaxy (see, e.g., [8,9]). Various explanations for this and similar discrepancies such as the "too big to fail" [10], "cusp vs. core" [11], and "satellite alignment" problems [12,13] were brought forward, with some of them attributed to feedback mechanisms in the baryonic sector that suppressed star formation in such small halos (see, e.g., [14]), thus leaving them without any observable tracers in the observational surveys [15], or altering the DM profiles within the halos [16–21]. Others turned to alternative models for the DM to account for the lower amount of small subhalos (see, e.g., [22,23])) or deviations of their expected properties [24–26]. The latter pathway is not only well motivated, as the properties of DM have yet remained largely a mystery, but in return, this also allows us to use the study of galaxies and their structural properties as effective probes into the very nature of the elusive DM particle.

One natural deviation from the collisionless CDM in the standard model is the assumption of the existence of interactions between DM and the standard model (SM) particles we know about, in particular, photons or neutrinos [27–29]. This does not only affect, as we show in this article, the formation of DM structures on small scales, but also provides an explanation for the exact relic abundance of DM, $\Omega_{cdm}h^2 = 0.12011$, found in the Universe today [1]. With such interactions, DM was in full thermal equilibrium with SM particles at sufficiently early times and then annihilated into SM particles until the DM decoupled from the standard sector as the Universe expanded and cooled down. The cross-section needed to retain the observed abundance of DM is surprisingly close to the one expected from the interaction via the weak force in the SM, thus coining the name "weakly interacting massive particles (WIMP) miracle". Beyond-SM theories provide a variety of WIMP DM candidates such as the minimal SUSY standard model with the neutralino and sneutrino and their electroweak scale interactions [30] or the minimal Universal extra dimension model of the Kaluza–Klein (KK) theory with the first excitation mode of the gauge field as the lightest KK-particle [31]. When it comes to the interaction partner, the usefulness of baryons is limited due to their relatively low abundance in the Universe at any time and the existing constrains on the cross-section with DM from direct detection experiments. Similarly, charged leptons, whose potential coupling with "leptophilic" DM was initially proposed as an explanation for an excess positron flux from outer space, as well as DM direct detection signals [32], are constrained by, e.g., the lack of observations of such interactions in collider experiments [33]. On the other hand, relativistic neutrinos and photons can be found in high abundance in the radiation-dominated era of the early Universe and particle-physics experiments, e.g., particle colliders, providing only very few constraints on their potential interaction with DM. Additionally, the cross-section considered in this work is sufficiently low that, e.g., the scattering rate of observable cosmic photons on DM halos is negligible as the mean free path of a photon even within the high-density inner regions of large DM halos is still many orders of a magnitude larger than these regions themselves.

In our work, we do not pick a specific model, but simply work within an effective theory, i.e., an effective interaction term between some unspecified, otherwise sterile DM particles and our SM particles of choice, photons and neutrinos in the Lagrangian. We will refer to this model as *interacting dark matter* (IDM). Depending on the actual type/mass of the mediator in our "black box", this can lead to the momentum/velocity-dependence of our effective cross-sections but, for simplicity, we mainly focus on the following inn velocity-independent scenarios. For any given cross-section, the DM remains coupled to the radiation in the early Universe until the latter is diluted enough as the Universe expands for the DM to become decoupled. As a result of this coupling, primordial perturbations, and thus, the seeds of late-time structures, are suppressed within the DM below a certain scale. This is visible as a cut-off in the linear matter power spectrum. For a DM–radiation scattering cross-section of $\sigma/\sigma_{Th} = 2 \times 10^{-9}(m_{dm}/GeV)$ with σ_{Th}, the Thomson cross-section, and m_{dm} the DM mass, this characteristic scale is $\sim$100 kpc [34] and increases or decreases with the cross-section [28,35–40].

Returning to the premise of using the halo and subhalo population as a probe into the nature of DM, we can use this suppression and its consequences for the structure formation to find bounds on the interaction cross-section. Unfortunately, as previously mentioned, a more direct study of the halo population is difficult as the distribution of its visible tracers, i.e., stars and gas, is also subject to not fully-quantified astrophysical processes. Strong lensing may provide a way to determine the DM profile of larger halos [41], but the halos around the cut-off scale are orders of magnitude smaller. Indirect methods, on the other hand, namely the detection of the annihilation or decay products of DM particles, are highly dependent on the statistical and structural properties of the halo and subhalo population. For instance, the extragalactic γ-ray and neutrino signals due to DM annihilations, when estimated via the so-called halo model [42–44], depend mainly on the DM halo and subhalo structural properties, as well as their abundances (see, e.g., [45–49]). Clearly, the considered cosmological model is crucial for such DM searches as different predictions for structure formation on small scales imply different gamma-ray or neutrino signal estimations. Ultimately, this may translate into different constraints on the DM annihilation cross-section when compared to those obtained assuming the standard ΛCDM scenario. In [50], the isotropic extragalactic signals expected from DM annihilations into γ-rays and neutrinos were investigated for both IDM and ΛCDM models using only the main halo properties as extracted from DM-only simulations. In this work, we study the properties of the halo substructure in the same IDM scenario of [50], for which we now use a set of N-body, DM-only cosmological simulations with higher particle resolution. Due, mainly, to the tidal stripping effects on the subhalo population, describing the subhalo DM density profiles is not a trivial task (see, e.g., the discussion in [49]). Thanks to our higher resolution simulations, by taking a profile-independent approach, we study IDM halo and subhalo structural properties as a function of the distance to the host halo centre and subhalo mass for the first time. As we explain in this work, one such way to characterize such properties without assuming a given density profile is to consider in the analysis the peak circular velocity V_{max} and the radius at which this velocity is attained R_{max}. In previous works [34,50], halo properties were presented as a function of halo mass. In order to compare halo and subhalo properties, in this work, we also present these properties as a function of V_{max}. On the other hand, in [51], a study about the number of subhalos in a Milky-Way-sized halo was performed as a function of V_{max}. In this work, we present such analyses as a function of the distance to the host halo centre and subhalo mass.

The work is organized as follows. We briefly summarize the theory behind IDM in Section 2, followed by a description of our simulations in Section 3. For both IDM and ΛCDM models, in Section 4, we present our results for subhalo properties such as concentrations, abundances, and subhalo radial distributions within the host halos. We finally discuss these results and draw our conclusions in Section 5.

2. Interacting Dark Matter

In our effective theory of IDM, the interactions between DM and photons (or alternatively neutrinos) result in additional terms in the equations governing the evolution of the cosmic components (see, e.g., [52]),

$$\dot{\theta}_b = k^2\psi - \mathcal{H}\theta_b + c_s^2 k^2 \delta_b - R^{-1}\dot{\kappa}(\theta_b - \theta_\gamma), \quad (1)$$

$$\dot{\theta}_\gamma = k^2\psi + \left(\frac{1}{4}\delta_\gamma - \sigma_\gamma\right)k^2\delta_b - \dot{\kappa}(\theta_\gamma - \theta_b) - C_{\gamma-\text{DM}}, \quad (2)$$

$$\dot{\theta}_{\text{DM}} = k^2\psi - \mathcal{H}\theta_{\text{DM}} - C_{\text{DM}-\gamma}, \quad (3)$$

where ψ is the gravitational potential, $\mathcal{H} = aH$ is the conformal Hubble rate, c_s is the baryon sound speed, and δ, θ, and σ are the density, velocity divergence, and anisotropic stress potential, respectively, associated with the baryon (b), photon (γ), and DM fluid. For the electromagnetic interactions (EM) in the SM, the first two equations include terms with the Thomson scattering rate $\dot{\kappa} \equiv a\sigma_{\text{Th}}cn_e$, where c is the speed of light and n_e the density of free electrons (the scale factor a appears since the derivative is taken with respect to conformal time). The ratio of the baryon to photon density, $R \equiv (3/4)(\rho_b/\rho_\gamma)$, is a pre-factor to ensure momentum conservation. $C_{\text{DM}-\gamma}$ and $C_{\gamma-\text{DM}} = -S^{-1}C_{\text{DM}-\gamma}$ are the new interactions terms that have to be added to include interactions between DM and the cosmic photon background with $S \equiv (3/4)(\rho_{\text{DM}}/\rho_\gamma)$ as the scaling of the counter term in the momentum, and ρ_{DM} is the dark matter energy density. Analogous to the EM interaction,

$$C_{\text{DM}-\gamma} = \dot{\mu}(\theta_{\text{DM}} - \theta_\gamma) \quad (4)$$

depends on the new interaction rate $\dot{\mu} \equiv a\sigma_{\text{DM}-\gamma}cn_{\text{DM}}$. Here, $\sigma_{\text{DM}-\gamma}$ is the elastic scattering cross-section between DM and photons, while $n_{\text{DM}} = \rho_{\text{DM}}/m_{\text{DM}}$ is the DM number density. For the DM–neutrino interactions, similar modifications can be added. In [38], an implementation of these modified Euler equations for the CLASS Boltzmann solver was presented. We are using this work to calculate the linear evolution of the Universe up to the point (in this work, at redshift $z = 127$) where we switch to simulations to also cover the full non-linear evolution and resulting structure formation accurately (for more details, see also [51]).

3. Simulations

For this work, we calculated the non-linear evolution of the matter distribution using a suite of cosmological DM-only simulations. This includes both simulations of single-resolution periodic volumes of 100 Mpc, as well as zoom-in simulations, which focus on representative sub-volumes to improve the maximum resolution for a subset of the obtained DM structure samples.

We performed these simulations with the parallel tree-particle mesh N-body code, P-Gadget3 [53], for both a standard, collision-less CDM and a γCDM model with a cross-section $\sigma/\sigma_{\text{Th}} = 2 \times 10^{-9}(m_{\text{DM}}/\text{GeV})$. This value is (roughly) the upper bound obtained in previous works from satellite number counts of Milky-Way-sized halos [34,51]. In [54], a more conservative constraint was claimed using measurements of the ionization history of the Universe at several redshifts, results from N-body simulations, and recent estimates of the number of Milky Way satellite galaxies. However, the approach implemented can generate large uncertainties since the presence of low-mass subhalos in galactic halos, which simulations cannot resolve, and extrapolations are necessary to obtain the results. Note that whereas larger cross-sections would erase most of the observed substructure, smaller cross-sections would imply results in between CDM and IDM. The simulations begin at a redshift of $z = 127$ (the DM–radiation interaction rate is negligible at all times afterwards). For the initial conditions, we used the same cosmology (WMAP7), random phases, and second-order Lagrangian perturbation theory (2LPT) method [55] as the APOSTLE project [56] and our previous studies of the impact of IDM on galactic substructures [51]. After having performed the full-volume run for both standard CDM and

γCDM with a particle mass $m_{\rm Part} = 1.96 \times 10^8$ M$_\odot/h$ and a comoving softening length $l_{\rm soft} = 2.7$ kpc, we identified the DM structures within using the Rockstar halo finder [57]. All halo properties were determined for spherically-overdense regions with a density of 200-times the critical density of the Universe at present, ρ_c. With these results, a cubic sub-volume was chosen at $z = 0$ with a side length of 14 Mpc/h that reproduces the overall halo mass function on the mass scales covered by it. A 1 Mpc-wide margin was added, and the resulting volume traced back to the initial redshift. We checked that the sub-volume thus constructed was still convex in these Lagrangian coordinates. This ensured that the progenitors of the structures within the targeted region evolved well within the high-resolution region, when the resulting volume was re-run using a zooming technique [58] with $m_{\rm Part} = 4.85 \times 10^5$ M$_\odot/h$ and $l_{\rm soft} = 860$ pc in the targeted region.

Throughout this work, we use the term *Box* to refer to the full-volume simulation (100 Mpc) at $z = 0$ for each cosmology. The zoom re-simulations model four Local Groups (*LGs*). We filtered the results to pick only those halos that were well within the higher resolution region, namely inside a $\sim$2.1-Mpc/h radius at $z = 0$. This was done in order to avoid boundary affects, such has halos that consist partly of higher-mass particles, which are ignored here. The total number of halos and subhalos found in both Box and LGs simulations is given in Table 1, together with the most relevant parameters of these simulations.

Table 1. Most relevant parameters of Box and Local Groups (*LGs*) simulations, together with their corresponding halo and subhalo abundances. Columns 2–4 indicate the box size $L_{\rm sim}$, the particle mass $m_{\rm Part}$, and the comoving softening length $l_{\rm soft}$. The rest of columns provide the total number of subhalos $N_{\rm sub, IDM/CDM}$ and halos $N_{\rm h, IDM/CDM}$ for each cosmological model. We remind that there are 4 LGs in each case.

	$L_{\rm sim}$	$m_{\rm Part}$	$l_{\rm soft}$	$N_{\rm sub, IDM}$	$N_{\rm sub, CDM}$	$N_{\rm h, IDM}$	$N_{\rm h, CDM}$
Box	100 Mpc	1.96×10^8 M$_\odot/h$	2.7 kpc	17,481	27,973	125,704	197,208
LGs	15 Mpc/h	4.85×10^5 M$_\odot/h$	860 pc	1606	11,092	10,513	40,874

4. Results

As mentioned, IDM exhibits a linear matter spectrum different from the one of CDM [28,35–40]. The IDM matter power spectrum features a cut-off around a smooth scale of $\sim$100 kpc for the cross-section that we are considering in this work ($\sigma/\sigma_{\rm Th} = 2 \times 10^{-9}(m_{\rm DM}/{\rm GeV})$). Therefore, a suppression of the number of halos below the scale of those hosting dwarf galaxies was expected (i.e., for halo masses below $\sim 10^{10}$ M$_\odot/h$). In addition, such a linear matter power spectrum impacts the structural halo properties, such as shape, spin, density profile, and halo concentrations [34,50,51]. In this section, we show the results we found for halo and subhalo concentrations in our simulations, as well as subhalo abundances.

4.1. Halo Concentrations

We considered two different definitions for the concentration parameter. The first and more standard is $c_\Delta \equiv R_{\rm vir}/r_{-2}$, i.e., the ratio between the halo virial radius, $R_{\rm vir}$, and the radius r_{-2} at which the logarithmic slope of the DM density profile $\frac{d \log \rho}{d \log r} = -2$. The other definition has the advantage of being independent of the adopted DM density profile and the particular definition used for the virial radius [59–61]:

$$c_{\rm V} = \frac{\bar{\rho}(R_{\rm max})}{\rho_c} = 2 \left(\frac{V_{\rm max}}{H_0 R_{\rm max}} \right)^2, \tag{5}$$

where $\bar{\rho}$ is the mean physical density within $R_{\rm max}$ and H_0 the Hubble constant. At a given $V_{\rm max}$, the concentration provides an alternative measure of the characteristic density of a halo.

Assuming an NFW profile [62,63], the relation between c_V and c_Δ is given by [59]:

$$c_V = \left(\frac{c_\Delta}{2.163}\right)^3 \frac{f(R_{\max}/r_s)}{f(c_\Delta)} \Delta, \quad (6)$$

where $f(x) = \ln(1+x) - x/(1+x)$ and $r_s = r_{-2}$ is the scale radius. For spherical (untruncated) halos[1] with a virial mass M_Δ and virial radius R_Δ at redshift $z = 0$, we have:

$$M_\Delta = \frac{4\pi}{3} R_\Delta^3 \rho_c \Delta. \quad (7)$$

where ρ_c is the critical density of the Universe at present, Δ is the overdensity factor that defines the halos, and r_Δ is its virial radius.

Using our set of simulations, both Box and LGs for IDM and CDM models, we obtained the medians of c_V and c_Δ. The latter was found by applying the c_V-c_Δ relation of Equation (6) to the $c_V(V_{\max})$ values found for every halo in the simulations. We adopted $\Delta = 200$ as the value for the overdensity to define the halos. For Box, we applied a restriction on halo maximum circular velocity such that only halos with $V_{\max} > 60$ km/s were included; in the case of the LG dataset, this restriction was set at $V_{\max} > 10$ km/s. Both criteria were adopted in order to avoid resolution issues in the determination of c_V at the smallest scales resolved by the simulations. We grouped halos into bins of $V_{\max}$ and obtained the medians of c_V. For both the LGs and Box simulations, similar bin sizes were chosen to cover the entire $V_{\max}$ range, ~ 10 km/s $< V_{\max} < 10^3$ km/s. For each cosmology, we considered five bins in LGs and nine bins for Box simulations.

In Figure 1, we show halo concentration values and the corresponding 1σ standard deviation (left panel) as found in Box (blue) and the four LG (red) simulation runs. Left and right panels show, respectively, results for both median $c_V(V_{\max})$ and $c_{200}(M_{200})$ values, the latter in bins of the halo mass M_{200} (four for both Box and LG data), calculated using Equation (7) and covering a mass range of $\sim 10^8$ M$_\odot/h < M_{200} < 10^{14}$ M$_\odot/h$. In order to have truly isolated halos, the so-called "field halos", in our analysis, we only considered halos that do not have another massive neighbour (defined as more than half the mass of the halo under consideration) located within a distance of 1.5-times its virial radius, R_{200}. In order to compare IDM and CDM subhalo concentrations one-to-one, we also include in Figure 1 the corresponding CDM concentrations. First, it is worth noting that Figure 1 shows an excellent agreement between the concentration values found in both Box and LGs at the scale where the simulations overlap. Furthermore, as expected, both IDM and CDM yielded similar results at large halo masses, while we derived significantly lower median concentration values below halo masses $\sim 10^{11}$ M$_\odot/h$ in the case of IDM compared to CDM. Interestingly, this decrease of concentration values was similar to that found in WDM simulations, an effect that has been explained as being due to the delayed formation time of low-mass halos [64]. In addition, a similar analysis for c_{200} was performed in [34,50] where also the dependence on redshift was presented. Our results were in good agreement with such previous ones at $z = 0$. As we explained above, such results for the concentration-mass relation, $c_{200}(M_{200})$, were obtained from $c_V(V_{\max})$ (see Equation (6)). In this way, we double checked previous results for IDM halo concentrations where a NFW profile was assumed. At late times, interacting DM models become (effectively) non-collisional for the cross-section studied here, in the same way that the free-streaming in WDM models becomes negligible at low redshifts. Therefore, the observed lower IDM concentration values at small halo masses also originate from the later collapse of DM halos in these models.

---

[1] Which are not affected by tidal forces.

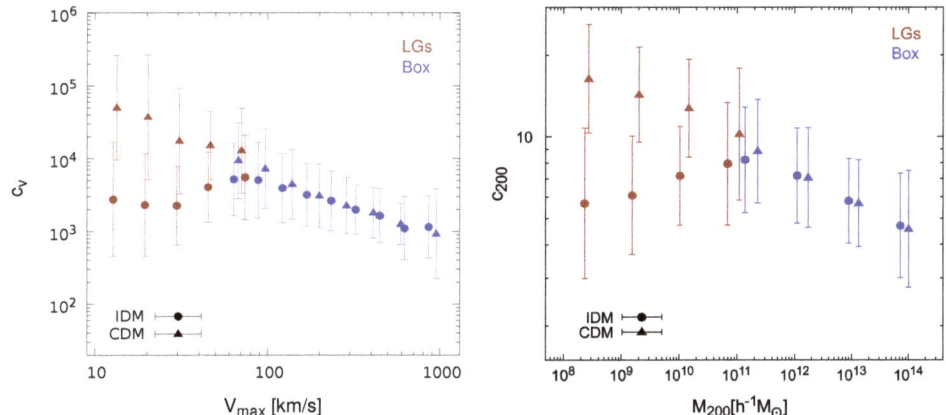

Figure 1. Median halo concentrations and 1σ errors as found in our set of simulations, Box (blue) and LGs (red), at $z = 0$. (**a**) Left panel: median c_V values as a function of $V_{\max}$. (**b**) Right panel: c_{200} as a function of M_{200}. In both panels, the circle symbols refer to the IDM simulations, whereas the triangles to CDM.

4.2. Subhalo Concentrations

The same analysis in $V_{\max}$ and subhalo mass, m_{200}, bins was performed for c_V and c_{200} subhalo concentrations, respectively. In this case for Box, eight bins were considered to cover the $V_{\max}$ range and five for m_{200}. We applied a restriction on subhalo maximum circular velocity such that only subhalos with $V_{\max} > 60$ km/s were included; for the *LGs*, this restriction was set at $V_{\max} > 10$ km/s considering just three bins for both $V_{\max}$ and m_{200} in order to obtain the median concentration values with good subhalo statistics. From the results of Box and LG simulations together, the $V_{\max}$ range covered was $10 < V_{\max} < 500$ km/s in each cosmology.

In the left panel of Figure 2, we depict median $c_V(V_{\max})$ values and corresponding 1σ errors as found in Box (blue) and the four LGs (red). The right panel shows the results for $c_{200}(m_{200})$. As in Figure 1, we also include the corresponding CDM concentrations. As can be seen, the medians of c_V (c_{200}) in both cosmologies were similar for $V_{\max} > 60$ km/s ($m_{200} > 2 \times 10^9$ M$_\odot$/h), while there was a significant departure between them at lower $V_{\max}$ (m_{200}) values. Unfortunately, the simulations had a limited mass resolution and subhalo statistics in that range, which translated into large 1σ errors, and as a consequence, our results were not conclusive. Yet, they provided a consistent picture of the subhalos' concentration behaviour at small $V_{\max}$ (m_{200}) values, IDM subhalos exhibiting lower concentrations than CDM subhalos in the mentioned $V_{\max} < 60$ km/s range.

Assuming a CDM framework, previous works have shown that the subhalo concentration depends not only on the mass of the subhalo, but also on the distance to the centre of its host halo [49,60,65]. In order to know if the same behaviour is found for IDM subhalos, Figure 3 depicts, for the LGs, the medians and 1σ errors of c_V (left panel) and c_{200} (right panel) as a function of the distance from the host halo centre in units of R_{200}. As before, we also include in the figure our results for the CDM case, which were in good agreement with the previous ones presented in [49]. The median IDM subhalo concentration increased towards the centre of the host halo more significantly than in the CDM case. Yet, for each considered radial bin, IDM concentrations were significantly and *consistently* lower than CDM ones. Such effects could be understood by studying in detail the properties of both CDM and IDM subhalos at infall. This study is beyond the scope of this paper and will be presented elsewhere. Again, large error bars prevented us from extracting firm conclusions, and thus, we will not propose any parametric fits to the data in this paper. However, this is an interesting qualitative result that points to a significantly different distribution of subhalo concentrations inside the host halo in the IDM scenario compared to CDM.

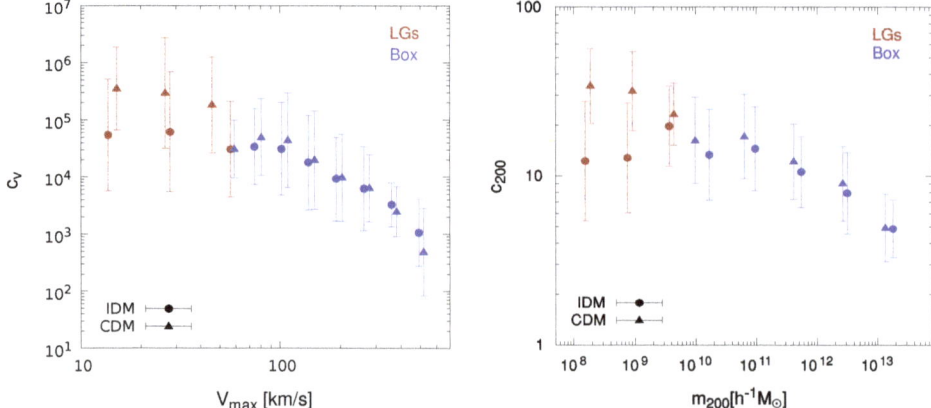

Figure 2. Median subhalo concentrations and 1σ errors as found in our set of simulations, Box (blue) and LGs (red), at $z = 0$. The circle symbols represent the results from the IDM simulations, whereas the triangle symbols correspond to the CDM results. (**a**) Left panel: the median c_V as a function of V_{max}. (**b**) Right panel: c_{200} as a function of m_{200} as obtained using Equations (6) and (7) for every subhalo in the simulations.

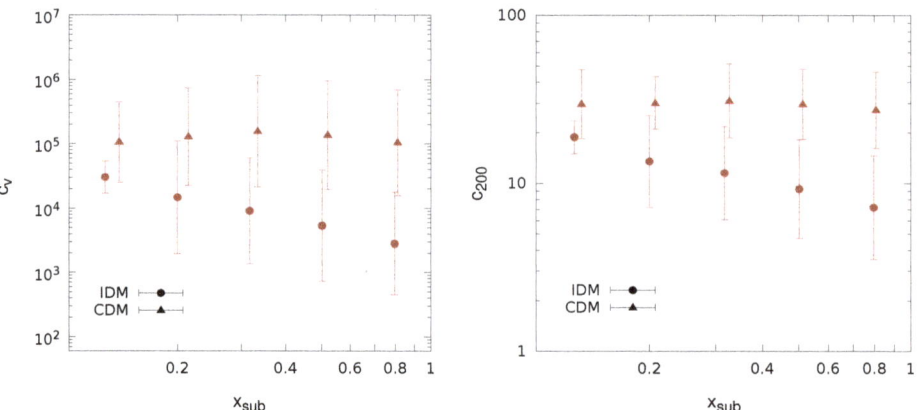

Figure 3. Median subhalo concentrations and 1σ errors as a function of x_{sub}, i.e., the distance to the centre of the host halo normalized to R_{200}. We show results for c_V (left) and c_{200} (right) as derived from our set of LG simulations.

In the standard CDM cosmological framework, it is well established from simulations that subhalos are more concentrated than field halos of the same mass [9,45,49,60,65–69]. It might not be the case in the IDM model; indeed, the mean subhalo concentration values (see Figure 2) fell within the values of halo concentrations studied in previous works for CDM. However, from Figure 1, we see that the IDM halos exhibited lower concentrations compared with the halo concentrations in CDM of the same mass, and then, differences were expected between the concentrations of subhalos and their hosts in the interacting models. In Figure 4, we shape such differences between halos and subhalos in the IDM scenario by comparing their median c_V (c_{200}) values and 1σ errors as a function of V_{max} (m_{200})

as found in our set of simulations. Analogously to what occurs in CDM, we obtained that also in IDM models, subhalos with mass $m_{200} < 10^{11}$ M$_\odot$/h tended to be more concentrated than their host halos. As in the previous cases above, a more quantitative statement about the observed trend is nevertheless not possible for the moment, given the relatively large uncertainties involved in our study.

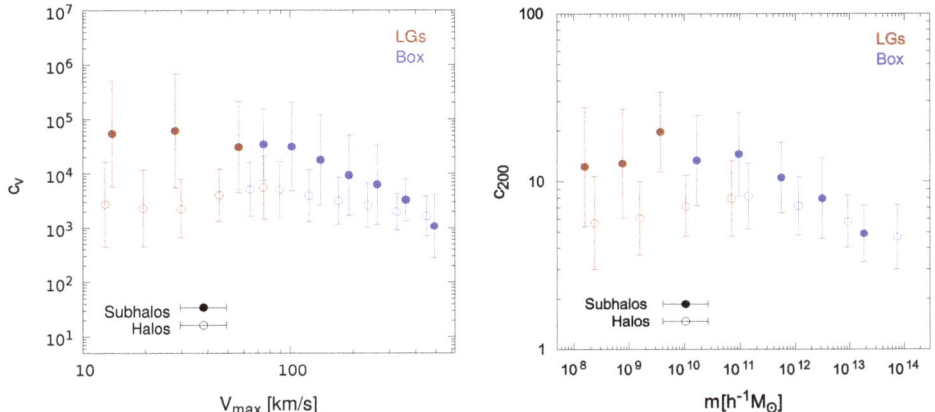

Figure 4. *Left panel*: Median halo (open circles) and subhalo (filled circles) c_V concentration values and corresponding 1σ errors, as a function of $V_{\max}$, as found in our set of simulations for interacting dark matter (IDM) at $z = 0$: Box (blue) and LGs (red). *Right panel*: the same, but for c_{200} as a function of m_{200}.

4.3. Subhalo Abundances

As mentioned, DM interactions lead to a matter power spectrum different from the one in CDM. This matter power spectrum features a cut-off around a smooth scale of ~ 100 kpc and therefore a suppression of the number of halos in the lower mass range. The impact of such an IDM initial matter power spectrum on the abundance of halos was studied in [34,50], where a comparison with the standard CDM result was also presented. A suppression of the number of low-mass halos with masses below $M_{200} \sim 10^{11} h^{-1}$ M$_\odot$ was found, which became particularly significant at the smallest considered halo masses. In this section, we will complement these previous studies by using our set of IDM simulations to obtain the first results for subhalo abundances. We will do so in a broad subhalo mass range, i.e., $[2 \times 10^6, 10^{12}]$ M$_\odot$/h.

In Figure 5, we show the cumulative number of subhalos, $N(> m_{200})$, as a function of subhalo mass, m_{200}, for both IDM and CDM scenarios and for both Box and LGs. Then, we consider all subhalos residing in halos with $M_h > 3 \times 10^{13}$ M$_\odot$/h for Box and 3×10^{11} M$_\odot$/h $< M_h < 1.4 \times 10^{12}$ M$_\odot$/h for LGs. These ranges allow us to have more than 30 subhalos per host in both cosmologies and both simulation sets. For each halo, we calculate the cumulative number of subhalos by adopting 100 subhalo mass bins and by finding the mean for each subhalo mass bin over all the main halos in the corresponding simulation. In the same Figure 5, we also show in solid lines the result of fitting the data with the following parametric expression:

$$N(> m_{200}) = \beta \, m_{200}^{\gamma} \qquad (8)$$

This fitting function follows previous works that calculated the cumulative subhalo mass function from N-body cosmological simulations and where the subhalo mass function was found to obey a power law $dN/dm \propto m_{200}^{-\alpha}$. [59]. Both the normalization factor, β, and the slopes, $\gamma = -\alpha + 1$, will depend on the adopted cosmological model. In Table 2, we report the best-fit values we found in our simulations for γ and β, both for the CDM and IDM scenarios. Using the LG data, the fits worked well for the subhalo mass range $[0.59, 3.39] \times 10^{10}$ $M_\odot/h$ and $[1.19, 9.66] \times 10^{10}$ $M_\odot/h$ for Box, in both cosmological models.

Table 2. Best-fit parameters and χ^2 values for the cumulative subhalo mass function given in Equation (8) according to our data. We show results for both IDM and CDM as obtained from our LG and Box simulations.

	γ^{LGs}	β^{LGs}	$\chi^{2,LGs}$	γ^{Box}	β^{Box}	$\chi^{2,Box}$
IDM	-0.71	6.04×10^6	0.27	-0.83	6.74×10^9	0.068
CDM	-0.88	7.22×10^8	0.19	-0.83	1.10×10^{10}	0.36

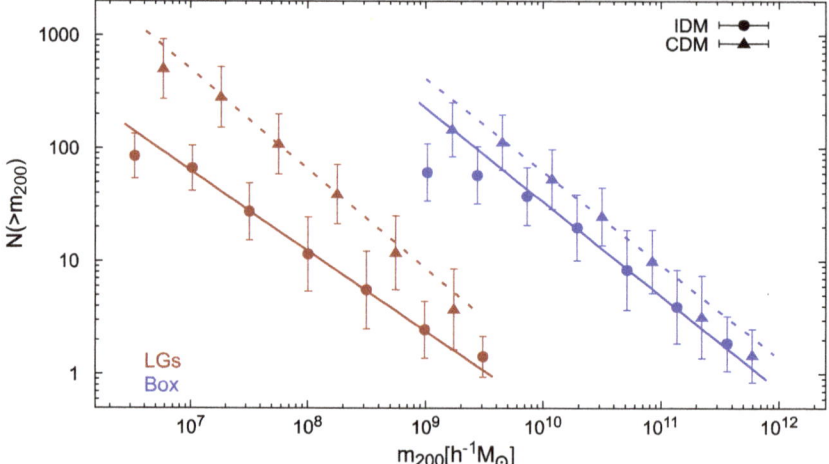

Figure 5. Cumulative number of subhalos, $N(> m_{200})$, as a function of subhalo mass, m_{200}, in the case of IDM (circle symbols) and CDM (triangles) as obtained from Box (blue) and LG (red) simulations at $z = 0$. We also show the corresponding fits using Equation (8) with the best-fit parameters reported in Table 2 (solid coloured lines).

As can be seen in Figure 5 and in Table 2, in the case of the LGs, the normalization of the cumulative subhalo mass function in the IDM case was significantly lower than that of CDM subhalos. More precisely, we found that mean $N(> m_{200})$ values for IDM subhalos were almost a factor ~ 10 lower than those of CDM for subhalos in the range 10^7 $M_\odot/h < m_{200} < 10^8$ $M_\odot/h$, this factor decreasing towards large subhalo masses. In Box, which covers comparatively larger halo masses, the differences among the two considered cosmologies were not statistically significant any more. Indeed, all these results were as expected. As discussed above, the particular differences between the IDM and CDM initial matter power spectra led to a suppression of smaller structures in the former case with respect to the latter, an effect that must become more evident in the LGs compared to Box, as the former simulations resolved smaller subhalo masses.

Finally, we also studied the radial dependence of the number of subhalos in the IDM case and compared it to the more standard CDM subhalo radial distribution. We did so only for the LGs, since high-resolution simulations are necessary to perform this kind of analysis. Indeed, we checked that the statistics in the Box simulation was not sufficient to perform the work properly. Figure 6 depicts mean values and corresponding 1σ errors of the number density as a function of the distance from the centre of the host halo (in units of its R_{200}) for halos with $[0.5–1] \times 10^{12}$ $M_\odot/h$. As can be seen, the radial number density of IDM subhalos increased towards the centre of the host halo as in the CDM case, but is significantly lower than the latter at all host radii.

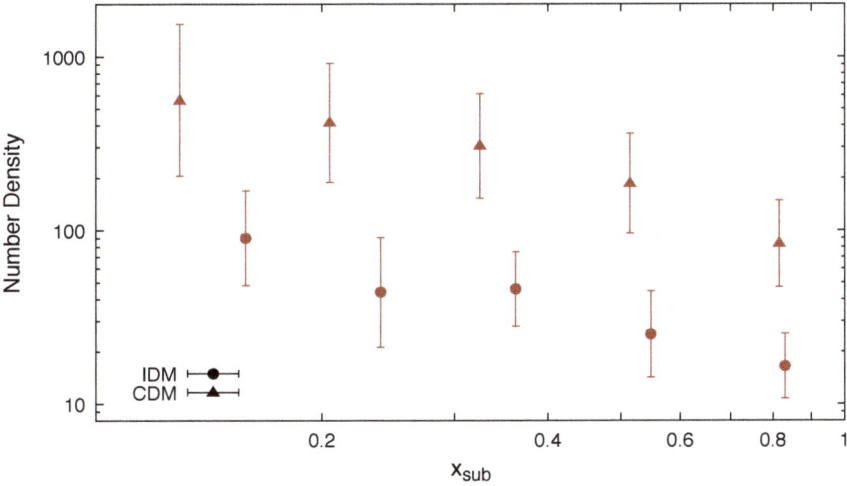

Figure 6. Number density of subhalos as a function of distance to the host halo centre, $x_{sub} = r_{sub}/R_{200}$. We show results for both IDM (circle symbols) and CDM (triangles). Both cases refer to the LG simulation set; see the text for details.

5. Summary and Discussion

We investigated DM subhalo properties in models where the linear matter power spectrum is suppressed at small scales due to DM interactions with radiation (photons or neutrinos). We did so by making use of N-body cosmological simulations, which are known to be a crucial tool to study the properties of DM structures. More precisely, we used data from our own set of simulations, described in Section 3. The runs were performed in both the standard CDM paradigm and in the IDM scenario, where the latter assumed interactions of DM with photons.[2] This allowed us to compare DM halo and subhalo properties as found in both cosmologies. Since the main impact of the DM-photon interactions on structure formation occurs mainly at small scales, we used data not only from a large simulation box (100 Mpc), but also high-resolution zoom-in simulations of four local groups.

First, in Sections 4.1 and 4.2, we studied, respectively, halo and subhalo concentrations as a function of halo/subhalo mass (and, alternatively, V_{max}). Both for halos and subhalos, we observed a significant reduction of the concentrations in the lower mass range (or, alternatively, small V_{max} values). Our result for halos confirmed the findings of previous works, e.g., [34,50], while this was the first time that the concentration of IDM subhalos was studied. This decrease of concentration values was expected and originated from the later collapse of low-mass DM halos and subhalos in IDM cosmologies, similarly to that observed in WDM simulations [64].

In Section 4.2, we studied subhalo concentrations as a function of the subhalo distance to the host halo centre. As in the CDM framework, we found that the median subhalo concentration values increased towards the innermost regions of the host for subhalos of the same mass. Yet, we obtained significantly lower median concentrations in the IDM case with respect to CDM at all radii (see Figure 3). Limitations in the number of subhalos prevented us from quantifying this effect more in detail; thus, it seemed robust to present in our data clearly. New N-body cosmological simulations with improved resolution will be needed in order to perform a more exhaustive analysis in this direction.

---

[2] We do not include the case of DM-neutrino interactions, yet the results are expected to be similar to those presented in this work; see the discussions, e.g., in [34,51].

In addition, when comparing our results for IDM halos and subhalos of the same mass, we concluded that in these IDM models, the subhalos were more concentrated than field halos (see Figure 4), similarly to what has been found for CDM, e.g., [49].

Finally, we also presented in Section 4.3 our results for subhalos abundances as a function of the distance to the host halo centre and subhalo mass. Our results were in agreement with expectations for IDM models, namely we found a significantly smaller number of subhalos in IDM with respect to that observed in our CDM simulations. However, not only the normalization of the cumulative subhalo mass function decreased (up to a factor ~ 10 at the smallest resolved subhalo scales), also its slope was substantially lower in IDM ($\gamma = -0.71$ versus $\gamma = -0.88$ for CDM in the approximated range $10^7\ M_\odot/h < m_{200} < 10^9\ M_\odot/h$; see Figure 5 and Table 2). As expected from theory, these differences among both cosmologies were not observed in the larger Box simulation. The radial distribution of subhalos within host halos exhibited a similar trend: there were fewer subhalos in IDM compared to CDM. Yet, we did not find appreciable differences in behaviour, i.e., the functional form of both radial distributions was similar.

In addition to the obvious interest in structure formation and the study of halo and subhalo properties, we note that our work has a direct application to studies aimed at the indirect detection of DM, namely the detection of the annihilation or decay products of DM particles. For instance, the extragalactic γ-ray and neutrino emission due DM annihilations depends mainly on the DM halos and subhalo properties (see, e.g., [45,46,48,49]). Another example is the so-called *subhalo boost*: subhalos are expected to boost the DM signal of their host halos significantly, e.g., [47,49]. This subhalo boost is very sensitive to the details of both subhalo concentration and subhalo abundance. Overall, from our results, we conclude that the role of halo substructure in DM searches will be less important in IDM scenarios than in CDM, given the fact that both the subhalo concentrations and abundances are lower in the former compared to the latter. Yet, it will not be negligible, as we also found in our IDM simulations larger concentrations for subhalos with respect to field halos of the same mass. Although this work represents an important step in addressing this and related issues, a quantitative study of the precise role of IDM subhalos for DM searches is left for future work: the IDM cosmological model mainly impacts low mass structures; thus, it will be necessary to have higher resolution simulations than those used in this work in order to do so. Likewise, for a full analysis of IDM halo and subhalo properties, it will be also necessary to run IDM simulations adopting other values of the cross-section of DM interactions.

Author Contributions: Conceptualization, Á.M.; Data curation, Á.M., J.A.S., A.A.-S. and S.A.C.; Formal analysis, Á.M.; Funding acquisition, M.A.S.-C. and S.A.C.; Investigation, Á.M., J.A.S., A.A.-S. and M.G.A.; Methodology, Á.M. and M.A.S.-C.; Project administration, Á.M.; Resources, J.A.S.; Software, J.A.S.; Supervision, Á.M. and M.A.S.-C.; Validation, Á.M.; Writing—original draft, Á.M.; Writing—review & editing, J.A.S., M.A.S.-C. and M.G.A.

Funding: A.M., A.A.-S., and M.A.S.-C. are supported by the *Atracción de Talento* Contract No. 2016-T1/TIC-1542 granted by the Comunidad de Madrid in Spain. They also acknowledge the support of the Spanish Agencia Estatal de Investigación through the grants PGC2018-095161-B-I00, IFT Centro de Excelencia Severo Ochoa SEV-2016-0597, and Red Consolider MultiDark FPA2017-90566-REDC. A.M. also thanks the Institute of Astrophysics and Space Sciences of Portugal, where part of this work was done and the partial support of the RAICES Argentinian program. We made use of the DiRACData Centric system at Durham University, operated by the ICC on behalf of the STFCDiRAC HPC Facility (www.dirac.ac.uk). This equipment was funded by BIS National E-infrastructure Capital Grant ST/K00042X/1, STFCCapital Grant ST/H008519/1, STFC DiRAC Operations Grant ST/K003267/1, and Durham University. DiRAC is part of the National E-Infrastructure. Furthermore, numerical computations were also done on the Sciama High Performance Compute (HPC) cluster, which is supported by the ICG, SEPNet, and the University of Portsmouth. SAC acknowledges funding from Consejo Nacional de Investigaciones Científicas y Técnicas (CONICET, PIP-0387), Agencia Nacional de Promoción Científica y Tecnológica (ANPCyT, PICT-2013-0317), and Universidad Nacional de La Plata (G11-124; G11-150), Argentina.

Conflicts of Interest: The authors declare no conflict of interest. The funders had no role in the design of the study; in the collection, analyses, or interpretation of data; in the writing of the manuscript, or in the decision to publish the results.

References

1. Aghanim, N.; Akrami, Y.; Ashdown, M.; Aumont, J.; Baccigalupi, C.; Ballardini, M.; Banday, A.J.; Barreiro, R.B.; Bartolo, N.; Basak, S.; et al. Planck 2018 results. VI. Cosmological parameters. *arXiv* **2018**, arXiv:astro-ph.CO/1807.06209.
2. Frenk, C.S.; White, S.D.M. Dark matter and cosmic structure. *Ann. Phys.* **2012**, *524*, 507–534. [CrossRef]
3. Netterfield, C.B.; Ade, P.A.R.; Bock, J.J.; Bond, J.R.; Borrill, J.; Boscaleri, A.; Coble, K.; Contaldi, C.R.; Crill, B.P.; de Bernardis, P. A measurement by Boomerang of multiple peaks in the angular power spectrum of the cosmic microwave background. *Astrophys. J.* **2002**, *571*, 604–614. [CrossRef]
4. Hinshaw, G.; Barnes, C.; Bennett, C.L.; Greason, M.R.; Halpern, M.; Hill, R.S.; Jarosik, N.; Kogut, A.; Limon, M.; Meyer, S.S.; et al. First year Wilkinson Microwave Anisotropy Probe (WMAP) observations: Data processing methods and systematic errors limits. *Astrophys. J. Suppl.* **2003**, *148*, 63. [CrossRef]
5. Riess, A.G.; Filippenko, A.V.; Challis, P.; Clocchiatti, A.; Diercks, A.; Garnavich, P.M.; Gilliland, R.L.; Hogan, C.J.; Jha, S.; Kirshner, R.P.; et al. Observational evidence from supernovae for an accelerating universe and a cosmological constant. *Astron. J.* **1998**, *116*, 1009–1038. [CrossRef]
6. Davis, T.M.; Mörtsell, E.; Sollerman, J.; Becker, A.C.; Blondin, S.; Challis, P.; Clocchiatti, A.; Filippenko, A.V.; Foley, R.J.; Garnavich, P.M.; et al. Scrutinizing Exotic Cosmological Models Using ESSENCE Supernova Data Combined with Other Cosmological Probes. *Astrophys. J.* **2007**, *666*, 716–725. [CrossRef]
7. Hamuy, M. The acceleration of the Universe in the light of supernovae - The key role of the Cerro Tololo Inter-American Observatory. *arXiv* **2013**, arXiv:astro-ph.CO/1311.5099.
8. Klypin, A.A.; Kravtsov, A.V.; Valenzuela, O.; Prada, F. Where are the missing Galactic satellites? *Astrophys. J.* **1999**, *522*, 82–92. [CrossRef]
9. Moore, B.; Ghigna, S.; Governato, F.; Lake, G.; Quinn, T.; Stadel, J.; Tozzi, P. Dark matter substructure within galactic halos. *Astrophys. J.* **1999**, *524*, L19–L22. [CrossRef]
10. Boylan-Kolchin, M.; Bullock, J.S.; Kaplinghat, M. Too big to fail? The puzzling darkness of massive Milky Way subhaloes. *Mon. Not. R. Astron. Soc.* **2011**, *415*, L40–L44. [CrossRef]
11. Dubinski, J.; Carlberg, R.G. The structure of cold dark matter halos. *Astrophys. J.* **1991**, *378*, 496–503. [CrossRef]
12. Goetz, M.; Sommer-Larsen, J. Galaxy formation: Warm dark matter, missing satellites, and the angular momentum problem. *Astrophys. Space Sci.* **2003**, *284*, 341–344. [CrossRef]
13. Vogelsberger, M.; Zavala, J.; Loeb, A. Subhaloes in Self-Interacting Galactic Dark Matter Haloes. *Mon. Not. R. Astron. Soc.* **2012**, *423*, 3740. [CrossRef]
14. Sawala, T.; Frenk, C.S.; Fattahi, A.; Navarro, J.F.; Bower, R.G.; Crain, R.A.; Vecchia, C.D.; Furlong, M.; Helly, J.C.; Jenkins, A.; et al. The APOSTLE simulations: Solutions to the Local Group's cosmic puzzles. *arXiv* **2015**, arXiv:1511.01098.
15. Kim, S.Y.; Peter, A.H.G.; Hargis, J.R. Missing Satellites Problem: Completeness Corrections to the Number of Satellite Galaxies in the Milky Way are Consistent with Cold Dark Matter Predictions. *Phys. Rev. Lett.* **2018**, *121*, 211302. [CrossRef]
16. Renaud, F.; Bournaud, F.; Emsellem, E.; Elmegreen, B.; Teyssier, R.; Alves, J.; Chapon, D.; Combes, F.; Dekel, A.; Gabor, J.; et al. A sub-parsec resolution simulation of the Milky Way: Global structure of the interstellar medium and properties of molecular clouds. *Mon. Not. R. Astron. Soc.* **2013**, *436*, 1836–1851. [CrossRef]
17. Rosdahl, J.; Schaye, J.; Dubois, Y.; Kimm, T.; Teyssier, R. Snap, crackle, pop: Sub-grid supernova feedback in AMR simulations of disc galaxies. *Mon. Not. R. Astron. Soc.* **2016**, *466*, 11–33. [CrossRef]
18. Pillepich, A.; Springel, V.; Nelson, D.; Genel, S.; Naiman, J.; Pakmor, R.; Hernquist, L.; Torrey, P.; Vogelsberger, M.; Weinberger, R.; et al. Simulating galaxy formation with the IllustrisTNG model. *Mon. Not. R. Astron. Soc.* **2017**, *473*, 4077–4106. [CrossRef]
19. Springel, V.; Pakmor, R.; Pillepich, A.; Weinberger, R.; Nelson, D.; Hernquist, L.; Vogelsberger, M.; Genel, S.; Torrey, P.; Marinacci, F.; et al. First results from the IllustrisTNG simulations: Matter and galaxy clustering. *Mon. Not. R. Astron. Soc.* **2017**, *475*, 676–698. [CrossRef]
20. Read, J.I.; Gilmore, G. Mass loss from dwarf spheroidal galaxies: The origins of shallow dark matter cores and exponential surface brightness profiles. *Mon. Not. R. Astron. Soc.* **2005**, *356*, 107–124. [CrossRef]
21. Navarro, J.F.; Eke, V.R.; Frenk, C.S. The cores of dwarf galaxy haloes. *Mon. Not. R. Astron. Soc.* **1996**, *283*, L72–L78. [CrossRef]

22. Bode, P.; Ostriker, J.P.; Turok, N. Halo formation in warm dark matter models. *Astrophys. J.* **2001**, *556*, 93–107. [CrossRef]
23. Bose, S.; Hellwing, W.A.; Frenk, C.S.; Jenkins, A.; Lovell, M.R.; Helly, J.C.; Li, B.; Gonzalez-Perez, V.; Gao, L. Substructure and galaxy formation in the Copernicus Complexio warm dark matter simulations. *Mon. Not. R. Astron. Soc.* **2016**, *464*, 4520–4533. [CrossRef]
24. Rocha, M.; Peter, A.H.; Bullock, J.S.; Kaplinghat, M.; Garrison-Kimmel, S.; Onorbe, J.; Moustakas, L.A. Cosmological simulations with self-interacting dark matter—I. Constant-density cores and substructure. *Mon. Not. R. Astron. Soc.* **2013**, *430*, 81–104. [CrossRef]
25. Vogelsberger, M.; Zavala, J.; Simpson, C.; Jenkins, A. Dwarf galaxies in CDM and SIDM with baryons: Observational probes of the nature of dark matter. *Mon. Not. R. Astron. Soc.* **2014**, *444*, 3684–3698. [CrossRef]
26. Bernal, N.; Cosme, C.; Tenkanen, T. Phenomenology of self-interacting dark matter in a matter-dominated universe. *Eur. Phys. J. C* **2019**, *79*, 99. [CrossRef]
27. Boehm, C.; Fayet, P.; Schaeffer, R. Constraining dark matter candidates from structure formation. *Phys. Lett. B* **2001**, *518*, 8–14. [CrossRef]
28. Boehm, C.; Riazuelo, A.; Hansen, S.H.; Schaeffer, R. Interacting dark matter disguised as warm dark matter. *Phys. Rev. D* **2002**, *66*, 083505. [CrossRef]
29. Boehm, C.; Schaeffer, R. Constraints on dark matter interactions from structure formation: Damping lengths. *Astron. Astrophys.* **2005**, *438*, 419–442. [CrossRef]
30. Chung, D.; Everett, L.; Kane, G.; King, S.; Lykken, J.; Wang, L.T. The soft supersymmetry-breaking Lagrangian: Theory and applications. *Phys. Rep.* **2005**, *407*, 1–203. [CrossRef]
31. Servant, G.; Tait, T.M. Is the lightest Kaluza–Klein particle a viable dark matter candidate? *Nucl. Phys. B* **2003**, *650*, 391–419. [CrossRef]
32. Fox, P.J.; Poppitz, E. Leptophilic dark matter. *Phys. Rev. D* **2009**, *79*, 083528. [CrossRef]
33. Fox, P.J.; Harnik, R.; Kopp, J.; Tsai, Y. LEP shines light on dark matter. *Phys. Rev. D* **2011**, *84*, 014028. [CrossRef]
34. Schewtschenko, J.A.; Wilkinson, R.J.; Baugh, C.M.; Bœhm, C.; Pascoli, S. Dark matter–radiation interactions: The impact on dark matter haloes. *Mon. Not. R. Astron. Soc.* **2015**, *449*, 3587–3596. [CrossRef]
35. Sigurdson, K.; Doran, M.; Kurylov, A.; Caldwell, R.R.; Kamionkowski, M. Dark-matter electric and magnetic dipole moments. *Phys. Rev. D* **2004**, *70*, 083501. [CrossRef]
36. Mangano, G.; Melchiorri, A.; Serra, P.; Cooray, A.; Kamionkowski, M. Cosmological bounds on dark matter-neutrino interactions. *Phys. Rev. D* **2006**, *74*, 043517. [CrossRef]
37. Serra, P.; Zalamea, F.; Cooray, A.; Mangano, G.; Melchiorri, A. Constraints on neutrino—Dark matter interactions from cosmic microwave background and large scale structure data. *Phys. Rev. D* **2010**, *81*, 043507. [CrossRef]
38. Wilkinson, R.J.; Lesgourgues, J.; Boehm, C. Using the CMB angular power spectrum to study Dark Matter-photon interactions. *J. Cosmol. Astropart. Phys.* **2014**, *1404*, 026. [CrossRef]
39. Wilkinson, R.J.; Boehm, C.; Lesgourgues, J. Constraining Dark Matter-Neutrino Interactions using the CMB and Large-Scale Structure. *J. Cosmol. Astropart. Phys.* **2014**, *1405*, 011. [CrossRef]
40. Cyr-Racine, F.Y.; de Putter, R.; Raccanelli, A.; Sigurdson, K. Constraints on Large-Scale Dark Acoustic Oscillations from Cosmology. *Phys. Rev.* **2014**, *D89*, 063517. [CrossRef]
41. Collett, T.E.; Buckley-Geer, E.; Lin, H.; Bacon, D.; Nichol, R.C.; Nord, B.; Morice-Atkinson, X.; Amara, A.; Birrer, S.; Kuropatkin, N.; et al. Core or Cusps: The Central Dark Matter Profile of a Strong Lensing Cluster with a Bright Central Image at Redshift 1. *Astrophys. J.* **2017**, *843*, 148. [CrossRef]
42. Neyman, J.; Scott, E.L. A Theory of the Spatial Distribution of Galaxies. *Astrophys. J.* **1952**, *116*, 144. [CrossRef]
43. Scherrer, R.J.; Bertschinger, E. Statistics of primordial density perturbations from discrete seed masses. *Astrophys. J.* **1991**, *381*, 349–360. [CrossRef]
44. Cooray, A.; Sheth, R. Halo models of large scale structure. *Phys. Rep.* **2002**, *372*, 1–129. [CrossRef]
45. Ullio, P.; Bergström, L.; Edsjö, J.; Lacey, C. Cosmological dark matter annihilations into γ rays: A closer look. *Phys. Rev. D* **2002**, *66*, 123502. [CrossRef]
46. Moliné, Á.; Ibarra, A.; Palomares-Ruiz, S. Future sensitivity of neutrino telescopes to dark matter annihilations from the cosmic diffuse neutrino signal. *J. Cosmol. Astropart. Phys.* **2015**, *2015*, 005. [CrossRef]
47. Sánchez-Conde, M.A.; Prada, F. The flattening of the concentration–mass relation towards low halo masses and its implications for the annihilation signal boost. *Mon. Not. R. Astron. Soc.* **2014**, *442*, 2271–2277. [CrossRef]
48. Fermi LAT Collaboration. Limits on dark matter annihilation signals from the Fermi LAT 4-year measurement of the isotropic gamma-ray background. *J. Cosmol. Astropart. Phys.* **2015**, *2015*, 008. [CrossRef]

49. Moliné, A.; Sánchez-Conde, M.A.; Palomares-Ruiz, S.; Prada, F. Characterization of subhalo structural properties and implications for dark matter annihilation signals. *Mon. Not. R. Astron. Soc.* **2017**, *466*, 4974–4990. [CrossRef]
50. Moliné, Á.; Schewtschenko, J.A.; Palomares-Ruiz, S.; Bœhm, C.; Baugh, C.M. Isotropic extragalactic flux from dark matter annihilations: Lessons from interacting dark matter scenarios. *J. Cosmol. Astropart. Phys.* **2016**, *2016*, 069. [CrossRef]
51. Boehm, C.; Schewtschenko, J.A.; Wilkinson, R.J.; Baugh, C.M.; Pascoli, S. Using the Milky Way satellites to study interactions between cold dark matter and radiation. *Mon. Not. R. Astron. Soc. Lett.* **2014**, *445*, L31–L35. [CrossRef]
52. Ma, C.P.; Bertschinger, E. Cosmological Perturbation Theory in the Synchronous and Conformal Newtonian Gauges. *Astrophys. J.* **1995**, *455*, 7. [CrossRef]
53. Springel, V. The Cosmological simulation code GADGET-2. *Mon. Not. R. Astron. Soc.* **2005**, *364*, 1105–1134. [CrossRef]
54. Escudero, M.; Lopez-Honorez, L.; Mena, O.; Palomares-Ruiz, S.; Villanueva-Domingo, P. A fresh look into the interacting dark matter scenario. *J. Cosmol. Astropart. Phys.* **2018**, *2018*, 007. [CrossRef]
55. Jenkins, A. Second-order Lagrangian perturbation theory initial conditions for resimulations. *Mon. Not. R. Astron. Soc.* **2010**, *403*, 1859–1872. [CrossRef]
56. Navarro, J.F.; Oman, K.A.; Fattahi, A.; Sawala, T.; Jenkins, A.; Frenk, C.S.; Schaller, M.; Furlong, M.; Theuns, T.; Crain, R.A.; et al. The APOSTLE project: Local Group kinematic mass constraints and simulation candidate selection. *Mon. Not. R. Astron. Soc.* **2016**, *457*, 844–856. [CrossRef]
57. Behroozi, P.S.; Wechsler, R.H.; Wu, H.Y. The Rockstar Phase-Space Temporal Halo Finder and the Velocity Offsets of Cluster Cores. *Astrophys. J.* **2012**, *762*, 109. [CrossRef]
58. Oñorbe, J.; Garrison-Kimmel, S.; Maller, A.H.; Bullock, J.S.; Rocha, M.; Hahn, O. How to zoom: Bias, contamination and Lagrange volumes in multimass cosmological simulations. *Mon. Not. R. Astron. Soc.* **2014**, *437*, 1894–1908. [CrossRef]
59. Diemand, J.; Kuhlen, M.; Madau, P. Formation and evolution of galaxy dark matter halos and their substructure. *Astrophys. J.* **2007**, *667*, 859–877. [CrossRef]
60. Diemand, J.; Kuhlen, M.; Madau, P.; Zemp, M.; Moore, B.; Potter, D.; Stadel, J. Clumps and streams in the local dark matter distribution. *Nature* **2008**, *454*, 735–738. [CrossRef]
61. Springel, V.; Wang, J.; Vogelsberger, M.; Ludlow, A.; Jenkins, A.; Helmi, A.; Navarro, J.F.; Frenk, C.S.; White, S.D.M. The Aquarius Project: The subhalos of galactic halos. *Mon. Not. R. Astron. Soc.* **2008**, *391*, 1685–1711. [CrossRef]
62. Navarro, J.F.; Frenk, C.S.; White, S.D. The Structure of cold dark matter halos. *Astrophys. J.* **1996**, *462*, 563–575. [CrossRef]
63. Navarro, J.F.; Frenk, C.S.; White, S.D. A Universal density profile from hierarchical clustering. *Astrophys. J.* **1997**, *490*, 493–508. [CrossRef]
64. Lovell, M.R.; Eke, V.; Frenk, C.S.; Gao, L.; Jenkins, A.; Theuns, T.; Wang, J.; White, S.D.M.; Boyarsky, A.; Ruchayskiy, O. The haloes of bright satellite galaxies in a warm dark matter universe. *Mon. Not. R. Astron. Soc.* **2012**, *420*, 2318–2324. [CrossRef]
65. Pieri, L.; Lavalle, J.; Bertone, G.; Branchini, E. Implications of High-Resolution Simulations on Indirect Dark Matter Searches. *Phys. Rev. D* **2011**, *83*, 023518. [CrossRef]
66. Ghigna, S.; Moore, B.; Governato, F.; Lake, G.; Quinn, T.R.; Stadel, J. Density profiles and substructure of dark matter halos. Converging results at ultra-high numerical resolution. *Astrophys. J.* **2000**, *544*, 616. [CrossRef]
67. Bullock, J.S.; Kolatt, T.S.; Sigad, Y.; Somerville, R.S.; Kravtsov, A.V.; Klypin, A.A.; Primack, J.R.; Dekel, A. Profiles of dark haloes. Evolution, scatter, and environment. *Mon. Not. R. Astron. Soc.* **2001**, *321*, 559–575. [CrossRef]
68. Diemand, J.; Moore, B. The structure and evolution of cold dark matter halos. *Adv. Sci. Lett.* **2011**, *4*, 297–310. [CrossRef]
69. Bartels, R.; Ando, S. Boosting the annihilation boost: Tidal effects on dark matter subhalos and consistent luminosity modeling. *Phys. Rev. D* **2015**, *92*, 123508. [CrossRef]

© 2019 by the authors. Licensee MDPI, Basel, Switzerland. This article is an open access article distributed under the terms and conditions of the Creative Commons Attribution (CC BY) license (http://creativecommons.org/licenses/by/4.0/).

MDPI
St. Alban-Anlage 66
4052 Basel
Switzerland
Tel. +41 61 683 77 34
Fax +41 61 302 89 18
www.mdpi.com

Galaxies Editorial Office
E-mail: galaxies@mdpi.com
www.mdpi.com/journal/galaxies

www.ingramcontent.com/pod-product-compliance
Lightning Source LLC
LaVergne TN
LVHW070403100526
838202LV00014B/1376